Lecture Notes in Computer Sci

Commenced Publication in 1973
Founding and Former Series Editors:
Gerhard Goos, Juris Hartmanis, and Jan van Leeuwen

Paola Flocchini   Leszek Gąsieniec (Eds.)

Structural Information and Communication Complexity

13th International Colloquium, SIROCCO 2006
Chester, UK, July 2-5, 2006
Proceedings

 Springer

Volume Editors

Paola Flocchini
University of Ottawa
School of Information Technology and Engineering (SITE)
Ottawa, ON, K1N 6N5, Canada
E-mail: flocchin@site.uottawa.ca

Leszek Gąsieniec
The University of Liverpool
Department of Computer Science
Liverpool, L69 7ZF, UK
E-mail: leszek@csc.liv.ac.uk

Library of Congress Control Number: 2006927796

CR Subject Classification (1998): F.2, C.2, G.2, E.1

LNCS Sublibrary: SL 1 – Theoretical Computer Science and General Issues

ISSN   0302-9743
ISBN-10   3-540-35474-3 Springer Berlin Heidelberg New York
ISBN-13   978-3-540-35474-1 Springer Berlin Heidelberg New York

Springer is a part of Springer Science+Business Media

springer.com

© Springer-Verlag Berlin Heidelberg 2006
Printed in Germany

Typesetting: Camera-ready by author, data conversion by Scientific Publishing Services, Chennai, India
Printed on acid-free paper   SPIN: 11780823   06/3142   5 4 3 2 1 0

Preface

The Colloquium on Structural Information and Communication Complexity (SIROCCO) is an annual meeting focused on the relationship between algorithmic aspects of computing and communication. Over its 13 years of existence, SIROCCO has become an acknowledged forum bringing together specialists interested in the fundamental principles underlying the interplay between information, communication and computing. SIROCCO covers topics such as distributed computing, high-speed networks, interconnection networks, mobile computing, optical computing, parallel computing, sensor networks, wireless networks, autonomous robots, and related areas.

SIROCCO 2006 was the 13th in this series, held in Chester, UK, July 3–5, 2006. Previous SIROCCO colloquia took place in Ottawa (1994), Olympia (1995), Siena (1996), Ascona (1997), Amalfi (1998), Lacanau-Océan (1999), L'Aquila (2000), Val de Nuria (2001), Andros (2002), Umeå (2003), Smolenice Castle (2004), and Mont Saint-Michel (2005).

In the tradition of previous occasions, this year's SIROCCO was a lively event, encouraging the emergence of new research areas (related to distributed computing in a broad sense) and the dissemination of original ideas. This was achieved by dedicating ample time for informal discussions and open problem sessions in addition to regular conference activities.

The 68 contributions submitted to SIROCCO 2006 were subject to a thorough refereeing process and 24 high-quality submissions were selected for publication. We would like to thank the authors of all the submitted papers. The excellent quality of the final program is also due to the dedicated and careful work of the Program Committee members. Our gratitude extends to the numerous subreferees for their valuable help.

We also thank the invited speakers: Hagit Attiya (Technion), Danny Krizanc (Wesleyan), and Roger Wattenhofer (ETH) for accepting our invitation to share their insights on new developments in their areas of interest, and for providing such entertaining talks.

We would like to express our sincere gratitude to the conference Chair David Peleg (Weizmann) for his enthusiasm and invaluable consultations, as well as the local organizing team, in particular Christoph Ambühl, Catherine Atherton, Alexey Fishkin, Dave Shield and Prudence Wong. Finally we would like to thank EPSRC and the University of Liverpool for their support.

These proceedings include all the accepted papers revised according to the feedback provided by the Program Committee, as well as the paper versions of the three invited talks. We hope you will enjoy them as much as we did at the conference.

July 2006

Paola Flocchini and Leszek Gąsieniec
Program Committee Co-chairs

Preface

Organization

Program Committee

Paolo Boldi, Milan
Shlomi Dolev, Ben-Gurion
Thomas Erlebach, Leicester
Paola Flocchini, Ottawa (Co-chair)
Fedor Fomin, Bergen
Pierre Fraigniaud, Paris Sud
Leszek Gąsieniec, Liverpool (Co-chair)
Christos Kaklamanis, Patras
Idit Keidar, Technion
Ralf Klasing, Bordeaux
Evangelos Kranakis, Carleton

Andrzej Lingas, Lund
Flaminia Luccio, Trieste
Andrzej Pelc, Québec
Giuseppe Prencipe, Pisa
Kirk Pruhs, Pittsburgh
Tomasz Radzik, Kings College
Violet R. Syrotiuk, Arizona State
Roger Wattenhofer, ETH
Masafumi Yamashita, Kyushu
Shmuel Zaks, Technion

Organizing Committee

Christoph Ambühl, Liverpool
Catherine Atherton, Liverpool
Alexei Fishkin, Liverpool

Leszek Gąsieniec, Liverpool (Chair)
Dave Shield, Liverpool
Prudence Wong, Liverpool

Steering Committee

Paola Flocchini, Ottawa
Leszek Gąsieniec, Liverpool
Christos Kaklamanis, Patras
Lefteris Kirousis, Patras
Rastislav Královič, Bratislava
Evangelos Kranakis, Carleton
Danny Krizanc, Wesleyan

Bernard Mans, Macquarie
Andrzej Pelc, Québec
David Peleg, Weizmann (Chair)
Michel Raynal, Rennes
Nicola Santoro, Ottawa
Paul Spirakis, Patras

Referees

Luca Allulli
Christoph Ambühl
Gal Badishi
René Beier

Amos Beimel
Sivan Bercovici
Edward Bortnikov
Olga Brukman

Nicolas Burri
Ioannis Caragiannis
Carlos Castillo
Stefano Chessa
David Coudert
Shantanu Das
Stefan Dobrev
Michael Elkin
Leah Epstein
Alexey Fishkin
Hen Fitoussi
Michele Flammini
Roland Flury
Vincenzo Gervasi
Olga Goussevskaia
Yinnon Haviv
Jesper Jansson
Panagiotis Kanellopoulos
Ronen Kat
Dariusz Kowalski
Rastislav Kralovic
Danny Krizanc
Michael Kuhn
Łukasz Kuszner
Christos Levcopoulos
Thomas Locher
Violetta Lonati
Zvi Lotker
Fabrizio Luccio

Euripides Markou
Morten Mjelde
Manuela Montangero
Luca Moscardelli
Thomas Moscibroda
Alfredo Navarra
Linda Pagli
Evi Papaioannou
Paolo Penna
Mia Persson
Alessandro Provetti
Geppino Pucci
Pascal von Rickenbach
Fabiano Sarracco
Stefan Schmid
Hadas Shachnai
Alexander Shraer
Francesco Silvestri
Savio Tse
Nir Tzachar
Chi-Hung Tzeng
Ugo Vaccaro
Sebastiano Vigna
Martin Wahlen
Mirjam Wattenhofer
Yves Weber
Prudence Wong
Qin Xin

Sponsoring Institutions

The Engineering and Physical Sciences Research Council (EPSRC)

Table of Contents

Mobile Agent Rendezvous: A Survey

Evangelos Kranakis[1], Danny Krizanc[2], and Sergio Rajsbaum[3]

[1] School of Computer Science, Carleton University, Ottawa, ON, Canada
[2] Department of Mathematics and Computer Science, Wesleyan University,
Middletown, Connecticut 06459, USA
[3] Instituto de Matemáticas, Universidad Nacional Autónoma de México (UNAM),
Ciudad Universitaria, D. F. 04510, Mexico

Abstract. Recent results on the problem of mobile agent rendezvous on distributed networks are surveyed with an emphasis on outlining the various approaches taken by researchers in the theoretical computer science community.

1 Introduction

Consider the following problem originally proposed by Alpern [1] (as quoted in [3]):

> Two astronauts land on a spherical body that is much larger than the detection radius (within which they can see each other). The body does not have fixed orientation in space, nor does it have an axis of rotation, so that no common notion of position or direction is available to the astronauts for coordination. Given unit walking speeds for both astronauts, how should they move about so as to minimize the expected meeting time T (before they come within the detection radius)?

This is just one version of a problem that has been studied under many guises under most of which it is referred to as *rendezvous*. In all settings, a set of agents are placed in a domain and are required to all meet at the same place and time within the domain, i.e., rendezvous. The settings differ mainly in the types and properties of the agents and the types and properties of domains. Besides the astronaut example above, rendezvous has been studied in settings as various as ships at sea, mother and child at a mall and autonomous robots on a hilly terrain.

Recently the theoretical computer science community has taken up the challenge of the problem of rendezvous for autonomous software agents moving through a distributed network. Requiring such agents to meet in order to synchronize, share information, divide up duties, etc. would seem to be a natural fundamental operation useful as a subroutine in more complicated applications such as web-crawling, peer-to-peer lookup, meeting scheduling, etc. In this paper, we provide a short (perhaps biased) survey of recent work done on this version of the problem. The research done in other settings is extensive and many of

P. Flocchini and L. Gąsieniec (Eds.): SIROCCO 2006, LNCS 4056, pp. 1–9, 2006.

the solutions can be applied here. But it is often the case that the models used and the concerns studied are sufficiently different as to require new approaches. Having said that, one ignores the earlier work at one's peril. For an excellent discussion of mainly continuous domains with randomized agents see the book by Alpern and Gal [3]. For work on robot rendezvous consider [26] and [28] as possible starting points.

2   The Model

The definition of rendezvous must begin with establishing the properties of the agents that will rendezvous and the domain in which rendezvous will occur. The model below captures the essence of what has been termed *the theory of mobile agent computing*.

2.1   Mobile Agents

We are interested in modeling a set of software entities that act more or less autonomously from their originator and have the ability to move from node to node in a distributed network maintaining some sort of state with the nodes of the network providing some amount of (possibly longterm) storage and computational support. Either explicitly or implicitly such a mobile (software) agent has most often been modeled using a finite automaton consisting of a set of states and a transition function. The transition function takes as input the agent's current state as well as possibly the state of the node it resides in and outputs a new agent state, possible modifications to the current node's state and a possible move to another node. In some instances we consider probabilistic automata which have available a source of randomness that is used as part of their input. Such agents are referred to as *randomized* agents.

An important property to consider is whether or not the agents are distinguishable, i.e., if they have distinct labels or identities. Agents without identities are referred to as *anonymous* agents. Anonymous agents are limited to running precisely the same program, i.e., they are identical finite automata. As the identity is assumed to be part of the starting state of the automaton, agents with identities have the potential to run different programs.

The knowledge the agent has about the network it is on and about the other agents can make a difference in the solvability and efficiency of rendezvous. For example, knowledge of the size of the network or its topology or the number of and identities of the other agents may be used as part of the program for rendezvous. If available to the agents, this information is assumed to be part of its starting state. (One could imagine situations where the information is made available by the nodes of the network and not necessarily encoded in the agent.)

Other properties that may be considered in mobile agent computing include whether or not the agents have the ability to "clone" themselves, whether or not they have the ability to "merge" upon meeting (sometimes referred to as "sticky" agents) or whether or not they can send self-generated messages. At this point, most of the research on rendezvous ignores these properties and they will not be discussed below.

2.2 Distributed Networks

The model of a distributed network is essentially inherited directly from the theory of distributed computing. We model the network by a graph whose vertices comprise the computing nodes and edges correspond to communication links.

The nodes of the network may or may not have distinct identities. In an *anonymous* network the nodes have no identities. In particular this means that an agent can not distinguish two nodes except perhaps by their degree. The outgoing edges of a node are usually thought of as distinguishable but an important distinction is made between a globally consistent edge-labelling versus a locally independent edge-labelling. A simple example is the case of a ring where clockwise and counterclockwise edges are marked consistently around the ring in one case, and the edges are arbitrarily - say by an adversary - marked 1 and 2 in the other case. If the labelling satisfies certain coding properties it is called a *sense of direction* [13]. Sense of direction has turned out to greatly effect the solvability and efficiency of solution of a number of problems in distributed computing and has been shown to be important in rendezvous as well.

Networks are also classified by how they deal with time. In a synchronous network there exists a global clock available to all nodes. This global clock is inherited by the agents. In particular it is usually assumed that in a single step an agent arrives at a node, performs some calculation, and exits the node and that all agents are performing these tasks "in sync". In an asynchronous network such a global clock is not available. The speed with which an agent computes or moves between nodes, while guaranteed to be finite, is not a priori determined.

Finally we have to consider the resources provided by the nodes to the agents. All nodes are assumed to provide enough space to store the agent temporarily and computing power for it to perform its tasks. (The case of malicious nodes refusing agents or even worse destroying agents - so-called *blackholes* - is also sometimes considered.) Beyond these basic services one considers nodes that might provide some form of long-term storage, i.e., state that is left behind when the agent leaves. In the rendezvous problem the idea of leaving an indistinguishable mark or *token* at a node (introduced in [5]) has been studied. More accommodating nodes might provide a *whiteboard* for agents to write messages to be left for themselves or for other agents.

3 The Rendezvous Problem

Given a particular agent model (e.g., deterministic, anonymous agents with knowledge they are on a ring of size n) and network model (e.g., anonymous, synchronous with tokens) a set of k agents distributed arbitrarily over the nodes of the network are said to *rendezvous* if after running their programs after some finite time they all occupy the same node of the network at the same time. It is generally assumed that two agents occupying the same node can recognize this fact (though in many instances this fact is not required for rendezvous to occur). As stated, rendezvous is assumed to occur at nodes. In some instances one considers the possibility of rendezvous on an edge, i.e., if both agents use the same edge (in opposite directions) at the same time. (For physical robots this makes sense. For software agents this perhaps is not so realistic but sometimes necessary to allow for the possibility of rendezvous at all - especially in instances where the network lacks a sense of direction.)

The first question one asks for an instance of rendezvous is whether or not it is solvable. There are many situations where it is not possible to rendezvous at all. This will depend upon both the properties of the agents (deterministic or randomized, anonymous or with identities, knowledge of the size of the network or not, etc.) and the network (synchronous or asynchronous, anonymous or with identities, tokens available or not, etc.). The solvability is also a function of the starting positions chosen for the agents. For example, if the agents start at the same node and can recognize this fact, rendezvous is possible in this instance. Given a situation where some starting positions are not solvable (i.e., rendezvous is not possible) but others are, we distinguish between algorithms that are guaranteed to finish for all starting positions, with successful rendezvous when possible but otherwise recognizing that rendezvous is impossible, versus algorithms that are only guaranteed to halt when rendezvous is possible. Algorithms of the former type are said to solve *rendezvous with detection*. (The distinction is perhaps analogous to Turing machines deciding versus accepting a language.)

For solvable instances of rendezvous one is interested in comparing the efficiency of different solutions. Much of the research focuses on the time required to rendezvous. In the synchronous setting the time is measured via the global clock. (In some situations, it makes a difference if the agents begin their rendezvous procedure at the same time or there is possible delay between start times.) In the asynchronous setting we adapt the standard time measures from the distributed computing model. Also of interest is the size of the program required by the agents to solve the problem. This is referred to as the memory requirement of the agents and is considered to be proportional to the base two logarithm of the number of states required by the finite state machine encoding the agent.

As is often the case, researchers are interested in examining the extremes in order to get an understanding of the limits a problem imposes. Over time it has become clear that for rendezvous symmetry (of the agents and the network) plays a central role in determining its solvability and the efficiency of its solutions. As such we divide our discussion below into the asymmetric and symmetric cases. For simplicity we restrict ourselves to the case of just two agents in most of the discussion below.

4 Asymmetric Rendezvous

Asymmetry in a rendezvous problem may arise from either the network or the agents.

4.1 Network Asymmetry

A network is asymmetric if it has one or more uniquely distinguishable vertices. A simple example is the case of a network where all of the nodes have unique identities chosen from a subset of some totally ordered set such as the integers. In this case, the node labelled with the smallest identity (for example) is unique and may be used as a meeting point for a rendezvous algorithm. Uniqueness need not be conferred using node labels. For example, in a network where there is a unique node of degree one, it may be used as a focal point.

If a "map" of the graph with an agent's starting position marked on it is available to the agents then the problem of rendezvous is easily solved by just traversing the path to an agreed upon unique node. Algorithms that use an agreed upon meeting place are referred to by Alpern and Gal [3] as FOCAL strategies. In the case where the graph is not available in advance but the agents know that a focal point exists (e.g., they know the nodes are uniquely labelled and therefore there exists a unique minimum label node) this strategy reduces to the problem of graph traversal or graph exploration whereby all of the nodes (sometimes edges) of the graph are to be visited by an agent. This has been extensively studied in a number of contexts, e.g., [9, 25]. Much of the work in this area has looked at improving the efficiency (time or memory) for restricted classes of graphs, e.g., trees [11]. A closely related problem is that of robot exploration of an unknown environment with obstacles which can often be modeled using graphs. See for example [6].

4.2   Agent Asymmetry

By agent asymmetry one generally means the agents have unique identities that allow them to act differently depending upon their values. In the simplest scenario of two agents, the agent with the smaller value could decide to wait at its starting position for the other agent to find it by exploring the graph as above. Alpern and Gal [3] refer to this as the Wait For Mommy (WFM) strategy and they show it to be optimal under certain conditions.

WFM depends upon the fact that the agents know in advance the identities associated with the other agents. In some situations this may be an unrealistic assumption. Yu and Yang [29] were the first to consider this problem. Under the assumption that the algorithm designer may assign the identities to the agents (as well as the existence of distinct whiteboards for each agent), they show that rendezvous may be achieved deterministically on a synchronous network in $O(nl)$ steps where n is the size of the network and l is the size of the identities assigned. The perhaps more interesting case where an adversary assigns the labels was first considered in [10]. Extensions to this work including showing rendezvous on an arbitrary graph is possible in time polynomial in n and l and that there exist graphs requiring $\Omega(n^2)$ time for rendezvous are described in [17, 18]. The case of an asynchronous network is considered in [8] where a (nonpolynomial) upper bound is set for rendezvous in arbitrary graphs (assuming the agents have an upper bound on the size of the graph). Improvements (in some cases optimal) for the case of the ring network are discussed in each of the above papers.

5   Symmetric Rendezvous

In the case of symmetric rendezvous, both the (generally synchronous) network and the agents are assumed to be anonymous. Further one considers classes of networks that in the worst case contain highly symmetric networks that do not submit to a FOCAL strategy. As might be expected some mechanism is required to break symmetry in order for rendezvous to be possible. The use of randomization and of tokens to break symmetry have both been studied extensively.

5.1   Randomized Rendezvous

Many authors have observed that rendezvous may be solved by anonymous agents on an anonymous network by having the agents perform a random walk. The expected time to rendezvous is then a (polynomial) function of the (size of the) network and is directly related to the cover time of the network. (See [24] for definitions relating to random walks.)

For example, it is straightforward to show that two agents performing a symmetric random walk on ring of size n will rendezvous within expected $O(n^2)$ time. This expected time can be improved by considering the following strategy (for a ring with sense of direction). Repeat the following until rendezvous is achieved: flip a (fair) coin and walk $n/2$ steps to the right if the result is heads, $n/2$ steps to the left if the result is tails. If the two agents choose different directions (which they do with probability $1/2$) then they will rendezvous (at least on an edge if not at a node). It is easy to see that expected time until rendezvous is $O(n)$. Alpern refers to this strategy as Coin Half Tour and studies it in detail in [2]. Note that the agents are required to count up to n and thus seem to require $O(\log n)$ bits of memory to perform this algorithm (whereas the straightforward random walk requires only a constant number of states to implement). It can be shown that $O(\log \log n)$ bits are sufficient for achieving linear expected rendezvous time[23].

5.2   Rendezvous Using Tokens

The idea of using tokens or marks to break symmetry for rendezvous was first suggested in [5] and expanded upon for the case of the ring in [27]. The first observation to make is that rendezvous is impossible for deterministic agents with tokens (or whiteboards) on an even size ring when the agents start at distance $n/2$ as the agents will remain in symmetric positions indefinitely. However, this is the only starting position for the agents for which rendezvous is impossible. This leads one to consider algorithms for rendezvous with detection where rendezvous is achieved when possible and otherwise the agents detect they are in an impossible to rendezvous situation. In this case, a simple algorithm suffices (described here for the oriented case). Each agent marks their starting position with a token. They then travel once around the ring counting the distances between their tokens. If the two distances are the same, they halt declaring rendezvous impossible. If they are different they agree to meet (for example) in the middle of the shorter side.

Again, one observes that the algorithm as stated requires $O(\log n)$ bits of memory for each agent in order to keep track of the distances. Interestingly enough this can be reduced to $O(\log \log n)$ bits and this can be shown to be tight for unidirectional algorithms [20]. If we are allowed two movable tokens (i.e., the indistinguishable marks can be erased and written with up to two marks stored per node) then rendezvous with detection becomes possible with an agent with constant size memory [21].

Multi-agent rendezvous, i.e., more than two agents, on the ring is considered in [14, 16], the second reference establishing optimal memory bounds for the problem. Two agent rendezvous on the torus is studied in [21] where tradeoffs between memory and the number of (movable) tokens used are given. Finally [15] considers a model in which tokens may disappear or fail over time.

5.3 Whiteboards and Blackholes

The use of whiteboards to achieve multiagent rendezvous on an arbitrary anonymous asynchronous network is studied in [4] where the problem is shown to be equivalent to that of leader election. In [12] rendezvous on an asynchronous ring with whiteboards in the presence of blackholes is considered.

6 Related Problems

As was mentioned above the rendezvous problem is intimately related under certain conditions to that of graph exploration [9, 25]. Besides this connection there are a number of "search" problems that are at least tangentially related to rendezvous in as much as techniques and/or models considered by researchers on these problems may be of some use to those working on rendezvous. Some examples are described below.

Imagine that the police have cornered a fugitive in a complicated warren of tunnels and rooms. They know that she is situated at one of the nodes and would like to capture her by searching the rooms. Their main concern is to decide what is the least number of police personnel necessary in order to be certain the fugitive does not evade capture. This number, referred to as the *search number* of a graph, was introduced by Megiddo et al. [22] and has since been studied by many authors.

Two friends check into an n room hotel. The hotel clerk is off duty and each doesn't know the room number of the other. They start calling rooms in order to find each other. What is the least number of calls they must make before at least one finds the other? This question was formalized and generalized as *mutual search* by Buhrman et al.[7]

A friend has given you the name but not the address of a restaurant in downtown Manhattan. You start searching for the restaurant by asking passersby which direction the restaurant is in. Their answers are sometimes erroneous, even contradictory. What is the best strategy to use in order to find the restaurant quickly? This problem was formalized in [19].

7   Conclusions

Rendezvous is a natural and fundamental problem for mobile software agents traveling through a distributed network. This makes it a natural choice as a model problem to study in the development of the theory of mobile agent computing much like leader election and consensus form model problems for the traditional study of the theory of distributed algorithms. We have attempted to bring together the current state of our knowledge of the study of rendezvous in the theoretical computer science context. A lot of work remains to be done.

Acknowledgements

Research of the first author was supported in part by NSERC (Natural Sciences and Engineering Research Council of Canada) and MITACS (Mathematics of Information Technology and Complex Systems) grants. Research of the third author was supported in part by PAPIIT-UNAM.

References

1. S. Alpern, Hide and Seek Games, Seminar, Institut für Höhere Studien, Wien, July 1976.
2. S. Alpern, The Rendezvous Search Problem, *SIAM Journal of Control and Optimization*, 33, pp. 673-683, 1995.
3. S. Alpern and S. Gal, *The Theory of Search Games and Rendezvous*, Kluwer Academic Publishers, Norwell, Massachusetts, 2003.
4. L. Barriere, P. Flocchini, P. Fraigniaud, and N. Santoro, Election and Rendezvous of Anonymous Mobile Agents in Anonymous Networks with Sense of Direction, *Proceedings of the 9th Sirocco*, pp. 17-32, 2003.
5. V. Baston and S. Gal, Rendezvous Search When Marks Are Left at the Starting Points, *Naval Research Logistics*, 47, No. 6, pp. 722-731, 2001.
6. A. Blum, P. Raghavan and B. Schieber, Navigating in Unfamiliar Geometric Terrain, *SIAM Journal on Computing* 26 (1997), 110-137.
7. H. Buhrman, M. Franklin, J. Garay, J. Hoepman, J. Tromp and P. Vitanyi, Mutual Search, *Journal of the ACM* 46 (1999), 517-536.
8. G. De Marco, L Gargano, E. Kranakis, D. Krizanc, A. Pelc and U. Vacaro, Asynchronous deterministic rendezvous in graphs, *Proc. 30th MFCS*, 2005, 271-282.
9. X. Deng and C. H. Papadimitriou, Exploring an Unknown Graph, *Journal of Graph Theory* 32 (1999), 265-297.
10. A. Dessmark, P. Fraigniaud, and A. Pelc, Deterministic Rendezvous in Graphs, *Proc. 11th ESA*, 184-195, 2003.
11. K. Diks, P. Fraigniaud, E. Kranakis, A. Pelc, Tree Exploration with Little Memory, *Journal of Algorithms*, 51 (2004) 38-63.
12. S. Dobrev, P. Flocchini, G. Prencipe, and N. Santoro, Multiple Agents Rendezvous in a Ring in spite of a Black Hole, *Proc. Symposium on Principles of Distributed Systems (OPODIS '03)*, LNCS 3144, pp. 34-46, 2004.
13. P. Flocchini, B. Mans and N. Santoro, Sense of direction: definition, properties and classes, *Networks* 32 (1998), 29-53.

14. P. Flocchini, E. Kranakis, D. Krizanc, N. Santoro, and C. Sawchuk, Multiple Mobile Agent Rendezvous in the Ring, *Proc. LATIN 2004*, LNCS 2976, pp. 599-608, 2004.
15. P. Flocchini, E. Kranakis, D. Krizanc, F. Luccio, N. Santoro and C. Sawchuk, Mobile Agent Rendezvous When Tokens Fail, *Proc. of 11th Sirocco*, 2004.
16. L. Gasieniec, E. Kranakis, D. Krizanc, X. Zhang, Optimal Memory Rendezvous of Anonymous Mobile Agents in a Uni-directional Ring, *Proc. of 32nd SOFSEM 2006*, to appear.
17. D. Kowalski and A. Pelc, Polynomial deterministic rendezvous in arbitrary graphs, *Proc. 15th ISAAC*, 2004.
18. D. Kowalski and A. Malinowski, How to meet in an anonymous network, *Proc. 13th Sirocco*, 2006.
19. E. Kranakis, and D. Krizanc, Searching with Uncertainty. *Proc. of 6th Sirocco*, 1999.
20. E. Kranakis, D. Krizanc, N. Santoro, and C. Sawchuk, Mobile Agent Rendezvous Search Problem in the Ring, *Proc. International Conference on Distributed Computing Systems (ICDCS)*, pp. 592-599, 2003.
21. E. Kranakis, D. Krizanc, E. Markou, Mobile Agent Rendezvous in a Synchronous Torus. *Proc. of LATIN*, 653-664, 2006.
22. N. Megiddo, S. Hakimi, M. Garey, D. Johnson and C. Papadimitriou. The Complexity of Searching a Graph, *Journal of the ACM* 35 (1988), 18-44.
23. P. Morin, personal communication.
24. R. Motwani and P. Raghavan, *Randomized Algorithms*, Cambridge University Press, New York, 1995.
25. P. Panaite and A. Pelc, Exploring Unknown Undirected Graphs, *Journal of Algorithms* 33 (1999), 281-295.
26. N. Roy and G. Dudek, Collaborative robot exploration and rendezvous: Algorithms, performance bounds and observations, *Autonomous Robots* 11 (2001), 117-136.
27. C. Sawchuk, *Mobile Agent Rendezvous in the Ring*, PhD thesis, Carleton University, School of Computer Science, Ottawa, Canada, 2004.
28. I. Suzuki and M. Yamashita, Distributed anonymous mobile robots: Formation of geometric patterns, *SIAM Journal of Computing* 28 (1999), 1347-1363.
29. X. Yu and M. Yung, Agent Rendezvous: A Dynamic Symmetry-Breaking Problem, *Proceedings of ICALP*, LNCS 1099, 610-621, 1996.

Adapting to Point Contention
with Long-Lived Safe Agreement
(Extended Abstract)

Hagit Attiya

Department of Computer Science
Technion
hagit@cs.technion.ac.il

Abstract. Algorithms with step complexity that depends only on the *point contention*—the number of simultaneously active processes—are very attractive for distributed systems with varying degree of concurrency. Designing shared-memory algorithms that adapt to point contention, using only read and write operations, is however, a challenging task.

The paper specifies the *long-lived safe agreement* object, extending an object of Borowsky et al. [1], and describes an implementation whose step complexity is adaptive to point contention. Then, we illustrate how this object is used to solve other problems, like *renaming* and *information collection*, in an adaptive manner.

1   Introduction

In order to coordinate the actions of a distributed application, processes must obtain up-to-date information from each other. In a typical *wait-free* algorithm, which guarantees that a process completes an operation within a finite number of its own steps, information is collected by reading from an array indexed with process' identifiers. If a distributed algorithm is designed to accommodate a large number of processes, this scheme is an over-kill when only a few processes simultaneously participate in the algorithm: many entries are read from the array although they contain irrelevant information about processes not wishing to coordinate.

The best performance is achieved when the step complexity of an operation is a function only of its *point contention*, namely, the maximal number of processes *simultaneously* executing the algorithm concurrently with it. In this way, an operation is delayed only when many processes are simultaneously active. Note that an algorithm whose step step complexity is *adaptive to point contention* is necessarily wait-free.

For example, if an algorithm is adaptive to point contention, then the step complexity of operation *op* in Figure 1 is constant since at most three processes simultaneously participate at each point during its interval. This holds although a large number of processes are active throughout *op*, and many processes are simultaneously active just before it starts.

P. Flocchini and L. Gąsieniec (Eds.): SIROCCO 2006, LNCS 4056, pp. 10–23, 2006.
© Springer-Verlag Berlin Heidelberg 2006

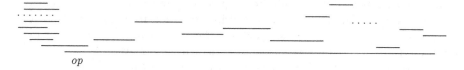

op

Fig. 1. An execution example

This article is devoted to explaining the design of algorithms that adapt to point contention using only read and write operations. Our exposition is based on an adaptive implementation of a long-lived safe agreement object.

The *safe agreement* object, originally defined by Borowsky et al. [1], allows processes to propose information and to agree on an identical value. In the BG simulation, the safe agreement object allows processes to agree on each step of a simulated process. Both the specification and the implementation of the object in [1] are neither wait-free (and hence, it is non-adaptive) nor long-lived.

We extend the specification to support adaptive and long-lived properties and present an implementation with $O(k)$ step complexity. Here and below, k denotes the point contention during an operation. This implementation is based on work by Attiya and Zach [2], which in turn follows ideas of Attiya and Fouren [3], as well as Afek et al. [4] and Inoue et al. [5].

The article also describes how safe agreement objects are used to solve the renaming problem and to implement two information collection objects, called *gather* and *collect*.

Renaming [6, 7] allows a process to obtain a *new name*—a positive integer, bounded by a function of the current number of active processes—and to release it afterwards. We adjust a renaming algorithm of Attiya and Fouren [3] to use long-lived safe agreement objects; in the algorithm, processes obtain names in a range of size $O(k^2)$.

Gather and collect objects allow processes to *store* information and to *retrieve* previously-stored information; they differ in the exact properties they provide. Gather and collect objects are commonly used as modular building blocks for efficient adaptive algorithms. We describe an implementation of a gather object, with $O(k^2)$ step complexity for storing information and $O(k)$ step complexity for retrieving information. A collect object is easily implemented from a gather object, with $O(k^2)$ step complexity for storing and retrieving information. The algorithms and their presentation combine ideas from [2, 3, 8, 9].

2   Model of Computation

We use a standard asynchronous shared-memory model of computation, following, e.g., [10]. A system consists of n processes, $p_1, \ldots, p_n$, and a set of registers that are accessed by read and write operations.

An *event* is a computation step by a single process; in an event, a process determines the operation to perform according to its local state, and determines its next local state according to the value returned by the operation.

An *execution* is a (finite or infinite) sequence of events; each event occurs at one process, which applies a read or a write operation to a single register and changes its state according to its algorithm. We assume an *asynchronous* model of computation, where process steps are arbitrarily interleaved.

An invocation of a high-level operation by a process causes the execution of the appropriate algorithm. The *execution interval* of an operation op_i by process p_i is the subsequence of the execution between the first event of p_i in op_i and the last event of p_i in op_i. We denote $op_i \rightarrow op_j$ if the execution interval of op_i precedes the execution interval of the operation op_j by process p_j; namely, the last event of p_i in op_i appears before the first event of p_j in op_j. If neither $op_i \rightarrow op_j$ nor $op_j \rightarrow op_i$, we say that op_i and op_j are *overlapping*.

Let α' be a finite prefix of an execution α; process p_i is *active* at the end of α' if α' includes an invocation of an operation op by p_i without the matching return.

The *point contention* at the end of α', denoted $pointCont(\alpha')$, is the number of active processes at the end of α'. Consider a finite interval β of an execution α; we can write $\alpha = \alpha_1 \beta \alpha_2$. The point contention during β is the maximum contention in prefixes $\alpha_1 \beta'$ of $\alpha_1 \beta$. We abuse notation and denote it by $pointCont(\beta)$, as well; that is

$$pointCont(\beta) = \max\{pointCont(\alpha_1 \beta') : \alpha_1 \beta' \text{ is a prefix of } \alpha_1 \beta \} \ .$$

If $pointCont(\beta) = k$, then k processes are simultaneously active at some point during β.

Another measure of concurrency is the *interval contention* during an interval β, denoted $intCont(\beta)$, counting the number of distinct processes that are active at some point during β. Clearly, $pointCont(\beta) \leq intCont(\beta)$.

3   The Adaptive Safe Agreement Object

We start by specifying and implementing the *one-shot* safe agreement object and later extend it to be long-lived.

The original *safe agreement* object [1] allows processes to propose values and to agree on a *single* value; in our safe agreement object, processes agree on an identical *set* of values. Clearly, a single value can be deduced from this set by choosing some predefined value, e.g., the minimum.

In the original specification, a process may wait indefinitely until the agreement value is decided, making it impossible to have a wait-free implementation. This means that there is no adaptive implementation, either. To admit adaptive implementations, we decouple the proposal of a value from the reading of the agreed set, and allow the reading to return an empty set, indicating that decision was not reached yet.

3.1   The One-Shot Object

The one-shot object provides two operations, propose and read. A process invokes propose at most once, but can invoke read to query the object several times.

A propose(*info*) operation tries to store *info* into the object; if it succeeds, it returns **true**, otherwise it returns **false**. A read operation returns a set of values, possibly *empty*.

A one-shot safe agreement object must provide the following properties:

Validity: Any value returned was previously proposed.
Agreement: All non-empty return sets are identical.

A process p_i *accesses* safe agreement object, if it calls a propose operation. A process p_i is *inside* a safe agreement object, if it gets **true** from a propose operation. A process is a *candidate* if its proposed value appears in the non-empty return set of values.

For example, assume processes p_0, p_1, p_2 and p_3 access a safe agreement object (we ignore the specific values proposed), and that p_3 returned **false**, while all other processes returned **true**, and thus, are inside the object. In later read operations, p_0 returns $\emptyset$, while p_1, p_2 and p_3 return $\{p_0, p_1, p_2\}$. In this case, the candidates are p_0, p_1, and p_2.

The *liveness* property, defined next, implies that at least one of the processes that get inside the safe agreement object is guaranteed to recognize itself as a candidate. These processes are called *winners*, and are necessarily also candidates. In the above example, the winners are p_1 and p_2, since they return a non-empty set containing themselves.

Liveness: In any execution, at least one of the processes accessing the object becomes a winner.
Concurrency: If a process gets inside a safe agreement object, and does not win, then some other process is inside the object concurrently.

The adaptive implementation of a one-shot safe agreement object is based on the *sieve* object of Attiya and Fouren [3]. We use the following data structures:

- A Boolean variable *inside*, initially **false**, indicates whether some process is already inside the object.
- An array $R[1 \ldots n]$ of views; all views are initially empty. $R[p_i]$ contains the view obtained by process p_i in this object.

We also associate a procedure for one-shot *atomic snapshot* [11] with the object, called osSnap. A process calls the procedure with *info* and obtains a *view*, namely a set of process id's and their information, which must include the process itself. The views returned by the procedure must be *comparable* by containment.

The pseudo-code appears in Algorithm 1.

In a propose operation, a process first checks if it is among the first processes to propose values. It succeeds only if the safe agreement object is empty, and *inside* is **false**. In this case, the process indicates that the safe agreement object is no longer empty, and then obtains a snapshot view, which it stores in its entry in R. The process returns with **true** from propose. If the process does not get

---

Algorithm 1. Adaptive one-shot safe agreement object: code for process p_i

---

data types:
   view : vector of $\langle id, info \rangle$
shared variables:
   inside : Boolean, initially **false**
   $R[1 \ldots n]$: array of views, initially $\emptyset$
local variables:
   V, W : view

Boolean procedure propose(*info*)
1: if (**not** *inside*) then
2:    *inside* = **true**   // notify that a process is inside the object
3:    $V = \mathsf{osSnap}(info)$ // propose *info*; return view of $\langle id, info \rangle$
4:    $R[id_i] = V$ // save the obtained view
5:    return(**true**)   // object is open
6: else return(**false**)   // object is not open

view procedure read()   // returns the subset of proposed values
7: $V = R[id_i]$
8: $W = min\{R[id_j] \mid \langle id_j, * \rangle \in V$ and $R[id_j] \neq \emptyset\}$ // min by containment
9: if $\forall \langle id_j, * \rangle \in W, R[id_j] \supseteq W$ then
10:   return(W)
11: else return($\emptyset$)

---

inside the object (i.e., *inside* is **true**), then some concurrent process is already inside the object, and the process returns with **false** from propose.

In a read operation, process p_i tries to distinguish the minimal (by containment) snapshot view. This is done by considering the minimal view written by any process in the view obtained by p_i; if one of these processes has not written its view yet, due to the asynchrony of the system, then the read is inconclusive and returns an empty set. Otherwise, the process returns the minimal view it finds. As we shall see in the first lemma below, this suffices to guarantee that all processes return the same set.

The *validity* property of the safe agreement object trivially holds by the code (see [3]).

We argue that all non-empty sets are identical.

Lemma 1. *Algorithm 1. satisfies the* agreement *property.*

Proof. Let V be the minimal view (by containment) obtained in osSnap, by some process p_k. We argue that any non-empty view W returned in an invocation of read by some process p_i is equal to V. Assume, by way of contradiction, that $W \neq V$. Since W is non-empty, it was obtained in some invocation of osSnap; therefore, W and V are comparable. By the minimality of V, $V \subset W$.

Since $p_k \in V \subset W$, p_i checks $R[id_k]$. If $R[id_k]$ is empty then p_i clearly fails the test; otherwise, $R[id_k] = V$ and since $V \subset W$, p_i also fails the test. In both cases, p_i returns $\emptyset$, which is contradiction.   □

Next, we show that at least one of the processes accessing a safe agreement object becomes a winner, when it calls read after returning from propose.

Lemma 2. *Algorithm 1. satisfies the liveness property.*

Proof. Let W_c be the set of candidates in the safe agreement object. W_c is not empty since some processes access the object. Let p_i be the last process in W_c to write its view in propose; clearly, p_i calls read after all processes in W_c write their views in propose.

Let V_i be the view obtained by p_i from osSnap. Since W_c is not empty, some process p_c obtains W_c, that is, the minimal (by containment) view, from osSnap. Note that p_c and p_i could be the same process.

Since W_c and V_i are comparable and since W_c is minimal, $p_c \in W_c \subseteq V_i$. By assumption, p_c writes its view to $R[id_c]$ before p_i calls read. Therefore, p_i reads W_c from $R[id_c]$, and sets V_i to W_c, by its minimality.

By assumption, p_i calls read after every process $p_j \in V_i = W_c$ wrote its view V_j to $R[id_j]$. By the minimality of W_c, $V_i = W_c \subseteq V_j$, and therefore, p_i evaluates the if condition to hold and obtains W_c. Thus, p_i finds that it is a winner in the safe agreement object. □

A process that gets inside the safe agreement object first reads **false** from *inside* and then sets *inside* to **true**. This fact can be used to prove that processes are *simultaneously* inside a safe agreement object.

Lemma 3. *The intervals of all processes that get inside a safe agreement object are overlapping.*

The next lemma shows that if process p_i gets inside safe agreement object and does not win, then some process p_j with an overlapping interval is a candidate in the object.

Lemma 4. *Algorithm 1. satisfies the concurrency property.*

Sketch of proof. If p_i gets **true** from open then p_i writes **true** to *inside*. By Lemma 3, the intervals of all processes that write **true** to *inside* overlap. By Lemma 2, at least one of them is a winner (and hence, a candidate) the safe agreement object, and the lemma follows. □

To evaluate the step complexity of the algorithm, suppose that p_i accesses safe agreement object in interval β_i, and let k be the point contention during β_i.

If A is the set of processes that are inside the object throughout the execution (including p_i itself), then Lemma 3 implies that processes in A access the object concurrently, and hence, $|A| \leq pointCont(\beta_i) = k$. Thus, at most k processes invoke the one-shot snapshot operation, which dominates the step complexity of propose.

Attiya and Fouren [12] present a one-shot snapshot algorithm, whose step complexity is $O(K log K)$, where K is the total number of processes that ever invoke it (also known as the *total contention*). As we have just argued, in our case we have that $K = k$, the point contention, and thus, the step complexity of the one-shot snapshot algorithm is $O(k \log k)$.

Inoue et al. [5] observed that the one-shot snapshot can be replaced with a *partial atomic snapshot* that guarantees that if one or more processes access the object concurrently, at least one process obtains a snapshot (the others may obtain an empty view). They also present a partial atomic snapshot algorithm with $O(k)$ step complexity, implying that the step complexity of propose is $O(k)$.

Since the maximal size of a view is k, the step complexity of read is $O(k)$. Therefore, a process performs $O(k)$ operations in each invocation of the one-shot safe agreement procedures.

This yields the following theorem:

Theorem 1. *Algorithm 1. implements an adaptive one-shot safe agreement object with $O(k)$ step complexity, where k is the point contention during the operation's execution interval.*

3.2   The Long-Lived Object

The long-lived safe agreement object has an infinite number of *generations*. Its interface is similar to that of the one-shot object, except that a *generation number* is added as a parameter to the operations. The generation number is visible from outside the object, in order to simplify the specification of the long-lived object, and to facilitate applications that use the generation number, e.g., the timestamps algorithm of [3].

Specifically, a long-lived safe agreement object supports three operations: propose(*info*) tries to store information in the current generation; if it succeeds, it returns ⟨**true**,c⟩, otherwise, it returns ⟨**false**,c⟩. read(c) returns either a non-empty set of values or an empty set. release(c) leaves generation c and activates the next generation $c + 1$, if possible.

A generation starts when the first invocation of a propose operation that returns ⟨∗,c⟩ occurs, and ends when the last process that got inside this generation leaves the generation, by invoking release(c). I_c denotes the execution interval of generation c.

We assume that there is at most one invocation of propose by each process for each generation, which means that after calling propose that returns a generation number c, the process must call release(c), before calling propose again.

The properties of the one-shot object (validity, agreement, liveness and concurrency) must hold for every generation of the long-lived safe agreement object. We also require the following property:

Synchronization: The generation number is a non-decreasing counter and it is incremented by one during interval I_c. Moreover, processes get inside generation c only after all candidates leave the smaller generations and the generation number is set to c (see Figure 2).

A long-lived safe agreement object is implemented using an unbounded array of one-shot safe agreement objects, each corresponding to a single generation. An

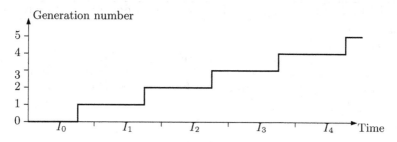

Fig. 2. Generation numbers are nondecreasing

integer variable, *count*, indicates the current generation and points to the current one-shot object. There is also an array $done[1 \ldots n][0 \ldots \infty]$ of Boolean variables, all initially **false**. The entry $done[p_i][c]$ indicates whether process p_i is done with the c'th one-shot safe agreement object. Boolean variables $allDone[0 \ldots \infty]$, all initially **false**, indicate whether all candidates of a specific generation are done.

The pseudo-code appears in Algorithm 2.

As in the one-shot object, a process proposes its value by checking if it is among the first processes to access the current generation of the safe agreement object. The process succeeds only if this generation is *empty*, that is, no other process is already inside it. If the process does not succeed to get inside the current generation, then some concurrent process is already inside the current generation.

The key idea in proving the correctness of the algorithm is that different generations of the same safe agreement object are accessed in disjoint intervals (this is the synchronization property of the object). This property allows correct handling of the non-decreasing counter, and implies that the long-lived safe agreement object inherits the properties of the one-shot object.

Process p_i is inside generation c of a long-lived safe agreement object since it gets **true** from a propose operation of generation c and until it leaves the generation. In other words, process p_i is inside generation c when it is inside the one-shot object $S[c]$ corresponding to generation c.

Careful inspection of the code, and agreement on the set of candidates show that a safe agreement object is released when all candidates are done.

Lemma 5. *If allDone[c] is set to* **true***, then all candidates of S[c] invoked* release*(c).*

This lemma is the key for showing, by induction, that a process is inside generation c only after all candidates leave smaller generations.

Lemma 6. *If process p_i is inside generation c of a safe agreement object, then all the candidates already called* release *of the smaller generations, $1, \ldots, c-1$.*

Sketch of proof. We concentrate on the induction step, assuming that the lemma holds for generation c. Only a winner p_j in generation c writes $c+1$ to *count*,

---

Algorithm 2. Adaptive long-lived safe agreement object: code for process p_i

---

shared variables:
  $count$: integer, initially 1
  $S[0 \ldots \infty]$: infinite array of one-shot safe agreement objects
  $done[1 \ldots n][0 \ldots \infty]$: two-dimensional array of Boolean variables, all initially **false**
  $allDone[0 \ldots \infty]$: infinite array of Boolean variables, all initially **false**
local variables:
  V, W : view

$\langle Boolean, integer \rangle$ procedure propose($info$)
1: if ($all\text{-}done[count - 1]$ and   // all candidates of the previous generation are done
              $S[count]$.propose($info$)) // and no process is inside the current generation
2:     return($\langle$**true**, $count\rangle$)   // generation $count$ is open
3: else release($count$)   // generation c is not open
4:     return($\langle$**false**, $count\rangle$)

view procedure read(c)   // returns the candidates of generation c
5: return ($S[c]$.read())

void procedure release(c)   // release generation c
6: $done[id_i][c] = $**true**   // indicate that p_i is done in generation c
7: $W = S[c]$.read()   // re-calculate the set of candidates
8: if ($p_i \in W$) then $count = c+1$ // p_i is a winner in generation c
9: if ($W \neq \emptyset$ and $\forall p_j \in W, done[p_j] == $**true**) then   // all candidates are done
10:    $allDone[c] = $**true**

---

implying that p_j is inside generation c. By the induction hypothesis, all candidates already called release of generation $1, \ldots, c-1$. Since p_i is inside generation $c + 1$ it reads **true** from $S[c].allDone$. Lemma 5 implies that all candidates of the one-shot object $S[c]$ already called release on this object.   □

Recall that I_c is the execution interval of generation c, starting with the first invocation of a propose operation in generation c and ending when the last process that got inside generation c leaves. The next lemma proves that the data structures of different generations of the same long-lived safe agreement object are modified in disjoint intervals.

Lemma 7. *For every $c \geq 1$, I_c starts after all previous intervals $I_1, \ldots, I_{c-1}$ end.*

Sketch of proof. We concentrate on the induction step, assuming that the lemma holds for for I_c. By definition, I_{c+1} starts with first invocation of a propose operation in generation $c + 1$. Lemma 6 implies that if this invocation occurs, then all the candidates already called release of the smaller generations. In particular, all candidates already called release of generation c, and by the induction hypothesis all previous intervals $I_1, \ldots, I_{c-1}$ end.

Since all candidates of generation(c) already called release, I_c also ended, implying that I_{c+1} starts after all previous intervals end.   □

The next lemma shows that the generation number is non-decreasing value, and it is incremented by one during interval I_c.

Lemma 8. *Algorithm 2. satisfies the* synchronization *property.*

Sketch of proof. By the code, only a winner p_w of generation c of a long-lived agreement object writes $c + 1$ to *count*. Since p_w is a winner, it previously writes **true** to *inside* of generation c, after I_c begins. Hence, p_w writes $c + 1$ to *count* after I_c begins, and before it leaves generation c. Lemma 5 implies that a one-shot safe agreement object is released after all candidates leave the object, and hence, p_w writes $c + 1$ to *count* before I_c ends.

Lemma 7 implies that processes get inside generation c only during interval I_c, after *count* is set to c. □

Thus, distinct one-shot objects, associated with different generations, are accessed in disjoint intervals, and the step complexity of accessing the object is dominated by the step complexity of the one-shot object.

Theorem 2. *Algorithm 2. implements an adaptive long-lived safe agreement object with $O(k)$ step complexity, where k is the point contention during the operation's execution interval.*

4   Application I: Renaming

In the (long-lived) renaming problem [6, 7] processes repeatedly *get* and *release* names from a small range. Ideally, the size of the name range should depend only on the current contention.

Long-lived safe agreement objects can be used in a simple renaming algorithm: Place n long-lived safe agreement objects, $S_{LL}[1], \ldots, S_{LL}[n]$, in a row. To get a name, a process p_i accesses the objects sequentially, applying **propose** to safe agreement object s; if it obtains $\langle$**true**,c$\rangle$, then p_i invokes **read**(c) to get the set of candidates W. Process p_i returns as its new name the pair composed of s and its rank in W. If p_i obtains $\langle$**false**,c$\rangle$, then it releases $S_{LL}[s]$ and continues to the next object, $S_{LL}[s + 1]$. In the latter case, we say that p_i *skips* the s'th safe agreement object.

To release a name, process p_i calls **release** on the safe agreement object where it obtained a name.

The concurrency properties of the long-lived safe agreement object can easily be used to prove that a process wins within k' iterations, where k' is the interval contention while p_i is getting a name.

It is more surprising—and harder to prove!—that in fact, process p_i wins within $2k - 1$ iterations, where k is the point contention while p_i is getting a name. This is done using an interesting *potential* method, which considers two sets of *simultaneously* active processes and shows that at least one of them is increased by 1 when p_i skips a safe agreement object.

The sets are indexed with an integer number $s = 1, \ldots, n$. The first set, denoted A_s, contains all processes that *access* the ℓ'th safe agreement object,

$1 \leq \ell \leq s$. The second set, denote W_s, contains all processes that *win* the ℓ'th safe agreement object, $1 \leq \ell \leq s$. (A_s and W_s are not disjoint.)

The exact definitions of A_s and W_s depend on specific execution intervals, and are used in proving that if process p_i skips the s'th safe agreement object, then $|W_s| + |A_s| \geq s + 1$ (for appropriately chosen intervals). The detailed definitions and statements, as well as the proofs, appear in [3, Section 4].

Since W_s and A_s contain simultaneously active processes, and since p_i is not in W_s, we have that $|W_s| + |A_s| \leq (k-1) + k$. This implies that p_i skips at most $2k - 2$ safe agreement objects, and the first component of the new name is $\leq 2k - 1$. The second component of the new name is equal to the rank of the process in the obtained view, and thus it is $\leq k$. It follows that the size of the name space is in $O(k^2)$.

Moreover, the step complexity of getting a name is $O(k)$ times the step complexity of proposing and reading in a safe agreement object, since p_i accesses at most $2k - 1$ sieves. The step complexity of releasing a name is proportional to the step complexity of releasing a safe agreement object.

Since both complexities can be bound by $O(k)$ (see the previous section), it follows that the algorithm solves long-lived $O(k^2)$-renaming with $O(k^2)$ step complexity.

5   Application II: Collecting Information

We discuss two objects, gather and collect, which allow a process to store its value in a shared memory, or to retrieve the values stored in the shared memory.

The *gather* object provides two operations: a put operation stores its parameter in the object, while a gather operation returns the set of values stored in the object before or possibly during the operation. (This is called the *validity* property.)

The *collect* object provides a collect operation that inherits the validity property of the gather operation, and further guarantees the *regularity* property. Namely, later collect operations are at least as updated as previous ones.

5.1   Implementing a Gather Object

To implement a gather object, we again put n long-lived safe agreement objects, $S_{LL}[1], \ldots, S_{LL}[n]$, in a row. With each safe agreement object we associate a set containing the values of processes that were candidates in this object. In order to store its information, a process iteratively tries to win in the safe agreement objects.

In the first part of a put operation, a process goes through the row of safe agreement objects, until it wins one of them. If it wins safe agreement object $S_{LL}[s]$, the process merges the values of all candidates of the current generation to the set associated with $S_{LL}[s]$. The synchronization and agreement properties of the safe agreement object guarantee that this set contains the values of all candidates in all generations of $S_{LL}[s]$.

In principle, in a **gather** operation, the process merely has to go through the safe agreement objects and read the associated sets of values. This might cause a problem, however, if a gather operation is performed after store operations with high contention: A put operation encountering high contention may store a new value in an entry with large index, and a subsequent **gather** operation will have to access many entries, even if its point contention is low.

This is solved by a technique called *bubble-up*, in which processes propagate new results to the beginning of the row. This allows a **gather** operation to find the new values at the beginning of the row, when contention is low. Bubble-up was presented by Afek et al. [8] and later used in [9, 3, 2].

After storing its value, the **put** operation bubbles the result to the top of the array. The process goes from entry s up to entry 1, and for each entry $s' = s, \ldots, 1$, it recursively gathers the values stored beyond entry s', and stores the result in its private register associated with entry s'. (See the description by Stupp [13, Section 7.1].)

In a **gather** operation, a process starts from the beginning of the row and reads until it find a value that was bubbled up by one of the processes. The number of entries accessed is linear in the point contention during the operation: if the contention is low, then the process reads the value near the beginning of the row; if the contention is high, then the process is allowed to go further in the row.

The key to showing that the gather object is correctly implemented is to show that the result array contains the most recent values of all candidates of all generations of the corresponding safe agreement object. The synchronization property of the safe agreement object is used to show that this property holds, although several processes (winners of the same safe agreement object) may concurrently write to the result entry. (See [3, Lemma 6.1].)

Next we explain how bubbling up restricts the step complexity of the **gather** operation. Assume that a **gather** operation by process p_i reads *skips* over entry s. The key to the complexity analysis is to show that a process skips an entry only if some concurrent process is bubbling up through this entry. By careful inspection of the intervals, this is used to show that p_i skips at most $3k$ entries, where k is the point contention.

The same potential argument used for renaming can be used to show that in its first part, a **put** operation accesses at most $2k - 1$ safe agreement objects. Together with the linear step complexity (in point contention) of operations on a safe agreement object, this implies that the step complexity of a **put** operation is $O(k^2)$.

5.2   Implementing a Collect Object

When a **gather** operation returns a value v for process p, it is still possible that a later **gather** operation returns a value that was written by process p *before* v. This can happen, for example, when the two **gather** operations are concurrent with the update operation of process p.

The regularity property of the collect object disallows this behavior. In order to guarantee this property, the **collect** procedure first calls **gather** and then calls put to store the result it has obtained. This increases the step complexity of a collect operation to be $O(k^2)$.

6   Summary

We tried to provide a closer look at the algorithmic ideas that are employed in order to have step complexity that adapts to point contention.

This article is not a comprehensive survey of the recent research on adaptive algorithms. The issues not covered here include the space complexity of adaptive algorithms [14,15], using stronger base objects [16,17], guaranteeing weaker types of adaptivity [12,18,19], and adaptive algorithms for other problems, e.g., mutual exclusion [20, 21, 22].

References

1. Borowsky, E., Gafni, E., Lynch, N., Rajsbaum, S.: The BG distributed simulation algorithm. Distributed Computing **14**(3) (2001) 127–146
2. Attiya, H., Zach, I.: Fully adaptive algorithms for atomic and immediate snapshots. www.cs.technion.ac.il/~hagit/pubs/AZ03.pdf (2003)
3. Attiya, H., Fouren, A.: Algorithms adaptive to point contention. Journal of the ACM **50**(4) (2003) 444–468
4. Afek, Y., Attiya, H., Fouren, A., Stupp, G., Touitou, D.: Adaptive long-lived renaming using bounded memory. www.cs.technion.ac.il/~hagit/pubs/AAFST99disc.ps.gz (1999)
5. Inoue, M., Umetani, S., Masuzawa, T., Fujiwara, H.: Adaptive long-lived $O(k^2)$-renaming with $O(k^2)$ steps. In: Proceedings of the 15th International Conference on Distributed Computing, Berlin, Springer-Verlag (2001) 123–135
6. Attiya, H., Bar-Noy, A., Dolev, D., Peleg, D., Reischuk, R.: Renaming in an asynchronous environment. Journal of the ACM **37**(3) (1990) 524–548
7. Moir, M., Anderson, J.H.: Wait-free algorithms for fast, long-lived renaming. Science of Computer Programming **25**(1) (1995) 1–39
8. Afek, Y., Stupp, G., Touitou, D.: Long-lived and adaptive collect with applications. In: Proceedings of the 40th IEEE Symposium on Foundations of Computer Science, Phoenix, IEEE Computer Society Press (1999) 262–272
9. Afek, Y., Stupp, G., Touitou, D.: Long-lived and adaptive atomic snapshot and immediate snapshot. In: Proceedings of the 19th Annual ACM Symposium on Principles of Distributed Computing, New-York, ACM Press (2000) 71–80
10. Herlihy, M.: Wait-free synchronization. ACM Transactions on Programming Languages and Systems **13**(1) (1991) 124–149
11. Afek, Y., Attiya, H., Dolev, D., Gafni, E., Merritt, M., Shavit, N.: Atomic snapshots of shared memory. Journal of the ACM **40**(4) (1993) 873–890
12. Attiya, H., Fouren, A.: Adaptive and efficient algorithms for lattice agreement and renaming. SIAM Journal on Computing **31**(2) (2001) 642–664
13. Stupp, G.: Long Lived and Adaptive Shared Memory Implementations. PhD thesis, Department of Computer Science, Tel-Aviv University (2001)
14. Afek, Y., Boxer, P., Touitou, D.: Bounds on the shared memory requirements for long-lived adaptive objects. In: Proceedings of the 19th Annual ACM Symposium on Principles of Distributed Computing, New-York, ACM Press (2000) 81–89

15. Attiya, H., Fich, F., Kaplan, Y.: Lower bounds for adaptive collect and related problems. In: Proceedings of the 23rd Annual ACM Symposium on Principles of Distributed Computing. (2004) 60–69
16. Afek, Y., Dauber, D., Touitou, D.: Wait-free made fast. In: Proceedings of the 27th ACM Symposium on Theory of Computing, New-York, ACM Press (1995) 538–547
17. Herlihy, M., Luchangco, V., Moir, M.: Space- and time-adaptive non-blocking algorithms. In: Electronic Notes in Theoretical Computer Science. Volume 78., Elsevier (2003)
18. Afek, Y., Stupp, G., Touitou, D.: Long-lived adaptive splitter and applications. Distributed Computing 15(2) (2002) 67–86
19. Attiya, H., Fouren, A., Gafni, E.: An adaptive collect algorithm with applications. Distributed Computing 15(2) (2002) 87–96
20. Anderson, J., Kim, Y.J.: Adaptive mutual exclusion with local spinning. In: Proceedings of the 14th International Conference on Distributed Computing. (2000)
21. Attiya, H., Bortnikov, V.: Adaptive and efficient mutual exclusion. Distributed Computing 15(3) (2002) 177–189
22. Choy, M., Singh, A.K.: Adaptive solutions to the mutual exclusion problem. Distributed Computing 8(1) (1994) 1–17

Sensor Networks: Distributed Algorithms Reloaded – or Revolutions?

Roger Wattenhofer

Computer Engineering and Networks Laboratory
ETH Zurich, 8092 Zurich, Switzerland
wattenhofer@tik.ee.ethz.ch

Abstract. This paper wants to motivate the distributed algorithms community to study sensor networks. We discuss why sensor networks are distributed algorithms, and why they are not.

1 Introduction

Wireless sensor networks currently exhibit an incredible research momentum. Computer scientists and engineers from all flavors are embracing the area. Sensor networks are adopted by researchers from hardware technology to operating systems, from antenna design to middleware, from graph theory to computational geometry. Information and communication theorists study fundamental scaling laws such as the capacity of a sensor network. Networking researchers propose new protocols for all layers of the stack. And for the database community, a sensor network essentially is – a database.

The distributed algorithms community should join this big interdisciplinary party! Distributed algorithms are central since – in a first approximation – a sensor network can be modeled as a message passing graph. Hence there is hope that distributed algorithms can be either directly used for or at least adapted to sensor networks.

In the last twenty years, distributed network algorithms have been a thriving theoretical research subject. So far however with limited influence on practice. Sensor networks may be a foremost application area of this vivid theory. Unlike other natural application areas such as the Internet or peer-to-peer/overlay networks, sensor networks are less prone to side effects such as selfish behavior of individual nodes, as generally the whole network is owned by a single entity.[1]

So, can we directly apply our distributed algorithms instruments when developing algorithms for sensor networks? In other words, are sensor networks nothing but distributed algorithms *reloaded*?! In this paper we study to what extent the wireless nature of sensor networks is changing the game. We identify and briefly discuss two modeling aspects for which we believe that sensor networks are fundamentally different from orthodox distributed algorithms.

---

[1] Interestingly, the other camp of the distributed computing community which deals less with loosely-coupled networks and more with tightly-coupled multiprocessors (a.k.a. shared memory systems) is currently experiencing a similar impetus from the application domain with forthcoming multicore architectures.

P. Flocchini and L. Gąsieniec (Eds.): SIROCCO 2006, LNCS 4056, pp. 24–28, 2006.

First, we need a model which reflects a typical topology of a sensor network. Traditionally, a sensor network is modeled as a *graph*, representing nodes by vertices and wireless links by edges. Geometry comes into play as the distribution of nodes in space, and the propagation range of wireless links, usually adhere to geometric constraints. Several models inspired by both graph theory and geometry are possible; what model is right depends on the question analyzed. A media access study might need a detailed model capturing several low-level aspects. For instance, it has to be taken into account that a message may not be received correctly due to a near-by concurrent transmission. Hence, it is crucial that the model appropriately incorporates interference aspects. However, for a transport layer study, a much simpler model which assumes random transmission errors might be sufficient. In a recent survey [12], a whole zoo of models borrowing from both graph theory and geometry is presented, comprising classic models such as the unit disk graph or the signal-to-interference-plus-noise ratio, but also novel generalizations such as the bounded independence graph or the unit ball graph. These geometric graph models will probably influence the research on distributed network algorithms. For details, we refer the reader to [12].

Second, the very definition of a distributed algorithm is about to change when entering the sensor network domain. We believe that new algorithm types will emerge, and will influence the distributed algorithms community in the coming years. In the remainder of the paper, we briefly discuss possible directions of research.

2 Algorithms

The distributed algorithms community has never been shy of models. We study message passing and shared memory systems, synchronous and asynchronous algorithms, Byzantine and selfish nodes, self-stabilization and failure-detection, to only name a few of the most typical modeling facets. In fact, what is (im)possible and/or (in)efficient *in which model* of distributed computation often outranks the importance of solving this or that problem in a specific model. Still, when it comes to sensor networks it seems that our abundance of models is not enough.

Most algorithms for sensor networks proposed in literature are meant to be *executed by the sensor nodes* during the system's operation. For example, when a node receives a message, it performs some (simple and local) computation, and—depending on the computation's results—sends a new message to its neighbors. A node a priori only knows its own state. In order to learn more about the other nodes in the network, it has to communicate with its neighbors. By collaboration of the nodes, global operations such as (multi-hop) routing can be achieved. Since the activity is distributed among the nodes, these algorithms are called *distributed algorithms* [10]. Distributed algorithms raise many interesting research questions. For example: What can be computed in a distributed fashion, and what not? How efficient is a distributed algorithm compared to a corresponding global algorithm?

Every (global) algorithm can easily be turned into a distributed algorithm: Simply centrally collect the distributed state, compute a global solution, and distribute this solution. However, this simple routine is often unreasonably pricey. Since sending and receiving messages are expensive operations in wireless networks (e.g., medium access control, energy consumption), a reasonable distributed algorithm should minimize communication. This motivates the introduction of *localized algorithms* [5, 13].[2] A localized algorithm is a special case of a distributed algorithm.

Model 1 (Localized Algorithms). *In a k-localized algorithm, for some parameter k, each node is allowed to communicate at most k times with its neighbors. A node can decide to retard its right to communicate; for example, a node can wait to send messages until all its neighbors having larger identifiers have reached a certain state of their execution.*

In spite of the restricted communication model, localized algorithms can be slow. A node u might have to wait for a neighbor v, while node v in turn has to wait for its neighbor w, etc. Thus, as a matter of fact there can be a *linear chain of causality*, with only one node being active at any time. This yields a worst-case execution time of $\Theta(n)$, where n is the number of nodes.[3] If we do not want this linear running time, we need to resort to another model [8, 10].

Model 2 (Local Algorithms). *In a k-local algorithm, for some parameter k, each node can communicate at most k times with its neighbors. In contrast to k-localized algorithms nodes cannot delay their decisions. In particular, all nodes process k synchronized phases, and a node's operations in phase i may only depend on the information received during phases 1 to $i - 1$. The most efficient local algorithms are often randomized [7, 9]; that is, the number of rounds k can vary.*

Observe that in a k-local algorithm, nodes can only gather information about nodes in their k-neighborhood. In some local algorithms [7] the algorithm designer can choose an arbitrarily small constant k (at the cost of a lesser approximation ratio). This makes local algorithms particularly suited in scenarios where the nodes' environment changes frequently, as the algorithm can constantly adapt to the new circumstances. However, due to the synchronous phases, local algorithms may make greater demands on the media access sub-layer than localized algorithms. In particular, in unreliable wireless networks it seems to be costly to implement a media access control scheme that allows for synchronous rounds, as messages will be lost due to interference (conflicting concurrent transmissions) or mobility (even if the nodes themselves are not mobile, the environment is typically dynamic, temporarily enabling/disabling links).

Dealing with unreliability has always been a core interest of the distributed computing community. A powerful concept for coping with failures is *self-stabilization* [4]. Fortunately, using a simple trick [3], every local algorithm is immediately self-stabilizing. The trick works as follows (Section 4 of [3]): Every node keeps a log of every state transition it has taken until its current state; generally this boils down to memorizing the local variables of each step of the main loop. If each node constantly sends its current log to all neighbor nodes, each node can check and correct every transition it has made in the past. Assuming that all inputs are correct (variable initialization and random seeds are stored in the imperishable program memory, sensor information

---

[2] To the best of our knowledge nobody has ever bothered to formally define what a localized algorithm is. However, all papers we are aware of implicitly use a model similar to Model 1.

[3] And many localized algorithm do exhibit this linear worst-case.

can be re-checked) every fault due to memory or message corruption will be detected and corrected. For details we refer to [3].

Turning a k-local algorithm into a self-stabilizing algorithm with [3] blows up messages by a factor k (in the worst case); on the other hand we immediately get an algorithm which works on a sensor network as the hardest wireless problems (messages lost due to interference and mobility) are covered by the self-stabilization model. Also, in case of an error (such as a lost message), only the k-neighborhood of a node is affected.[4]

In practice, for some local algorithms the detour to self-stabilization may be costly, as the message overhead is prohibitive;[5] instead we need models that integrate interference. One solution is the so-called unstructured radio network model [1, 2, 6] where the algorithm designer has to implement her own medium access scheme from scratch.

Model 3 (Unstructured Radio Networks). *In the* unstructured radio network *model time is divided into slots. In each time slot, each node can decide whether to transmit, listen (or sleep). If two conflicting nodes transmit simultaneously, a potential receiver cannot decode any message. Nodes are distributed in an arbitrary (worst-case) multi-hop fashion, and may wake-up asynchronously (also worst-case).*

The unstructured radio network model may be classified further, for example depending to what extent collisions can be detected by a receiver. The unstructured radio network model seems to fit practice well, especially if teamed up with a sensible topology/interference model such as signal-to-interference-plus-noise ratio or bounded independence graph [12]. Clearly, the slotted-time assumption is a simplification, however as usual the difference between slotted and unslotted can easily be bounded [11].

Unfortunately, unstructured radio network algorithms tend to be quite technical, as even higher-layer algorithms need to specify media access. We believe that there is room for novel models with more coarse-grained assumptions how the media is accessed. One might for example imagine a model abstracting away from media access, where an adversary schedules transmissions. It seems that this model only makes sense if the adversary is restricted appropriately, that is, if there are fairness guarantees. For example, the adversary might have to schedule each node at least once every $\Theta(n)$ rounds. Moreover, one could imagine an adversary which delivers a message only to a subset of a node's neighbors, because the other neighbors experience collisions.

3   Conclusions

This paper has presented and compared a subjective selection of algorithmic models. For other modeling aspects, we refer to [12]. We want to emphasize that there is no optimal model, and that an engineer has to choose the model which reflects her needs best. Generally, we believe that for efficiency considerations, a slightly idealistic model can be fine. However, when it comes to issues such as correctness of an algorithm,

---

[4] In principle localized algorithms can also benefit from [3], however, errors are not restricted to a k-neighborhood but may propagate the whole network – we experience a troublesome *butterfly effect*.

[5] Currently the payload constant of a packet in TinyOS is 29 bytes.

it seems that a more pessimistic or conservative model should be preferred. In other words, a *robust* algorithm is also correct in a more general model than for which it has been studied or proven efficient.

Acknowledgments

We would like to thank Stefan Schmid and Thomas Moscibroda for valuable discussions.

References

1. N. Abramson. The ALOHA System. In *Computer-Communication Networks, Prentice Hall,* 1973.
2. N. Alon, A. Bar-Noy, N. Linial, and D. Peleg. A Lower Bound for Radio Broadcast. In *Journal of Computer and System Sciences*, 1991.
3. B. Awerbuch and G. Varghese. Distributed Program Checking: A Paradigm for Building Self-stabilizing Distributed Protocols. In *32nd Annual IEEE Symposium on Foundations of Computer Science (FOCS)*, 1991.
4. E. W. Dijkstra. Self-stabilizing Systems in Spite of Distributed Control. In *Communications of the ACM*, 1974.
5. D. Estrin, R. Govindan, J. S. Heidemann, and S. Kumar. Next Century Challenges: Scalable Coordination in Sensor Networks. In *Fifth Annual International Conference on Mobile Computing and Networking (MobiCom)*, 1999.
6. F. Kuhn, T. Moscibroda, and R. Wattenhofer. Initializing Newly Depoloyed Ad-hoc and Sensor Networks. In *10th Annual Intl. Conf. on Mobile Computing and Networking (MobiCom)*, 2004.
7. F. Kuhn, T. Moscibroda, and R. Wattenhofer. The Price of Being Near-Sighted. In *ACM-SIAM Symp. on Discrete Algorithms (SODA)*, 2006.
8. N. Linial. Distributive Graph Algorithms – Global Solutions from Local Data. In *28th Annual IEEE Symposium on Foundations of Computer Science (FOCS)*, 1987.
9. M. Luby. A Simple Parallel Algorithm for the Maximal Independent Set Problem. In *SIAM Journal on Computing*, 1986.
10. D. Peleg. *Distributed Computing: A Locality-sensitive Approach.* Society for Industrial and Applied Mathematics, Philadelphia, PA, USA, 2000.
11. L. G. Roberts. Aloha packet system with and without slots and capture. In *Computer Communication Review*, 1975.
12. S. Schmid and R. Wattenhofer. Algorithmic Models for Sensor Networks. In *14th International Workshop on Parallel and Distributed Real-Time Systems (WPDRTS), Island of Rhodes, Greece*, April 2006.
13. Y. Wang, X.-Y. Li, P.-J. Wan, and O. Frieder. Sparse Power Efficient Topology for Wireless Networks. *Journal of Parallel and Distributed Computing*, 2002.

Local Algorithms for Autonomous Robot Systems

Reuven Cohen[1] and David Peleg[2,*]

[1] Dept. of Electrical and Computer Eng., Boston University, Boston, MA, USA
`cohenr@bu.edu`
[2] Dept. of Computer Science, Weizmann Institute, Rehovot, Israel
`david.peleg@weizmann.ac.il`

Abstract. This paper studies local algorithms for autonomous robot systems, namely, algorithms that use only information of the positions of a bounded number of their nearest neighbors. The paper focuses on the spreading problem. It defines measures for the quality of spreading, presents a local algorithm for the one-dimensional spreading problem, prove its convergence to the equally spaced configuration and discusses its convergence rate in the synchronous and semi-synchronous settings. It then presents a local algorithm achieving the exact equally spaced configuration in finite time in the synchronous setting, and proves it is time optimal for local algorithms. Finally, the paper also proposes an algorithm for the two-dimensional case and presents simulation results of its effectiveness.

1 Introduction

1.1 Background and Motivation

Swarms of low cost robots provide an attractive alternative when facing various large-scale tasks in hazardous or hostile environments. Such systems can be made cheaper, more flexible and potentially resilient to malfunction. Indeed, interest in autonomous mobile robot systems arose in a large variety of contexts (see [3, 4, 11, 13, 14, 15, 16, 21] and the survey [5]).

Along with developments related to the physical engineering aspects of such robot systems, there have been recent research attempts geared at developing suitable algorithmics, particularly for handling the distributed coordination of multiple robots [2, 6, 17, 19, 20]. A number of computational models were proposed in the literature for multiple robot systems. In this paper we consider the model of [2, 19]. (An alternative, weaker, model is found in [6, 9, 18].) In this model, the robots are assumed to be identical and indistinguishable, lack means of communication, and operate in discrete cycles. Each robot may wake up at each of these cycles, and make a move according to the locations of the other robots in the environment. The moves are assumed to be instantaneous.

---

* Supported in part by a grant from the Israel Science Foundation.

P. Flocchini and L. Gąsieniec (Eds.): SIROCCO 2006, LNCS 4056, pp. 29–43, 2006.

The model also makes the assumption that the robots are *oblivious*, i.e., have no memory and can only act according to a calculation based only on the their last observation of the world. Oblivious algorithms have the advantage of being self-stabilizing (i.e., insensitive to transient errors and to the addition or removal of robots) and are also usually simple to design and implement.

Algorithms were developed in the literature for a variety of control problems for robot swarms [2, 6, 9, 11, 17, 18, 19, 20]. Most of those algorithms, however, require the robots to perform "global" calculations in each cycle, relying on the entire current configuration. For example, algorithms for the gathering problem, which requires the robots to gather at one point, typically instruct each robot to calculate the goal point to which it should move based on the exact locations of all the other robots in the current configuration. This calculation can be fairly simple and require only linear time (e.g., computing the average location of the robots), but in some cases involves a more complex computation, such as finding the smallest enclosing circle or the convex hull of the robots. Moreover, for very large swarms, and when each robot operates in fast short cycles, even a linear time computation per cycle may be too costly. A similar difficulty may arise when the task at hand involves coordinated movement in a certain direction while avoiding collisions, or evenly spreading the robot swarm in a given area.

The current paper addresses the issue of local and simple algorithms for control and coordination in robot swarms. It may be instructive to turn to the metaphor of insect swarms, and consider the way such problems are managed. It would appear that in a large swarm of bees, for instance, an individual bee would not calculate its next position based on the exact positions of all other bees. Rather, it is likely to decide its course of movement based on the positions and trajectories of a few nearby bees forming its immediate neighborhood. This local information is often sufficient to allow the individual insect to plan its movement so as to follow its swarm, avoid collisions and so on. Such policy would lead to a much "lighter" calculation, which can be carried out more frequently and effectively.

This analogy thus motivates the idea of exploring the behavior and performance of local, or "light-weight" algorithms for controlling a swarm of robots. Such algorithms have several obvious advantages over their more traditional "global" counterparts, in simplicity, computational complexity, energy consumption and stability. They also have the advantage of being applicable in a "limited visibility" model, where each robot sees only its close vicinity.

An enabling prerequisite is that each robot be equipped with a mechanism enabling it to efficiently obtain as input information about its locality, i.e., its immediate neighbors. (Clearly, if the robot's input device is designed so that its input consists a global picture of the configuration, and the robot's neighborhood can only be deciphered by going through this entire picture and performing complex calculations, then the robot might as well perform a global algorithm on its input). In particular, a robot equipped with a sonar may be able to first detect nearby objects. If the input to the robot is in the form of a visual image, then it may be necessary to have a scanning algorithm on the image, sweeping the image from the point of the robot outwards, thus hitting the nearby robots first. The scan can be terminated after identifying the immediate neighbors. Hereafter we will ignore this issue, and assume that a suitable input mechanism exists, efficiently providing the robot in each cycle with a concise and accurate description of its immediate neighborhood.

In this paper we explore the "light-weight" approach to robot swarm control through the concrete example of the *robot spreading* problem. In this problem, N robots are initially distributed in the plane, and the goal is to spread them "evenly" within the perimeters of a given region. To consider methods for spreading one should first define a criterion for the quality of spreading of a given robot distribution. For the one-dimensional case this is easy, as the best configuration is clearly the equal spacing arrangement on the line, and all other configurations can be compared to this one. Quantitative measures for the spreading quality are presented in Section 2. In higher dimensions even the definition of the spreading quality becomes more difficult, as different criteria may be devised. One may consider, to list but a few examples, the minimum distance over all robot pairs, the average distance between each robot and its nearest neighbor, or the time needed to reach the point most distant from any robot. The choice of a definition may also depend on the motivation for the application of the spreading algorithm. The definition we use here is the average over all robots of the minimum distance to the nearest "object",

$$d_{\mathrm{av}} = \frac{1}{N} \sum_i \min_{j \neq i} \{d_{i,j}\},$$

where the objects considered in taking the minimum are all other robots and all points of the perimeter of the region.

We discuss two timing models for the robot operations. In the $\mathcal{FSYNC}$ (synchronous) model, it is assumed that the robots operate in cycles, where all robots operate in each cycle, i.e. take a snapshot of their surroundings, run their algorithm on the snapshot, and move accordingly. In the $\mathcal{SSYNC}$ (semi-synchronous) model, the robots are again assumed to operate in cycles. However, not all robots are active in every cycle. The activation schedule is assumed to be determined by an adversary, but it is guaranteed that every robot is activated an infinite number of times.

When the robots share a common orientation and the model is synchronous the task of spreading via a *global* algorithm is quite easy. A simple solution can be obtained by agreeing on an ordering (such as top-to-bottom and then left-to-right) and deciding on each robot's final location accordingly. This solution can be applied even in asynchronous settings if the movements of each robot are restricted so as to preserve the ordering. When only local information is used, it is difficult to prevent the swarm from converging to a non-optimal configuration, which is locally optimal for almost all robots. This phenomenon is well known in physical systems, e.g., such "defects" are known to determine many of the crystal's properties, as well as in artificial intelligence search, where finding the optimal state is difficult when many local minima exist.

Another advantage of global algorithms over local ones is the convergence rate or finishing time of the algorithm. It is easily seen that an algorithm based on ordering in a synchronous setting, will achieve the final position in a single step. A local algorithm will require at least $O(N)$ steps for the information to reach each of the robots, as proven below in Section 3.

In this paper we present an oblivious local algorithm for spreading in one dimension. We prove its validity for the synchronous and semi-synchronous models and discuss its convergence rate. We also present a non-oblivious local algorithm for spreading in finite (and optimal) time, and discuss some of its shortcomings. We then describe a generalization of the first algorithm to two dimensions and present simulation results showing its behavior. We also discuss several other alternatives and their relative strengths and weaknesses.

A related problem was studied by Dijkstra in [8]. There, n units labeled $\{0, 1, \ldots, n-1\}$ are initially placed around a ring, and apply a rule whereby unit i moves to the middle point between units $i-1$ and $i+1$ (modulo n). It is shown therein that under this rule, the system might in certain cases oscillate and fail to converge, even in a synchronous setting. A discussion of a possible solution to this problem appears in [7].

1.2   The Model

The basic model studied in [2, 20] can be summarized as follows. The N robots execute a given algorithm in order to achieve a prespecified task. Time is divided to discrete events, where in each of these events each robot may or may not be active, under the condition that each robot is activated an infinite number of times. Whenever a robot is activated it takes a snapshot of its surroundings, which is used as the input for the algorithm executed by the robot. The output of the algorithm is a destination point, to which the robot moves instantaneously.

Following the common model in this area, the robots are assumed to be rather limited. To begin with, they have no means of directly communicating with each other. Moreover, they are assumed to be *oblivious* (or memoryless), namely, they cannot remember their previous states, their previous actions or the previous positions of the other robots. Hence the algorithm used in the robots' computations cannot rely on information from previous cycles, and its only input is the current configuration. As explained earlier, while this is admittedly an over-restrictive and unrealistic assumption, developing algorithms for the oblivious model has the advantages of self-stabilization (i.e., the ability to recover from any finite number of transient errors) and suitability to dynamical settings, where robots are added and removed during operation.

1.3   Preliminaries

We review some properties of discrete Fourier transforms to be used in our analysis later on.

The discrete sine transform: The discrete sine transform is appropriate for anti-symmetric functions or alternatively, functions fixed to zero at both edges, which can be made antisymmetric by an appropriate continuation.

Consider the $N-2$ vectors $\bar{v}_k$ of dimension $N-2$ defined componentwise by

$$(v_k)_i = \sqrt{\frac{2}{N-1}} \sin ki, \quad i = 1, 2, \ldots, N-2,$$

for k values from the range $K = \{k = \frac{m\pi}{N-1} \mid m = 1, \ldots, N-2\}$. We make use of the following known lemmas (proofs can be found in standard books on Fourier series or digital signal processing; see, e.g., [1]).

Lemma 1. *For $k, q \in K$, we have $\bar{v}_k \cdot \bar{v}_q = \delta_{k,q}$, where δ stands for Kroneker's delta, i.e., $\delta_{k,q} = 1$ if $k = q$ and 0 otherwise.*

Corollary 1. *The $N-2$ vectors $\bar{v}_k$ form an orthonormal basis to $\mathcal{R}^{N-2}$.*

For a sequence of reals η_i, $i = 0, \cdots, N-1$, satisfying $\eta_0 = \eta_{N-1} = 0$, and for $k \in K$, one can define the discrete Fourier transform of $\bar{\eta}$ as

$$\mu_k = \sqrt{\frac{2}{N-1}} \sum_{i=0}^{N-1} \eta_i \sin ki. \tag{1}$$

The inverse transform is given by the following lemma.

Lemma 2. $\eta_j = \sqrt{\dfrac{2}{N-1}} \displaystyle\sum_{k=\frac{\pi}{N-1}}^{\frac{(N-2)\pi}{N-1}} \mu_k \sin kj.$

The discrete cosine transform: The discrete cosine transform is widely used in digital signal processing. The version presented below is appropriate for symmetric functions, having a symmetry axis at $x = -1/2$.

Define the functions $f(i, k) = A(i, k) \cos \dfrac{k\hat{i}\pi}{m}$, for $i, k \in \{0, 1, \ldots, 2m-1\}$, where hereafter $\hat{i} = i + \frac{1}{2}$, and

$$A(i, k) = \begin{cases} \frac{1}{\sqrt{m}}, & k = 0, \\ \frac{1}{\sqrt{2m}}, & \text{otherwise.} \end{cases}$$

We bring the following without proof.

Lemma 3. *(1) $\sum_{i=0}^{2m-1} f(i, k)f(i, q) = \delta_{k,q}$,   (2) $\sum_{k=0}^{2m-1} f(i, k)f(j, k) = \delta_{i,j}$.*

Suppose now that η_i, $i = 0, \ldots, 2m-1$ is a sequence of numbers with $\eta_i = \eta_{2m-1-i}$, and define the transformed sequence $\mu_k = \sum_i \eta_i f(i, k)$ for $k = 0, \ldots, 2m-1$. Lemma 3 implies the following.

Corollary 2. $\eta_i = \sum_k \mu_k f(i, k).$

Let $\phi(j, k, q) = \displaystyle\sum_{i=0}^{2m-1} f(i, k)f(i+j, q).$

Lemma 4. $\phi(i, k, q) = \delta_{k,q} \cos \frac{jq\pi}{m}.$

2   A Local Spreading Algorithm in One Dimension

2.1   The Algorithm

A swarm of N robots are positioned on a line. The aim is to spread the robots along this line with equal spacing between each pair of adjacent robots, where the size of the occupied segment is determined by the positions of the leftmost and rightmost robots. One may assume, instead, that the leftmost and rightmost positions represent some perimeter marks rather than robots. Each robot uses its own coordinate system. However, since the algorithm, presented below, is linear, any coordinate system will give the same resulting destination. Therefore, we use an external, global coordinate system, on which the robots have no knowledge. We refer to the robots according to their order on the line, and denote the position of each robot i, $0 \leq i \leq N - 1$, at time t, in this global coordinate system by $R_i[t]$.

The local algorithm for spreading in constant distances operates as follows.

Algorithm Spread (Code for robot i)
If no other robot is seen on the left or on the right then do nothing.
Otherwise, move to the point $\frac{R_{i+1}+R_{i-1}}{2}$.

2.2   The $\mathcal{FSYNC}$ Model

We now turn to prove the convergence of Algorithm Spread. We choose the global coordinate system such that $R_0[t] = 0$ and $R_{N-1}[t] = 1$ for all t (since the external robots do not move). As the goal is to spread the robots uniformly, upon termination the ith robot should be placed in position $i/(N - 1)$. Define $\eta_i[t]$ as the *shift* of the ith robot's location at time t from its final designated position, namely,

$$\eta_i[t] \;=\; R_i[t] - \frac{i}{N - 1} \;.$$

As our progress measure we define the quantity

$$\psi[t] \;=\; \sum_{i=0}^{N-1} \eta_i^2[t] \;,$$

where by definition, $\eta_0[t] = \eta_{N-1}[t] = 0$ for all t.

By executing the algorithm, the position of a robot changes to

$$R_i[t + 1] \;=\; \frac{R_{i-1}[t] + R_{i+1}[t]}{2} \;,$$

hence for $1 \leq i \leq N - 2$, the shifts change with time as

$$\eta_i[t + 1] = \frac{\eta_{i+1}[t] + \eta_{i-1}[t]}{2}. \tag{2}$$

We now turn to prove our main lemma.

Lemma 5. *For N robots executing Algorithm **Spread** in the $\mathcal{FSYNC}$ model $\psi[t]$ is a decreasing function of t unless the robots are already equally spread (in which case it remains constantly zero).*

Proof. Eq. (2) gives the change in η in every time step. The value of ψ thus changes to

$$\psi[t+1] = \frac{1}{4} \sum_{i=1}^{N-2} (\eta_{i+1}[t] + \eta_{i-1}[t])^2 .$$

The decrease in ψ is therefore

$$\psi[t] - \psi[t+1] = \frac{1}{4}(\eta_1[t]^2 + \eta_{N-2}[t]^2) + \frac{1}{4}\sum_{i=1}^{N-2}(\eta_{i+1}[t] - \eta_{i-1}[t])^2 ,$$

which is a positive quantity, proving the lemma. ∎

Theorem 1. *For N robots executing Algorithm **Spread** in the $\mathcal{FSYNC}$ model:*
1. *Every $O(N)$ cycles, $\psi[t]$ is at least halved.*
2. *The robots converge to a point.*

Proof. The equations for the shift changes of the robots are given in Eq. (2) for all $0 < i < N - 1$. Denote by μ_k the Fourier (Sine) series as defined in Eq. (1). By Eq. (1) and Lemmas 2 and 1,

$$\psi[t] = \sum_i \eta_i^2[t] = \frac{2}{N-1} \sum_{i,k,q} \mu_k[t]\mu_q[t] \sin ki \sin qi$$

$$= \frac{2}{N-1} \sum_{k,q} \mu_k[t]\mu_q[t] \frac{N-1}{2} \delta_{k,q} = \sum_k \mu_k^2[t]. \qquad (3)$$

Similarly, applying the transform (1) to the linear equation array (2) and noting that $\eta_0[t] = \eta_{N-1}[t] = 0$ for all times, t, yields

$$\mu_k[t+1] = \sqrt{\frac{1}{2(N-1)}} \sum_{i=1}^{N-2} \sin ki(\eta_{i-1}[t] + \eta_{i+1}[t])$$

$$= \sqrt{\frac{1}{2(N-1)}} \sum_{j=0}^{N-3} \sin k(j+1)\eta_j[t] + \sqrt{\frac{1}{2(N-1)}} \sum_{j=2}^{N-1} \sin k(j-1)\eta_j[t]$$

$$= \sqrt{\frac{1}{2(N-1)}} \sum_{j=0}^{N-1} \sin k(j+1)\eta_j[t] + \sqrt{\frac{1}{2(N-1)}} \sum_{j=0}^{N-1} \sin k(j-1)\eta_j[t],$$

where the two terms added to each sum at the last line are zero as $\sin k \cdot (N-1) = \sin k \cdot 0 = 0$, and $\eta_0[t] = \eta_{N-1}[t] = 0$. Using the fact that $\sin k(i \pm 1) = \sin ki \cos k \pm \sin k \cos ki$, we get

$$\mu_k[t+1] = \sqrt{\frac{1}{2(N-1)}} \sum_{j=0}^{N-1} (\sin k(j+1) + \sin k(j-1))\eta_j[t]$$

$$= \sqrt{\frac{1}{2(N-1)}} \sum_{j=0}^{N-1} 2\sin kj \cos k\eta_j[t]$$

$$= \cos k \cdot \mu_k[t].$$

Hence, by Eq. (3),

$$\psi[t+1] \;=\; \sum_k \mu_k^2[t+1] \;=\; \sum_k (\cos k\mu_k[t])^2 \;=\; \sum_k \cos^2 k \cdot \mu_k^2[t]. \qquad (4)$$

The values of k are $k = \frac{\pi}{N-1}, \dots, \frac{\pi(N-2)}{N-1}$. The largest factor is $\cos \frac{\pi}{N-1} \approx 1 - \frac{c}{N}$ for constant $c > 0$, hence by (4) and using (3) again, we get that

$$\psi[t+1] \;\leq\; \left(1 - \frac{c}{N}\right)^2 \sum_k \mu_k^2[t] \;\leq\; \left(1 - \frac{c}{N}\right)^2 \psi[t].$$

Thus, ψ is at least halved every $\frac{\ln 2}{2c} N$ steps, proving the claim. ∎

2.3   The $\mathcal{SSYNC}$ Model

We now turn to the $\mathcal{SSYNC}$ model. In the $\mathcal{SSYNC}$ model we can no longer assume that all robots move at each time step. Therefore, when looking at the robots moving at some time step we can no longer assume that the robots at the boundary of the moving group are located at their final designated position. Thus, the subsequent analysis employs the cosine series rather than the sine series, since the cosines allow for a constant displacement term (because $\cos 0 \neq 0 = \sin 0$).

Our rational for proving convergence is as follows. We first introduce a non-decreasing quantity and prove its monotonicity. To prove that it decreases by a constant factor, we need to show that the presented quantity is related to the non-constant terms of the cosine series. This is done in Lemmas 7–9. Finally, in Theorem 2, we show that the non-constant terms are decreased by a constant factor on every round, proving convergence.

For $i = 1, \dots, N-1$, define the robot *gaps* as the quantities

$$\gamma_i[t] \;=\; \eta_i[t] - \eta_{i-1}[t].$$

For each robot *moving* in the tth step, the equation for its position is given in (2). For any time step t, the moving robots can be grouped into chains, each of which is bounded by two stationary robots. Since the chains have no effect on each other, each can be handled separately.

Consider one such chain consisting of $m-1$ moving robots, which we number $1, \dots, m-1$ regardless of their real numbering. The boundary stationary robots will be numbered 0 and m. The appropriate equations for the gaps between them are

$$\gamma_j[t+1] = \begin{cases} \frac{\gamma_1[t]+\gamma_2[t]}{2} , & j = 1, \\[2mm] \frac{\gamma_{j+1}[t]+\gamma_{j-1}[t]}{2} , & j = 2,\ldots,m-1, \\[2mm] \frac{\gamma_m[t]+\gamma_{m-1}[t]}{2} , & j = m. \end{cases}$$

To simplify this, we define virtual robots for every integer j outside the range $[0, N-1]$, and extend the definition of γ to every j outside $[0, N-1]$ by requiring that $\gamma_j = \gamma_{j+2m}$, and $\gamma_j = \gamma_{-j+1}$ for every j. Using this definition, $\gamma_0 = \gamma_1$ and $\gamma_{m+1} = \gamma_{-m+1} = \gamma_m$. Therefore the equations take a simpler form

$$\gamma_j[t+1] = \frac{\gamma_{j+1}[t] + \gamma_{j-1}[t]}{2}, \quad \text{for every } j . \tag{5}$$

Define the Fourier transform of γ_j to be ϵ_k for $k = 0,\ldots,2m-1$ satisfying

$$\gamma_j[t] = \sum_{n=0}^{2m-1} \epsilon_k[t] f(j,k) ; \qquad \epsilon_k[t] = \sum_{n=0}^{2m-1} \gamma_j[t] f(j,k)$$

We now define the quantity $\varphi[t] = \sum_{j=1}^{N-1} \gamma_j^2[t]$.

Lemma 6. *In the $\mathcal{SSYNC}$ model $\varphi[t]$ is a non-increasing function of t.*

Proof. Apply the transform to Eq. (5) to obtain

$$\epsilon_k[t+1] = \sum_{j=0}^{m-1} \gamma_j[t+1] f(j,k)$$

$$= \sum_{j=0}^{2m-1} \frac{f(j,k)}{2} \left(\sum_{p=0}^{2m-1} f(j+1,p)\epsilon_p[t] + \sum_{p=0}^{2m-1} f(j-1,p)\epsilon_p[t] \right)$$

$$= \sum_{p=0}^{2m-1} \cos\frac{p\pi}{m} \delta_{p,k} \epsilon_p[t] = \cos\frac{k\pi}{m} \epsilon_k[t] , \tag{6}$$

where the last line uses Lemma 4. Now, from the orthogonality of the functions $f(j,k)$ (Lemma 3) it follows that

$$\sum_{j=0}^{m-1} \gamma_j^2[t] = \frac{1}{2} \sum_{j=0}^{2m-1} \gamma_j^2[t] = \frac{1}{2} \sum_{k=0}^{2m-1} \epsilon_k^2[t].$$

By Eq. (6),

$$\sum_{k=0}^{2m-1} \epsilon_k^2[t+1] = \sum_{k=0}^{2m-1} \epsilon_k^2[t] \cos^2\frac{k\pi}{m} \leq \sum_{k=0}^{2m-1} \epsilon_k^2[t].$$

Since the sum of terms for the moving robots decreased, and the other terms were not affected, φ could not increase. ∎

For a chain of robots $0, \ldots, m$, define $\gamma_{\max}[t] = \max_{j \in \{1,\ldots,m-1\}} \gamma_j[t]$ and $\gamma_{\min}[t] = \min_{j \in \{1,\ldots,m-1\}} \gamma_j[t]$

Lemma 7. *For a chain of robots moving at time t, $[\gamma_{\min}[t+1], \gamma_{\max}[t+1]] \subseteq [\gamma_{\min}[t], \gamma_{\max}[t]]$.*

Proof. By Eq. (5) each new value is the average of two old ones. ∎

For the entire group of robots define Γ_i to be the ith smallest value of the γ_j's. $\Gamma_{\max}[t] = \Gamma_{N-1}[t] = \max_{j \in \{1,\ldots,N-1\}} \gamma_j[t]$ and $\Gamma_{\min}[t] = \Gamma_1[t] = \min_{j \in \{1,\ldots,N-1\}} \gamma_j[t]$.

Lemma 8. *There exists $0 < i < m$ such that $(\Gamma_i[t] - \Gamma_{i-1}[t])^2 \geq \frac{\varphi[t]}{N^3}$.*

Proof. By the definition, $\sum_{j=0}^{N-1} \gamma_j[t] = 0$, and therefore, $\Gamma_{\max}[t] \geq 0 \geq \Gamma_{\min}[t]$. Thus, for every i $\gamma_i[t] \leq \Gamma_{\max}[t] + \Gamma_{\min}[t]$ and $\varphi[t] \leq (N-1)(\Gamma_{\max}[t] + \Gamma_{\min}[t])$. Now, $\Gamma_{N-1}[t] - \Gamma_1[t] = \sum_{j=2}^{N-1} \Gamma_j[t] - \Gamma_{j-1}[t]$, so there must exist some i such that

$$\Gamma_i[t] - \Gamma_{i-1}[t] \geq \frac{\Gamma_{\max}[t] - \Gamma_{\min}[t]}{N-2} \geq \frac{\Gamma_{\max}[t] + \Gamma_{\min}[t]}{N-2} .$$

Therefore, $(\Gamma_i[t] - \Gamma_{i-1}[t])^2 \geq \frac{\varphi[t]}{N^3}$. ∎

Lemma 9. *For a chain of gaps, $0, \ldots, m-1$ with $\gamma_{\max}[t] - \gamma_{\min}[t] \geq c$, $\epsilon_0^2[t] + \frac{c^2}{2} \leq \sum_{k=0}^{2m-1} \epsilon_k^2[t]$.*

Proof. By definition $\epsilon_0^2[t] = \frac{1}{2m}(\sum \Gamma_i[t])^2 \leq \sum_i \gamma_i^2[t]$. Denote $\gamma_s[t] = \gamma_{\max}[t] = a + b$ and $\gamma_p[t] = \gamma_{\min}[t] = a - b$, with $b \geq c/2$. The sequence with $\gamma_s[t]$ and $\gamma_p[t]$ replaced by a will have the same $\epsilon_0[t]$ since the average remains the same. However

$$\sum_i \gamma_i^2[t] = \sum_{i \neq s,p} \gamma_i^2[t] + (a-b)^2 + (a+b)^2 = \sum_{i \neq s,p} \gamma_i^2[t] + 2a^2 + 2b^2 \geq \epsilon_0^2[t] + 2b^2.$$

Since $b \geq c/2$ the lemma follows. ∎

Theorem 2. *In the SSYNC model N robots executing Algorithm Spread will converge to a configuration with equal distances.*

Proof. For a time t take $\Delta[t] = \max_j(\Gamma_j[t] - \Gamma_{j-1}[t])$ and $g = \arg\max_j(\Gamma_j[t] - \Gamma_{j-1}[t])$. By Lemma 8, $\Delta^2[t] > \varphi[t]/N^3$. Define the sets of gaps $A = \{i \mid \gamma_i[t] \geq \Gamma_g[t]\}$ and $B = \{i \mid \gamma_i[t] \leq \Gamma_{g-1}[t]\}$. Suppose that $t' \geq t$ is the first time a robot i, such that one of its neighboring gaps is of set A and the other of set B makes a move. For as long as no robot sitting between a gap in A and a gap in B is activated no gap can leave either set, by Lemma 7 and therefore $\Gamma_g[t^*] - \Gamma_{g-1}[t^*] \geq \Gamma_g[t] - \Gamma_{g-1}[t]$ for $t \leq t^* \leq t'$.

Denote by $C = \{\ldots, i, i+1 \ldots\}$ the chain of gaps surrounding robot i that change at time t'. By Lemma 8 $|\Gamma_i[t'] - \Gamma_{i-1}[t']| \geq \frac{\sqrt{\varphi}[t']}{N^{3/2}}$ and therefore $\max(|\Gamma_i[t']|, |\Gamma_{i-1}[t']|) \geq \frac{\sqrt{\varphi}[t']}{2N^{3/2}}$, leading to $\max(\Gamma_i^2[t'], \Gamma_{i-1}^2[t']) \geq \frac{\varphi[t']}{4N^3}$. Now, $\sum_{i \in C} \gamma_i^2[t'] \geq \max(\Gamma_i^2[t'], \Gamma_{i-1}^2[t']) \geq \frac{\varphi[t']}{4N^3}$. Consider the change to $\sum_{i \in C} \gamma_i^2$ after step t'. Again, we number the gaps in C by $j = 0, \ldots, m-1$ and complete with virtual robots. We have

$$\sum_{j=0}^{2m-1} \gamma_j^2[t'+1] = \sum_{k=0}^{2m-1} \epsilon_k^2[t'+1] = \sum_{k=0}^{2m-1} \cos^2 \frac{k\pi}{m} \epsilon_k^2[t'].$$

For all $k > 0$, $\cos^2 \frac{k\pi}{m} \leq \cos^2 \frac{\pi}{m} = O\left(1 - \frac{1}{m^2}\right) \leq O\left(1 - \frac{1}{N^2}\right)$.

By Lemma 9 $\sum_{k \neq 0} \epsilon_k^2[t'] \geq \frac{\varphi[t']}{N^3}$ and by the above, whenever an appropriate robot makes a move the terms ϵ_k, $k \neq 0$ are decreased by $O(1/N^2)$, therefore, φ is decreased by $O(1/N^5)$ at timestep t'. By Lemma 6 $\varphi[t'] \leq \varphi[t]$. The theorem follows. ∎

3   An Exact Global Algorithm in One-Dimension

Assuming each robot knows the number of robots at each of its sides (i.e., robot $1 \leq i \leq N$ knows it is the ith robot in the line), it is possible to achieve the goal state after a finite number of steps using the following algorithm.

Algorithm Fast_Spread (Code for robot i)
1. $t \leftarrow 1$.
2. While $t < N - 2$, move to the point $\frac{R_{i+1}+R_{i-1}}{2}$, and set $t \leftarrow t + 1$.
3. If $t = N - 2$ then solve the linear equation array Eq. (7) and move to the point $(x_1[0] + x_N[0]) \frac{i-1}{N-1}$.

We assume each robots designates its coordinate center at the point where it starts the algorithm. We first define the equation array:

$$x_i[0] = 0 ; \qquad x_{i\pm1}[t] = \sum_{j=0}^{t} \binom{t}{j} \frac{x_{i\pm1+2j-t}[0]}{2^t} \tag{7}$$

We now show that Algorithm **Fast_Spread** guarantees reaching the wanted position after exactly $N - 2$ moves.

Lemma 10. *The equation array (7) for a fixed $1 < i < N$, in conjunction with the data of the robot's neighbors $x_{i\pm1}[t]$ at times $t = 0, 1, \ldots, N-3$ provides a unique solution for $x_1[0], \ldots, x_N[0]$.*

Proof. Look at the set of equations for $x_i[0]$, $x_{i-1}[t]$ for $t = 0, 1, \ldots, i-2$ and $x_{i+1}[t]$ for $t = 0, 1, \ldots, N-i-1$. This set includes exactly N equations in N unknowns. The equations are independent since for each time t_0 there is a nonzero coefficient that was zero for all times $t < t_0$. Therefore, a unique solution exists. ∎

Theorem 3. *In the $\mathcal{FSYNC}$ model N robots executing algorithm* `Fast_Spread` *will reach their exact final position after $N-2$ steps.*

Proof. By Lemma 10 each robot can deduce the position of all other robots after $N-3$ steps. Afterwards it can solve an array of linear equations (7) and deduce its final position, which it will assume in the last step.   ∎

Theorem 4. *In the SYN model the algorithm* `Fast_Spread` *achieves the fastest possible convergence to the final position.*

Proof. Since each robot can only see its nearest neighbors, and no communication is allowed, no information of the positions of the external robots can move more than one robot per move. At the beginning, each robot has information on its own position and its nearest neighbors. After the j step it has information on its $j+1$st nearest neighbors. Therefore, at least $N-3$ steps are needed for the 2nd and $N-2$st robots to get information on the position of the most distant robots, and another move to achieve their final position.   ∎

Notice, however, that the coefficient of $\eta_j[0]$ in the linear equations obtained by robot k is proportional to $2^{-|j-k|}$. Therefore, the information accuracy decays exponentially quickly with the distance, and thus is hardly usable in any reasonable model of finite accuracy robots.

4   Local Spreading in Two Dimensional Space

In two dimensions the task of spreading via a local algorithm becomes more complicated. The set of nearest neighbors is of undetermined size, as even with no articulation points, a robot may have all other robots at an equal distance. Furthermore, the boundaries of the region in which the robots may spread cannot be efficiently marked by robots in a local manner. To simplify the situation we assume that the robots are confined to the region $[0,1] \times [0,1]$, and that the walls of this region are visible by the robots and serve as detractors. Furthermore, we assume the robots share the same orientation, where the axes parallel the walls. The algorithm is based on each robot, i, dividing space into four quadrants, Q_0 to Q_3, according to the orientation (see Fig. 1). In our simulation the quadrant boundaries were taken to lay in an angle of $45°$ to the axes to simplify the treatment of walls. This, assumption, however, should not be relevant to the results. Objects situated on the dividing lines may be considered, for instance, to belong to the lower numbered quadrant.

To illustrate the behavior of local algorithms for achieving spreading in two dimensions, we introduce an algorithm, named `Spread_2d`, in the spirit of the 1-dimensional Algorithm `Spread` presented above, and present empirical results indicating that this algorithm converges to a good approximation of the equal spacing spreading. Let us remark that an inetresting alternative approach to the problem follows from reversing the gathering technique presented in [10], for a slightly different continuous model.

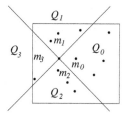

Fig. 1. The four quadrants of a robot's view. Here $m_q = m_2$

Algorithm Spread_2d (Code for robot i)
For $j = 0, \ldots, 3$ **do:**
  (a) $m_j \leftarrow$ coordinate of nearest robot or perimeter point in quadrant Q_j.
  (b) $d_j \leftarrow dist(i, m_j)$.
$q \leftarrow \arg\min_j \{d_j\}$; $d_{min} = \min_j \{d_j\}$; $d_{opp} = d_{3-q}$;
Move away from current location by $\frac{d_{min} - d_{opp}}{2d_{min}} m_q$.

Notice that in Algorithm Spread_2d an object may be either a robot or (the nearest point of) the perimeter.

To study the behavior of Algorithm Spread_2d simulations were conducted under various circumstances. We define the parameter

$$d_{av} = \frac{1}{N} \sum_i \min_{j \neq i} \{d_{i,j}\},$$

where the minimum is taken over all other robots and all points on the perimeter of the region. The behavior of this parameter, used as an indication of the spreading efficiency, was studied as a function of time. In an equally spaced square grid formation of the robots, the value of the parameter is $d_{av}^{opt} = \frac{1}{\sqrt{N}+1}$. This value can be used as an indication of the optimal spreading.

Fig. 2 presents the behavior of d_{av} under Algorithm Spread_2d as a function of time. The optimal value for this number of robots is $d_{av}^{opt} = 0.1$ and, as can be seen, the system saturates at a value close to the optimum. Fig. 3 presents the locations of the robots at the initial and final states of the algorithm application.

One may consider other generalizations of the one-dimensional algorithm to two dimensions. One such possible algorithm is based on moving to the average position of the four nearest robots (one in each quadrant). Our experiments show that this algorithm saturates quickly to a configuration which may sometimes be very far from the optimal. Another possible algorithm may rely on each robot calculating its Voronoi cell and moving to its center (according to some measure). This algorithm has the problem of failing the locality criterion, since in the worst case some of the robots may need to consider $\Omega(N)$ other robots in order to calculate their Voronoi cell. The analysis of this algorithm is also difficult, since the robots movement may change the entire structure of the Voronoi diagram such that the Voronoi neighbors of a robot may change from cycle to cycle. An approximation for this method may be obtained by using θ-graphs [12],

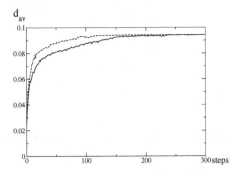

Fig. 2. Results of the simulation of algorithm `Spread_2d` with $N = 81$ robots. The dashed line is for the $\mathcal{FSYNC}$ timing model. The solid line is for the $\mathcal{SSYNC}$ timing model with each robot having probability $1/2$ for moving at each step.

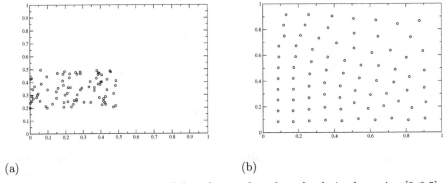

(a)   (b)

Fig. 3. (a) The initial location of the robots, selected randomly in the region $[0, 0.5] \times [0.2, 0.5]$. (b) The final locations of the robots, taken after 1000 steps.

which guarantees that each robot has exactly $2\pi/\theta$ neighbors, thus guaranteeing locality. The algorithm `Spread_2d` may be considered as such an approximation with $\theta = \pi/2$.

It should also be noted that Algorithm `Spread_2d` presented above, as well as any other algorithm, will fail to break symmetry when starting from some highly symmetrical configurations such as a line or a circle, which are very far from the optimal configuration. This problem can be overcome by adding randomness to the algorithm.

References

1. Ali N. Akansu and Richard A. Haddad. *Multiresolution Signal Decomposition.* Academic Press, San Diego, CA, USA, 1992.
2. H. Ando, I. Suzuki, and M. Yamashita. Formation and agreement problems for synchronous mobile robots with limited visibility. In *Proc. IEEE Symp. of Intelligent Control*, pages 453–460, August 1995.

3. T. Balch and R. Arkin. Behavior-based formation control for multi-robot teams. *IEEE Trans. on Robotics and Automation*, 14, December 1998.

4. G. Beni and S. Hackwood. Coherent swarm motion under distributed control. In *Proc. DARS'92*, pages 39–52, 1992.

5. Y.U. Cao, A.S. Fukunaga, and A.B. Kahng. Cooperative mobile robotics: Antecedents and directions. *Autonomous Robots*, 4(1):7–23, March 1997.

6. M. Cieliebak, P. Flocchini, G. Prencipe, and N. Santoro. Solving the robots gathering problem. In *Proc. 30th Int. Colloq. on Automata, Languages and Programming*, pages 1181–1196, 2003.

7. X. Defago and A. Konagaya. Circle formation for oblivious anonymous mobile robots with no common sense of orientation. In *Proc. 2nd ACM Workshop on Principles of Mobile Computing*, pages 97–104. ACM Press, 2002.

8. Edsger W. Dijkstra. *Selected Writings on Computing: A Personal Perspective*. Springer, New York, 1982. pages 34-35.

9. P. Flocchini, G. Prencipe, N. Santoro, and P. Widmayer. Hard tasks for weak robots: The role of common knowledge in pattern formation by autonomous mobile robots. In *Proc. 10th Int. Symp. on Algorithms and Computation*, 93–102, 1999.

10. N. Gordon, I.A. Wagner, and A.M. Bruckstein. Gathering multiple robotic a(ge)nts with limited sensing capabilities. In *Proc. 4th Int. Workshop on Ant Colony Optimization and Swarm Intelligence*, pages 142–153, September 2004.

11. D. Jung, G. Cheng, and A. Zelinsky. Experiments in realising cooperation between autonomous mobile robots. In *Proc. Int. Symp. on Experimental Robotics*, 1997.

12. J.M. Keil and C.A Gutwin. Classes of graphs which approximate the complete euclidean graph. *Discrete computational Geometry*, 7:13–28, 1992.

13. M.J. Mataric. *Interaction and Intelligent Behavior*. PhD thesis, MIT, 1994.

14. L.E. Parker. Designing control laws for cooperative agent teams. In *Proc. IEEE Conf. on Robotics and Automation*, pages 582–587, 1993.

15. L.E. Parker. On the design of behavior-based multi-robot teams. *J. of Advanced Robotics*, 10, 1996.

16. L.E. Parker, C. Touzet, and F. Fernandez. Techniques for learning in multi-robot teams. In T. Balch and L.E. Parker, editors, *Robot Teams: From Diversity to Polymorphism*. A. K. Peters, 2001.

17. G. Prencipe. CORDA: Distributed coordination of a set of atonomous mobile robots. In *Proc. 4th European Research Seminar on Advances in Distributed Systems*, pages 185–190, May 2001.

18. G. Prencipe. *Distributed Coordination of a Set of Atonomous Mobile Robots*. PhD thesis, Universita Degli Studi Di Pisa, 2002.

19. K. Sugihara and I. Suzuki. Distributed algorithms for formation of geometric patterns with many mobile robots. *J. of Robotic Systems*, 13(3):127–139, 1996.

20. I. Suzuki and M. Yamashita. Distributed anonymous mobile robots: Formation of geometric patterns. *SIAM J. on Computing*, 28:1347–1363, 1999.

21. I.A. Wagner and A.M. Bruckstein. From ants to a(ge)nts. *Annals of Mathematics and Artificial Intelligence*, 31, special issue on ant-robotics:1–5, 1996.

How to Meet in Anonymous Network

Dariusz R. Kowalski[1] and Adam Malinowski[2,*]

[1] Department of Computer Science
The University of Liverpool
Liverpool L69 7ZF, UK
[2] Instytut Informatyki, Uniwersytet Warszawski
Banacha 2, 02-097 Warszawa, Poland

Abstract. A set of k mobile agents with distinct identifiers and located in nodes of an unknown anonymous connected network, have to meet at some node. We show that this *gathering problem* is no harder than its special case for $k = 2$, called the *rendezvous problem*, and design deterministic protocols solving the rendezvous problem with arbitrary startups in rings and in general networks. The measure of performance is the number of steps since the startup of the last agent until the rendezvous is achieved.

For rings we design an oblivious protocol with cost $O(n \log \ell)$, where n is the size of the network and ℓ is the minimum label of participating agents. This result is asymptotically optimal due to the lower bound showed in [18].

For general networks we show a protocol with cost polynomial in n and $\log \ell$, independent of the maximum difference τ of startup times, which answers in affirmative the open question from [22].

Keywords: mobile agents, gathering, rendezvous, anonymous networks, distributed algorithms.

1 Introduction

We consider a *gathering problem* defined as follows: a set of k mobile agents originally located at arbitrary nodes of a network—modeled as an undirected connected graph—have to meet at some node. An important special case is the version for two agents, known as the *rendezvous problem*. If nodes of the graph are equipped with unique labels or the agents are allowed to save messages in nodes, then the problem gets significantly simpler and can be reduced to graph exploration. However, in many applications such facilities may not be available e.g. for technical or security reasons, which implies the need to design gathering protocols working in *anonymous* networks modeled by graphs with unlabeled nodes. We must assume however that ports at a node are *locally* distinguishable: otherwise two agents might be unable to meet even in a complete graph with four nodes K_4. Hence we make a natural assumption that all ports at a node are locally labeled $1, \ldots, d$, where d is the degree of the node (local labelings at different nodes do not have to be consistent in any way). Unless otherwise stated, we do not assume agents to know the topology of the graph or its size. We also assume that the agents have distinct identifiers—otherwise it is easy to see that *deterministic* gathering even in anonymous graph K_2 (a clique of two nodes) is impossible.

* Work supported by the KBN Grant 4T11C04425.

P. Flocchini and L. Gąsieniec (Eds.): SIROCCO 2006, LNCS 4056, pp. 44–58, 2006.
© Springer-Verlag Berlin Heidelberg 2006

1.1   Related Work

Gathering is a very natural problem and it was considered in various settings, see e.g., [12, 13, 19, 21, 24, 28]. Even for the ring topology, the gathering problem is important due to its application to self-stabilization problem where, starting from an arbitrary state of the system, the goal is to recover into the legal single-agent state (see e.g., [21, 25]).

Previous solutions considered either models with some additional information provided to agents, like sense of directions or a map of the network, or used randomization to break the symmetry of the anonymous network. A model similar to the one studied in this paper was considered in [29], but it still assumed some knowledge about the graph and the localization of agents. Randomized approach to gathering exploits random walks in graphs, which have been widely studied and applied in various contexts, such as graph traversing [2], on-line algorithms [15], etc.

The special case of the gathering problem — the rendezvous problem — has been also extensively studied (see [5] for references). It was introduced in [27] and continued in two directions: geometric space [8, 9, 10, 11, 20] and graphs [3, 6]. Most of the papers, e.g., [3, 4, 7, 10, 21] adopted the probabilistic scenario where either inputs are random or protocols use randomization (or both). Deterministic rendezvous with anonymous agents working in unlabeled graphs but equipped with tokens used to mark nodes was considered e.g. in [23]). The so called hunter-rabbit game, in which one agent has to catch the other, was also considered, see [1] for recent results and references. De Marco et al. [17] considered the rendezvous problem in asynchronous networks. Unlike in our model, they allowed a meeting of two agents passing the same edge but in the opposite directions.

Dessmark et al. [18] were the first to study the rendezvous problem in the model adopted in this paper. They designed protocols for trees, rings and general networks, both in simultaneous and arbitrary startup model, and established lower bounds on time complexity. We briefly describe their contribution focusing on the results corresponding strictly to the results presented in this paper. For simultaneous startup they proved that the optimal cost of rendezvous in any ring is $\Theta(D \log \ell)$, where D is the initial distance between agents. For arbitrary startup, they showed that $\Omega(n + D \log \ell)$ is a lower bound on the cost required for rendezvous on an n-node ring (since an adversary can make $D = \Theta(n)$, the worst-case lower bound—viewed as a function of n and ℓ—is $\Omega(n \log \ell)$). They designed a rendezvous protocol for rings with cost $O(\ell\tau + ln^2)$, where τ is the difference between startup times, if the agents do not know ring size n. They also proved that for arbitrary networks the problem is feasible.

Kowalski and Pelc [22] presented a deterministic rendezvous protocol with cost polynomial in n, τ and $\log \ell$ for arbitrary connected graphs, as well as a lower bound $\Omega(n^2)$ on the cost of rendezvous in some family of graphs.

1.2   Our Results

For the ring network topology we design an oblivious gathering protocol with cost $O(n \log \ell)$, which is asymptotically worst-case optimal due to the lower bound $\Omega(n \log \ell)$ showed in [18]. Note that this lower bound also holds for protocols which are adaptive and work in simultaneous startup model. In this sense we show that the problem for rings with and without arbitrary startup has the same difficulty, as well as for adaptive and non-adaptive protocols.

For general networks we show a gathering protocol with cost polynomial in $n, \log \ell$ and *not depending* on τ, which answers in affirmative the question stated in [22]. Our protocol (as well as the one from [22]) is partly non-constructive: it uses some combinatorial objects which are only proved to exist by the probabilistic method. Nevertheless the protocol is in fact deterministic, since the agents can find the required combinatorial objects locally by brute force. This may be quite expensive but our model described below counts only moves of the agents and does not care about local computation time.

1.3   The Model

The model we adopt is essentially the same as the one used in [18, 22], so we just give its brief description.

The network is modeled as a simple undirected connected graph with unlabeled nodes. The ports at each node (i.e. the edges incident with the node) are labeled $1, \ldots, d$, where d is the degree of the node, but this labeling is only local and it does not have to be consistent with labelings of this node's neighbors. The size of the network, denoted by n, is a-priori unknown to the agents.

There are k agents in the network, where k is a-priori not known to the agents (although for the rendezvous problem we assume that the information that $k = 2$ is provided a-priori). Agents have distinct identifiers (labels), which are *different integers* written as binary strings. Every agent knows only its own label, which is the sole input to a gathering protocol it executes. Agents move in synchronous steps. In a single step, an agent may either remain idle or move to an adjacent node. Agents can exchange their informations iff they are in the same node at the same time step. The initial location of agents is arbitrary (i.e. decided by an adversary). During the execution of the protocol, if an agent currently located at some node decides to traverse some yet unexplored incident edge, the actual edge is chosen by an adversary. The agent, however, learns the local port number by which it enters a node and the degree of the node. Agents actually meet only if they get to the same node in the same step, not if they cross each other along an edge.

We consider a general scenario with *arbitrary startup*, in which starting times of agents are arbitrarily decided by an adversary. Agents are not aware of the difference between starting times; each of them is created at its startup and begins executing its protocol and counting steps since then.

The *cost* of a gathering protocol is the worst-case number of steps *since the startup of the latest agent* until gathering is achieved, taken over all mentioned above adversary choices of initial locations, unexplored edges and startup times. Time of local computations performed by agents does not contribute to cost.

1.4   Notation and Preliminaries

Labels of agents are denoted by $L_1, L_2, \ldots, L_k$, in the order of the startup times: agent with label L_i appears not later than agent with label L_{i+1}. An agent knows only the value of its label L_i, not its relative position i. The smallest of the considered labels $L_1, \ldots, L_k$ is denoted by ℓ and the difference between startup times of agents L_1 and L_k is denoted by τ. The number of nodes in the graph is denoted by n, and it is not known to the agents.

For label L and integer r, let $f(L,r)$ denote the string obtained by replacing each 0 in the binary representation of L with $0^{4^r}1^{4^r}$ and each 1 in a binary representation of L with $1^{4^r}0^{4^r}$. Note that $f(L,r)$ is partitioned into homogenous (only 0s or only 1s) *blocks* of size 4^r, numbered from 1 to $2|L|$. We say that one block is *opposite* to another if one of them is a block of ones while the other one is a block of zeros. For technical reasons we assume that $|L| \geq 16$ (this can be achieved e.g. by adding 2^{15} to all labels). We use $f(L,r)$ to control the actions of agent L in stage r of our gathering/rendezvous protocols.

Let $C_r(L)$ denote the concatenation of strings for label L and stages $0, 1, \ldots r$, i.e.

$$C_r(L) = f(L,0) \sqcup f(L,1) \sqcup \cdots \sqcup f(L,r) \ .$$

Each step of our protocols will correspond to a single bit in the infinite control string $C_\infty(L)$, thus we will identify time segments during the execution of the protocol with substrings of $C_\infty(L)$ and say, e.g., that some segment S *is covered* by block P (which means that the string corresponding to S is contained in P).

The following lemma shows useful combinatorial properties of control strings (proof in the full version of the paper).

Lemma 1. *For any substring S of $C_r(L)$, with $|S| = 4^{r+1}$, S contains at least 4^r 0s and at least 4^r 1s.*

2   Gathering vs. Rendezvous Problem

In this section we show that each rendezvous protocol can be easily modified to obtain gathering protocol with asymptotically the same time complexity. Given the rendezvous protocol **R**, each agent applies this protocol with the following modification:

> after some agents meet in a node, they all continue rendezvous protocol **R** for the smallest label from them, as it would be no rendezvous (we say that the agent with the smallest label *sticks* all the other agents met in this node).

We call this protocol *sticky-***R** protocol.

Lemma 2. *Each sticky-***R** *protocol is a gathering protocol with the same asymptotic complexity as the original rendezvous protocol* **R** *for the same smallest label agent.*

Proof. Fix anonymous network G and starting nodes and times. Consider two executions: first is the worst-case execution of the sticky-**R** protocol for the smallest label ℓ, the second is the worst-case execution of the original rendezvous protocol **R** for the smallest label ℓ. It is clear that if the agent with label ℓ does not "stick" some other agent L in the gathering protocol then it also does not meet L in the rendezvous execution. □

3   Rendezvous in Rings

In this section we assume that agents know that the underlying graph is a ring, although they do not know its size. An agent arbitrarily chooses one direction in the ring as 'left' and the other as 'right' (note that these assignments may be different for different agents). The starting direction for each agent is arbitrary. We present an asymptotically optimal rendezvous protocol which, by Lemma 2, also solves the gathering problem. (A similar idea was applied independently in an asynchronous setting in [17].)

Protocol RING-WALK(L)
step_count $\leftarrow 0$
$\Delta \leftarrow 1$
for *stage* $\leftarrow 0, 1, \ldots$ **do**
   $T \leftarrow f(L, stage)$
   for $i \leftarrow 1, \ldots, |T|$ **do**
      if $T[i] = 1$ **then** move to the next node in the current direction
      else remain idle for one step
      step_count $\leftarrow$ *step_count*$+1$
      if *step_count*$= \Delta$ **then**
         change direction
         $\Delta \leftarrow 2\Delta$

Step counter determines when the direction of walk is changed and after how many steps Δ it will happen; the interval between ith and $(i+1)$th change of direction is called *period* $i+1$. Note that period i, for $i \geq 1$, takes $\Delta = 2^i$ steps. The current value of Δ at the beginning of a step is the *period size* of an agent in this step. It is straightforward to prove by induction on i that the period size in step k is exactly i for $2^i \leq k < 2^{i+1}$.

Now we prove that the RING-WALK protocol guarantees fast rendezvous.

Theorem 1. *Two agents performing* RING-WALK *protocol in a ring of size n meet after* $O(n \log \ell)$ *steps.*

Proof. Let $i_0 = \lceil \log_4 n \rceil$ and $b = 4^{i_0}$ (i_0 is the number of the first stage with block size $b \geq n$).

First we note the general property which we will use in this proof: during any time segment which consists of at most eight blocks in the same stage $i > 1$ of agent L_2, each agent can change direction at most once. The proof of this fact is as follows. Fix stage i of agent L_2 and segment S consisting of at most eight blocks of this stage. Since $|L_2| \geq 16$, agent L_2 executes at least

$$2|L_2|\frac{4^i - 1}{3} \geq 2 \cdot 4^{i+1}$$

steps by the end of stage $i-1$, which means that at the beginning of stage i its period size is at least $8 \cdot 4^i/2$. If agent L_2 changes its direction in S, its period size grows to at least $8 \cdot 4^i \geq |S|$, so it will not change again in S. The property for agent L_1 is straightforward, since its period size is always not smaller than the current period size of agent L_2.

Now we proceed with the proof of the theorem, which is divided into three cases.

Case 1: $|L_1| > |L_2|$.

Let S be the time segment which consists of the first four blocks in stage $i_0 + 2$ of agent L_2. Note that S has length $64b$. We divide S into four succesive subsegments S_1, S_2, S_3, S_4, each consisting of $16b$ steps.

Subcase 1a: at the beginning of segment S agent L_1 is in stage bigger than $i_0 + 2$ or in one of its last two blocks of stage $i_0 + 2$.

If agent L_1 is in stage bigger than $i_0 + 2$ at the beginning of segment S then it can change its block at most once in S, and so at least one block of agent L_2 in S is covered by an opposite block of agent L_1. This is also true when agent L_1 is in one of its last two blocks of stage $i_0 + 2$ at the beginning of S, because then the first block in stage $i_0 + 3$ of agent L_1 covers S_3 and S_4, so it is opposite to one of them (the third and the fourth blocks of a stage are always opposite to each other, so any block is opposite to one of them). Summing up, we proved that S contains a subsegment of size $16b$ in which one agent is idle and the other is moving. Since the period size of both agents is at least $8b$, the moving agent can change direction at most once in this subsegment, so it makes at least $8b > n$ consecutive moves in one direction. Hence the rendezvous is achieved in S.

Subcase 1b: at the beginning of segment S agent L_1 is in stage $i_0 + 2$, but neither in the last two blocks nor in the first two blocks.

Let S' consists of the last four blocks of stage $i_0 + 1$ of agent L_2. Note that S' has size $16b$ and ends just before S. Using similar argument as in Subcase 1a but applied to segment S', we conclude that the rendezvous occurs in S'. More precisely, since $|L_1| > |L_2|$ and at the end of S' agent L_1 is at least in the third block of stage $i_0 + 2$, we see that S' is covered by at most two blocks of size $16b$ from stage $i_0 + 2$ of agent L_1. It follows that one of four blocks of agent L_2 in S' is covered by one opposite block of agent L_1. In time segment corresponding to this block of size $4b$ the moving agent can change its direction at most once, hence it makes at least $2b > n$ consecutive moves in one direction and the rendezvous is completed.

Subcase 1c: at the beginning of segment S agent L_1 is either in its stage $i_0 + 1$ or in the first two blocks of its stage $i_0 + 2$.

Consider time segment S' consisting of the first $640b$ steps in stage $i_0 + 5$ of agent L_2. Since $|L_1| > |L_2|$ and at the beginning of S agent L_1 is at most in the second block of its stage $i_0 + 2$, during segment S' agent L_1 is at least in its stage $i_0 + 1$ (this is obvious since S precedes S') and at most in its stage $i_0 + 4$. Indeed, stages $i_0 + 2, i_0 + 3, i_0 + 4$ of agent L_1 take

$$2|L_1| \cdot (16b + 64b + 256b) \geq 2(|L_2| + 1) \cdot 336b = 2|L_2| \cdot 336b + 32b + 640b$$

steps, so even if we remove from stage $i_0 + 2$ its first two blocks of total length $32b$, the remaining part of stage $i_0 + 2$ together with stages $i_0 + 3, i_0 + 4$ covers S'. Segment S', which in fact is a part of the first block of stage $i_0 + 5$ of agent L_2, contains blocks of size at least $4b$ and at most $256b$ of agent L_1, so there is a subsegment of S' containing at least

$$\min\{4b, 640b - 2 \cdot 256b\} = 4b$$

consecutive zeroes and another subsegment of S' containing at least $4b$ consecutive ones of agent L_1. It follows that during one of these subsegments the agents are opposite, i.e.

one of them is idle and the other is moving. Since the considered subsegment has length $4b$ and is only a part of one block of agent L_2, the moving agent can change its direction at most once in this subsegment. Consequently during at least half of this subsegment the agent moves in one direction, while the other one is idle, hence the rendezvous is achieved.

Subcase 1d: at the beginning of segment S agent L_1 is in stage at most i_0.
Note that stage $i_0 + 1$ of agent L_1 lasts

$$2|L_1| \cdot 4b \geq 2(|L_2| + 1) \cdot 4b = 2|L_2| \cdot 4b + 8b \geq 136b > |S|$$

steps, since $|L_1| > |L_2| \geq 16$. It follows that during segment S agent L_1 is in stage at most $i_0 + 1$, while agent L_2 is in one of the first four blocks of its stage $i_0 + 2$. Consider the first block *of zeros* of agent L_2 within segment S. Its length is $16b$ and it is covered by blocks from stages up to $i_0 + 1$ of agent L_1, hence by Lemma 1 agent L_1 moves in at least $4b$ steps within this block. S contains only four blocks of one stage of agent L_2, so each agent may change direction at most once within S (hence also within any subsegment of S). It follows that agent L_1 makes at least $2b$ moving steps (perhaps separated with idle steps) without change of direction while agent L_2 is idle. Hence agent L_1 finds agent L_2 during the considered block in segment S.

To summarize Case 1, we proved in all subcases that the rendezvous happens by at most the end of stage $i_0 + 5$ of agent L_2, which is by step $\frac{4^{i_0+6}-1}{3} \cdot 2|L_2| = O(n \log \ell)$ after the startup of agent L_2.

Case 2: $|L_1| < |L_2|$.

Subcase 2a: $\tau \geq 16b|L_1|$.
Consider time segment S of length $16b$ which begins $4b$ steps after the startup of agent L_2. At the beginning of S agent L_1 is in stage at least $i_0 + 2$, hence S is covered by at most two blocks of agent L_1. During this time segment the period size Δ_1 of agent L_1 is at least $16b$ and agent L_2 changes direction twice: first after $4b$ steps and then after another $8b$ steps. It follows that S contains a subsegment S' of length at least $4b$ covered by a single block of agent L_1 and such that the directions of both agents agree in S'. Agent L_2 is in stage at most i_0 during S', hence, by Lemma 1, the rendezvous is achieved in this segment.

Subcase 2b: $\tau < 16b|L_1|$.
Consider the first block S *of zeroes* in stage $i_0 + 4$ of agent L_1. Note that at the beginning of segment S agent L_2 has period size at least

$$\frac{4^4 b - 1}{3} 2|L_1| - \tau \geq \frac{4^4 b - 1}{3} 2|L_1| - 16b|L_1| =$$

$$4^4 b|L_1| \cdot \left(\frac{2}{3} - \frac{1}{3 \cdot 4^4 b} - \frac{1}{4^2} \right) > 4^4 b/2 = |S|/2 ,$$

since segment S starts at least $\frac{4^4 b-1}{3} 2|L_1| - \tau$ steps after the startup of agent L_2. It follows that agent L_2 can change direction in S at most once—otherwise the length of S would be at least twice the period size at the beginning of S, a contradiction. Consequently, agent L_1 also may change its direction within segment S at most once (its period size is the same or bigger). Note also that the current stage of agent L_2 is never bigger than the one of agent L_1, hence at the end of segment S agent L_2 is in stage $i_0 + 4$ or less.

First assume that at the end of segment S agent L_2 is in stage at most $i_0 + 3$. Agent L_2 may change direction at most once in S, so, by Lemma 1, it makes at least $16b > n$ moving steps (perhaps separated with idle steps) without change of direction, while agent L_1 is idle, so the rendezvous happens in segment S.

Otherwise at the end of segment S agent L_2 is in stage $i_0 + 4$. We have

$$(f(L_2, i_0 + 3) + \tau) - f(L_1, i_0 + 3) > 2 \cdot 4^{i_0 + 3} ,$$

and two situations may happen:

- if S is the first block in stage $i_0 + 4$ of agent L_1 then during the first $2 \cdot 4^{i_0 + 3}$ steps of S agent L_2 is in its stage $i_0 + 3$. It follows that during these steps agent L_2 makes at least $4^3 b / 2 > n$ moves without change of direction.
- if S is the second block in stage $i_0 + 4$ of agent L_1 then consider the previous block S', which is the block of ones. During the first $2 \cdot 4^{i_0 + 3}$ steps of S' agent L_2 is in stage $i_0 + 3$. It follows that during at least $4^{i_0 + 3} / 2$ consecutive steps agent L_2 is idle while the earlier one is moving and changes direction at most once. Consequently it makes at least $4^{i_0 + 3} / 4 = 16b > n$ moves in one direction, so the rendezvous is achieved in S'.

Summarizing Case 2, in all subcases the rendezvous is achieved by the end of segment S, which in both cases is at most $20b + 2|L_1| \frac{4^{i_0 + 5} - 1}{3} = O(n \log \ell)$ steps after the startup of agent L_2.

Case 3: $|L_1| = |L_2|$.

Subcase 3a: $\tau \geq 8b$.

Consider time segment S consisting of the last eight blocks of stage i_0 of agent L_2. During S agent L_1 is already in stage at least $i_0 + 1$, and the period size Δ (for both agents) is at least $8b$, hence by Lemma 1 during this segment there are $b \geq n$ consecutive steps in which one agent moves *in one direction* while the other remains idle, so the rendezvous is achieved.

Subcase 3b: $\tau < 8b$.

Let j be the position of the first difference in code strings $f(L_1, 0)$ and $f(L_2, 0)$ of the agents and consider the j-th block of stage $i_0 + 2$ of both agents. The block size is $16b$ and the delay is less than $4b$, so the size of the maximal segment S covered by both these blocks is more than $12n$. The period size Δ (for both agents) is at least $16b$, hence, by Lemma 1, during segment S there are $b \geq n$ consecutive steps in which one agent moves *in one direction* while the other remains idle, so the rendezvous is achieved.

In view of the above theorem, the lower bound $\Omega(n \log \ell)$ for rendezvous obtained in [18], and Lemma 2, we have the following.

Corollary 1. RING-WALK *is an asymptotically optimal agent protocol for gathering in unknown rings.*

4   Gathering in Arbitrary Networks

As in the previous section, we design a protocol for rendezvous problem, which automatically works also for the general gathering problem.

4.1   Cover Walks and Network Exploration

In this subsection we present a generalization of the construction of deterministic cover walks [26], previously used e.g., in [22] in the context of the rendezvous problem.

A *walk* of length x in a graph is a sequence $(v_1, ..., v_x)$ of nodes such that node v_{i+1} is adjacent to v_i, for all $i < x$. A *cover walk* is a walk in which every node of the graph appears at least once. Given an unknown network, we consider a *markovian procedure* which produces a walk in the network, i.e. a procedure in which the next move depends only on the number of previous moves and the degree of the currently visited node (ports are numbered but destination nodes have no labels). The aim of this subsection is to design a deterministic markovian procedure called UCW (short for *Universal Cover Walk*), consisting of $\lambda(N)$ steps, where N is a given positive integer parameter and $\lambda(N)$ is polynomial in N, which satisfies the following condition for some function $\gamma_N(n)$ which is polynomial in n:

Property UniCoverWalk. for any initial node of an anonymous graph with $n \leq N$ nodes, and for any step number $a \leq \lambda(N) - \gamma_N(n) + 1$, the sequence of steps $a, a + 1, ..., a + \gamma_N(n) - 1$ of the procedure yields a cover walk in this graph.

Procedure UCW will be an important ingredient of our rendezvous protocol. Note that this procedure itself guarantees exploring in $\gamma_N(n)$ steps, which means that a single agent visits all n nodes of anonymous network in every interval of $\gamma_N(n)$ steps while executing this procedure.

The existence of cover walks guaranteeing exploration of any n-node graph, for a *known* parameter n, is a well known fact (cf. [26]). Note that the problem of explicit construction of cover walks is hard (cf. hardness of construction of a universal traversal sequence even for 3-regular graphs [14]). Our universal cover walk differs in two points from the previously considered cover walks:

- it works fast for all networks of any size $n \leq N$,
- it works even if we start in any point of the sequence and continue exploration from this point according to the following part of the sequence.

Now we proceed with the construction of procedure UCW. We do it in two steps: first we construct procedure ALMOSTUCW, which satisfies Property UniCoverWalk only for restricted range of parameters $n \leq N$, and then we use it as a building block in construction of procedure UCW.

Almost Universal Cover Walks. Let $\lambda^*(N)$ be a non-decreasing positive integer function, polynomial in N. For any positive integer N and any function $h_N : \{1, ..., \lambda^*(N)\} \times \{1, ..., N-1\} \longrightarrow \{1, ..., N-1\}$, such that $h_N(i, d) \leq d$, we define the following generic procedure ALMOSTUCW describing a walk of length $\lambda^*(N)$ in an anonymous n-node network G, for $n \leq N$, starting at arbitrary node v (location v is unknown to the agent executing the procedure, but a particular walk defined by this procedure depends on v).

Procedure ALMOSTUCW(N, h_N)
In step i such that $0 < i \leq \lambda^*(N)$, the agent, currently located at a node of degree d, moves to an adjacent node by port $h_N(i, d)$.

In order to instantiate the procedure ALMOSTUCW we have to define functions h_N for integer $N > 1$. First we consider a *random walk* of an agent in graph G, i.e. a walk in which the agent, currently located at a node of degree d, selects one port uniformly at random (independently with probability $1/d$) and exits the node through this port.

Let us define $\lambda^*(N) = \lceil 2\alpha N^5 \log N \rceil$, where $\alpha = 4/27$ is the constant coefficient in the upper bound on the expected length of the random cover walk [16]. Let $\gamma_N^*(n) = \lambda^*(n)$, for all $n \leq N$. The following lemma was proved in [22]:

Lemma 3. *A random walk of length $\gamma_N^*(n)$ starting at node v in a connected graph G with at most n nodes is a cover walk with probability at least $1 - 2^{-2n^2 \log n}$.* □

Now we prove that for some deterministic functions h_N procedure ALMOSTUCW behaves like a random walk.

Lemma 4. *For any positive integer N, there exists a function $h_N : \{1, ..., \lambda^*(N)\} \times \{1, ..., N-1\} \longrightarrow \{1, ..., N-1\}$, such that $h_N(i, d) \leq d$ and for any starting node of any connected n-node graph G, where $\sqrt{8 \log N} < n \leq N$ for $N > 4$ or $1 \leq n \leq N$ for $1 \leq N \leq 4$, the sequence of any $\gamma_N^*(n)$ consecutive steps $a, a+1, ..., a + \gamma_N^*(n) - 1$ during the execution of procedure ALMOSTUCW(N, h_N) produces a cover walk in this graph.*

Proof. First assume that $N > 4$. We can select parameter $n \leq N$ in at most N different ways, graph G (with labeled ports) in at most n^{n^2} ways, starting node v in at most n ways, and the first step a of the sequence in at most $\lambda^*(N)$ ways. Hence we can make a selection of a quadruple $\langle n, G, v, a \rangle$ in at most

$$n^{n^2} \cdot N^2 \cdot \lambda^*(N) \leq 2^{n^2 \log n + 8 \log N}$$

different ways.

Consider function h_N selected randomly as follows: values $h_N(i, d)$, over all possible i, d, are selected independently; $h_N(i, d) = j$ with probability $1/d$ for every $1 \leq j \leq d$. It is easy to see that procedure ALMOSTUCW(N, h_N) instantiated by this function h_N, started in node v and considered for consecutive steps in interval $[a, a + \gamma_N^*(n))$, generates a random walk on any graph G of at most n nodes.

Consider this random walk. By Lemma 3, the probability of the event 'there exists a connected n-node graph G and a starting node v in G, such that the random walk of length $\gamma_N^*(n)$ in graph G starting in v is not a cover walk' is at most

$$2^{-2n^2 \log n} \cdot 2^{n^2 \log n + 8 \log N} < 1 \,,$$

since $n > \sqrt{8 \log N}$. Using the probabilistic argument we prove the existence of the desired function, which completes the proof for case $N > 4$.

For $N \leq 4$ the proof is similar, but we obtain the existence of a function h_N yielding a cover walk for all $n \leq N$. Indeed, for $n \leq 2$ or $N \leq 2$ it is straighforward. For $3 \leq n \leq N \leq 4$ the number of possible configurations is at most

$$n^{n^2} \cdot N^2 \cdot \lambda^*(N) \leq 2^{n^2 \log n} \cdot \frac{4^9}{27} \, ,$$

and again the probability of the event 'there exists a connected n-node graph G and a starting node v in G, such that the random walk of length $\gamma_N^*(n)$ in graph G starting in v is not a cover walk' is at most

$$2^{-2n^2 \log n} \cdot 2^{n^2 \log n} \cdot \frac{4^9}{27} \leq 2^{-9 \log 3} \cdot \frac{4^9}{3^3} \leq \frac{4^9}{3^{12}} < 1 \, ,$$

and the probabilistic argument shows the existence of h_N.   □

Although an explicit construction of functions h_N is hard, the agents can find them locally by exhaustive search (which does not contribute to the cost of rendezvous protocol). Hence from now on we will assume that functions h_N satisfying Lemma 4 are fixed for all $N \geq 1$ and we will omit the second parameter in calls to procedure AL-MOSTUCW.

Universal Cover Walks. To overcome the constraints on n in Lemma 4 we modify procedure ALMOSTUCW(N) to obtain more flexible procedure UCW(N), which essentially after each step of the walk calls itself recursively for smaller argument, that is for which the previous procedure ALMOSTUCW did not guarantee the cover walk.

Procedure UCW(N)

Case $2 \leq N \leq 4$: Run procedure ALMOSTUCW(N)
Case $N > 4$: For each i, where $0 < i \leq \lambda^*(N)$ the agent, currently located at a node of degree d, does the following actions:
   – it recursively runs procedure UCW(x), where $x = \lfloor \sqrt{8 \log N} \rfloor$, and goes backward;
   – it moves to an adjacent node by port $h_N(i, d)$;
   – it recursively runs procedure UCW(x), where $x = \lfloor \sqrt{8 \log N} \rfloor$, and goes backward.

Let $\lambda(N)$ denote the number of steps in the procedure UCW(N).

Lemma 5. $\lambda(N) = O(N^5 \log^4 N)$.

Proof. We have the recurrence relation following directly from the pseudo-code of procedure UCW(N)

$$\lambda(N) = \lambda^*(N) \cdot (1 + 4\lambda(x))   \text{ for } N > 4 \, ,$$

where $x = \lfloor \sqrt{8 \log N} \rfloor$, and $\lambda(N) = O(1)$ for $N \leq 4$. Expanding this recurrence gives $\lambda(N) = O(N^5 \log^4 N)$.   □

Now we prove that procedure UCW satisfies Property UniCoverWalk for function

$$\gamma_N(n) = \min\{8(\lambda(n))^2, \lambda(N')\},$$

where N' is such that procedure $UCW(N')$ is called during the execution of procedure $UCW(N)$ and $\sqrt{8\log N'} < n \leq N'$. Note that $\gamma_N(n)$ is polynomial in n, since it is not bigger than $8(\lambda(n))^2$, which is polynomial in n by Lemma 5.

Lemma 6. *For any positive integer N, there exists a function $h_N : \{1,...,\lambda(N)\} \times \{1,...,N-1\} \longrightarrow \{1,...,N-1\}$, such that $h_N(i,d) \leq d$ and the procedure $UCW(N)$ produces a cover walk of length at most $\gamma_N(n)$ in any connected n-node graph G, for $1 \leq n \leq N$, and starting in any node v and in any step $a \leq \lambda(N) - \gamma_N(n) + 1$.*

Proof. For $N \leq 4$ it follows directly from Lemma 4, so we assume that $N > 4$.

Consider parameters n, N and a corresponding value N' as in the definition of function $\gamma_N(n)$. Fix time segment S of length $\gamma_N(n)$ in procedure $UCW(N)$. Note that during the execution of procedure $UCW(N)$ at most one step from the call of some procedure $\text{ALMOSTUCW}(N'')$ with $N'' > N'$, is performed between two consecutive calls of procedure $UCW(N')$. This property follows immediately from the observation that any two steps from a single call of procedure $\text{ALMOSTUCW}(N'')$, as well as any two steps from calls of procedure $\text{ALMOSTUCW}(N'')$ and procedure $\text{ALMOSTUCW}(N''')$, for $N''', N'' > N'$ are separated by at least one execution of procedure $UCW(N')$. We consider two cases.

Case 1: $\gamma_N(n) = \lambda(N')$.
In this case S contains either at least the first half or at least the second half of steps of procedure $UCW(N')$ (otherwise the length of S would be at most $\lambda(N')/2 - 1$ taken twice plus at most one step according to some call to procedure $\text{ALMOSTUCW}(N'')$, which would be less than $\lambda(N') = |S|$). In either case the whole procedure $\text{ALMOST-}UCW(N')$ is executed within teh considered segment of the call to procedure $UCW(N')$ (either in the first part or in return part), which by Lemma 4 and inequality $n \leq N'$ yields the existence of cover walk in S for any n-node network.

Case 2: $\gamma_N(n) < \lambda(N')$.
In this case $\gamma_N(n) = 8(\lambda(n))^2$. Similarly as in the previous case, it follows that S contains at least $4(\lambda(n))^2$ consecutive steps from a call to procedure $UCW(N')$ which take place either in the first or in the second half of this procedure (otherwise the length of S is at most $4(\lambda(n))^2 - 1$ taken twice plus additional at most one step according to some procedure $\text{ALMOSTUCW}(N'')$, for $N'' > N'$, which would result in a contradiction $|S| \leq 2 \cdot (4(\lambda(n))^2 - 1) + 1 = 8(\lambda(n))^2 - 1)$. In either case at least $4(\lambda(n))^2$ consecutive steps of procedure $UCW(N')$ are made. It follows that the number of consecutive steps of a call to procedure $\text{ALMOSTUCW}(N')$ which are performed in either such segment is at least

$$\frac{4(\lambda(n))^2}{4\lambda(\lfloor\sqrt{8\log N'}\rfloor)} \geq \lambda(n) \geq \lambda^*(n) = \gamma_N^*(n).$$

Since $\sqrt{8\log N'} < n \leq N'$, Lemma 4 yields the existence of cover walk in S for any n-node network. $\square$

4.2 Rendezvous

Procedure UCW is a building block of our rendezvous protocol. We start with describing the procedure $RV(L, j)$, for label L and upper bound j on the number of stages, which is the main ingredient of the protocol.

Procedure $RV(L, j)$
for $stage \leftarrow 1, 2, \ldots j$ **do**
    $T \leftarrow f(L, stage)$
    $N \leftarrow \max\{n : \lambda(n) \leq |L| \cdot 4^{stage}\}$
    $count \leftarrow 1$
    for $i \leftarrow 1, \ldots, |T|$ **do**
        if $T[i] = 1$ **then** perform step number $count$ in procedure $UCW(N)$
        else stay idle for one step
        $count \leftarrow count + 1 \mod \lambda(N)$

For label L, let $L|_j$ be the first j positions in the binary representation of L, where $16 \leq j \leq |L|$.

Protocol RENDEZVOUS(L)
for $epoch\ j \leftarrow 16, 17, \ldots |L| - 1$ **do**
    $RV(L|_j, \lceil \log_4 j \rceil)$
$RV(L, \infty)$

Note that the agent *moves* only when a step number $count$ from procedure UCW is done, otherwise it *idles*. The execution of procedure $RV(L|_j, \lceil \log_4 j \rceil)$, for $1 \leq j < |L|$, is called the *j-th epoch*, while the execution of procedure $RV(L, \infty)$ is called the $|L|$-*th epoch*. Stage r of epoch j is partitioned into *blocks*, each consisting of consecutive 4^r moving steps or consecutive 4^r idle steps. Since the smallest epoch number is 16, the length of a label in every call to procedure RV is at least 16.

Theorem 2. *Protocol* RENDEZVOUS *achieves rendezvous in time* $O(\log^3 \ell + (\lambda(n))^3)$.

Proof. The idea is similar to the proof of theorem 1. We show that after $O(\log^3 \ell + (\lambda(n))^3)$ steps since the startup of agent L_2 there is a segment of length $O(\log^3 \ell + (\lambda(n))^3)$ in which one agent is idle while the other one makes at least $\gamma_N(n)$ moving steps (perhaps separated with idle steps) during a single call to procedure $UCW(N)$ for some $N \geq n$, which by Lemma 6 guarantees the rendezvous in this segment. Due to the lack of space the detailed proof will appear in the full version of the paper.   □

Applying upper bound $O(n^5 \log^4 n)$ on $\lambda(n)$ from Lemma 5, as well as Lemma 2, we get the following results.

Corollary 2. *Protocol* RENDEZVOUS *achieves rendezvous/gathering in time* $O(\log^3 \ell + n^{15} \log^{12} n)$.

4.3 Conclusions

We considered the problem of deterministic rendezvous and gathering with arbitrary startup in anonymous networks. For rings we presected an optimal protocol, reaching the lower bound $\Theta(n \log \ell)$ from [18]. For arbitrary connected graphs we showed a deterministic rendezvous protocol polynomial in n and $\log \ell$, and independent of τ, which gives a positive answer to the question stated in [18, 22] about the existence of such a protocol.

The following problems seem to be an interesting challenge for future research:

Reducing complexity. Can delay-independent rendezvous/gathering in general networks be made practical by eliminating non-constructive ingredients from the protocol and/or lowering degrees of polynomials in complexity formula?

Simultaneous rendezvous/gathering and exploring. We say that a walk in a graph is T-*exploring* if during its any interval of length T it visits all nodes of the graph. Does there exist an optimal delay-independent rendezvous protocol in rings which is $O(n)$-exploring? Does there exist a delay-independent rendezvous/gathering protocol in general networks polynomial in n and ℓ (or better $\log \ell$) which is also T-exploring, where T is a function polynomial in n and ℓ (or better $\log \ell$)?

Dynamic and fault-tolerant settings. Although we did not address this subject here, it is clear that our protocol for rings is robust with respect to some dynamic changes of the network (inserting/deleting nodes) and non-permanent faults. What is the degree of this robustness? How about the case of general networks?

References

1. M. Adler, H. Racke, C. Sohler, N. Sivadasan, and B. Voecking, Randomized pursuit-evasion in graphs, Proc. 29th Int. Colloquium on Automata, Languages and Programming (ICALP'2002), 901-912.
2. R. Aleliunas, R.M. Karp, R.J. Lipton, L. Lovász, and C. Rackoff, Random walks, universal traversal sequences, and the complexity of maze problems, Proc. 20th Annual Symposium on Foundations of Computer Science (FOCS'1979), 218-223.
3. S. Alpern, The rendezvous search problem, SIAM J. on Control and Optimization 33 (1995), 673-683.
4. S. Alpern, Rendezvous search on labelled networks, Naval Research Logistics 49 (2002), 256-274.
5. S. Alpern, and S. Gal, The theory of search games and rendezvous. Int. Series in Operations research and Management Science, Kluwer Academic Publisher, 2002.
6. J. Alpern, V. Baston, and S. Essegaier, Rendezvous search on a graph, Journal of Applied Probability 36 (1999), 223-231.
7. E. Anderson, and R. Weber, The rendezvous problem on discrete locations, Journal of Applied Probability 28 (1990), 839-851.
8. E. Anderson, and S. Fekete, Asymmetric rendezvous on the plane, Proc. 14th Annual ACM Symp. on Computational Geometry, 1998.
9. E. Anderson, and S. Fekete, Two-dimensional rendezvous search, Oper. Research 49 (2001), 107-118.

10. V. Baston, and S. Gal, Rendezvous on the line when the players' initial distance is given by an unknown probability distribution, SIAM J. on Control and Optimization 36 (1998), 1880-1889.
11. V. Baston, and S. Gal, Rendezvous search when marks are left at the starting points, Naval Reaserch Logistics 48 (2001), 722-731.
12. N. H. Bshouty, L. Higham, and J. Warpechowska-Gruca, Meeting times of random walks on graphs, Information Processing Letters 69(5) (1999), 259-265.
13. M. Cielibak, P. Flocchini, G. Prencipe, and N. Santoro, Solving the robots gathering problem, Proc. 30th International Colloquium on Automata, Languages and Programming (ICALP'2003), LNCS 2719, 1181-1196.
14. S.A. Cook and P. McKenzie, Problems complete for deterministic logarithmic space, Journal of Algorithms 8 (5) (1987), 385-394.
15. D. Coppersmith,, P. Doyle, P. Raghavan, and M. Snir, Random walks on weighted graphs, and applications to on-line algorithms, Proc. 22nd Annual ACM Symp. on Theory of Computing (STOC'1990), 369-378.
16. D. Coppersmith, P. Tetali, and P. Winkler, Collisions among random walks on a graph, SIAM J. on Discrete Math. 6 (1993), 363-374.
17. G. De Marco, L. Gargano, E. Kranakis, D. Krizanc, A. Pelc, and U. Vaccaro, Asynchronous deterministic rendezvous in graphs, Proc. 30th Int. Symp. on Math. Found. of Comp. Science (MFCS'2005), LNCS 3618, 271-282.
18. A. Dessmark, P. Fraigniaud, and A. Pelc, Deterministic rendezvous in graphs, Proc. 11th European Symposium on Algorithms (ESA'2003), LNCS 2832, 184-195.
19. P. Flocchini, G. Prencipe, N. Santoro, and P. Widmayer, Gathering of asynchronous oblivious robots with limited visibility, Proc. 18th Ann. Symp. on Theoretical Aspects of Comp. Science (STACS'2001), LNCS 2010, 247-258.
20. S. Gal, Rendezvous search on the line, Operations Research 47 (1999), 974-976.
21. A. Israeli, and M. Jalfon, Token management schemes and random walks yield self stabilizing mutual exclusion, Proc. 9th ACM Symp. on Principles of Distributed Computing (PODC'1990), 119-131.
22. D. Kowalski, and A. Pelc, Polynomial deterministic rendezvous in arbitrary graphs, Proc. 15th Annual Symp. on Algorithms and Computation (ISAAC'2004), LNCS 3341, 644-656.
23. E. Kranakis, D. Krizanc, N. Santoro, and C. Sawchuk, Mobile agent rendezvous in a ring, Proc. 23rd Int. Conference on Distributed Computing Systems (ICDCS'2003), 592-599.
24. W. Lim, and S. Alpern, Minimax rendezvous on the line, SIAM J. on Control and Optimization 34 (1996), 1650-1665.
25. A.J. Mayer, R. Ostrovsky, and M. Yung, Self-stabilizing algorithms for synchronous unidirectional rings, Proc. 7th Annual ACM-SIAM Symposium on Discrete Algorithms (SODA'1996), 564-573.
26. Motwani, Raghawan, Randomized Algorithms, Cambridge University Press, 1995.
27. T. Schelling, The strategy of conflict, Oxford University Press, Oxford, 1960.
28. L. Thomas, Finding your kids when they are lost, Journal on Operational Res. Soc. 43 (1992), 637-639.
29. X. Yu, and M. Yung, Agent rendezvous: a dynamic symmetry-breaking problem, Proc. International Colloquium on Automata, Languages, and Programming (ICALP'1996), LNCS 1099, 610-621.

Setting Port Numbers
for Fast Graph Exploration

David Ilcinkas*

LRI, Université Paris-Sud, France
ilcinkas@lri.fr

Abstract. We consider the problem of periodic graph exploration by a finite automaton in which an automaton with a constant number of states has to explore all unknown anonymous graphs of arbitrary size and arbitrary maximum degree. In anonymous graphs, nodes are not labeled but edges are labeled in a local manner (called *local orientation*) so that the automaton is able to distinguish them. Precisely, the edges incident to a node v are given port numbers from 1 to d_v, where d_v is the degree of v.

Periodic graph exploration means visiting every node infinitely often. We are interested in the length of the period, i.e., the maximum number of edge traversals between two consecutive visits of any node by the automaton in the same state and entering the node by the same port. This problem is unsolvable if local orientations are set arbitrarily. Given this impossibility result, we address the following problem: what is the mimimum function $\pi(n)$ such that there exist an algorithm for setting the local orientation, and a finite automaton using it, such that the automaton explores all graphs of size n within the period $\pi(n)$?

The best result so far is the upper bound $\pi(n) \leq 10n$, by Dobrev et al. [SIROCCO 2005], using an automaton with no memory (i.e. only one state). In this paper we prove a better upper bound $\pi(n) \leq 4n$. Our automaton uses three states but performs periodic exploration independently of its starting position and initial state.

1 Introduction

The task of visiting all nodes is fundamental when searching for data in a network. The specific case of periodic exploration is particularly useful for network maintenance, where every node has to be regularly checked. In this paper we consider the task of periodic exploration, in which a mobile entity, or robot, has to periodically visit every node of an unknown graph.

We assume that the graph is anonymous, i.e., the nodes are unlabeled. Note that node labels would not help much the robot anyway because, as we will see later, it is modeled as a finite automaton, and thus is unable to store even a

---

* Supported by the project "PairAPair" of the ACI Masses de Données, the project "Fragile" of the ACI Sécurité et Informatique, and the project "Grand Large" of INRIA.

P. Flocchini and L. Gąsieniec (Eds.): SIROCCO 2006, LNCS 4056, pp. 59–69, 2006.

single node label. To enable the robot to distinguish the different edges incident to a node, edges at a node v are assigned port numbers in $\{1, \ldots, d_v\}$ in a one-to-one manner, where d_v is the degree of node v. Such port-numbering is called a *local orientation*.

The robot is modeled by a deterministic finite automaton. More precisely, we consider Mealy automata. A Mealy automaton has a transition function f and a finite number of states. If the automaton enters a node of degree d through port i, in state s, then it switches to state s' and exits the node through port i', with $f(s, i, d) = (s', i')$. Since the transition function takes as input a port number, we say that the automaton is on an edge e towards the extremity v of e or, in short, is on (e, v). Such a pair is called a *position*.

We consider the problem of periodic graph exploration where the finite automaton has to explore any unknown anonymous connected graph of arbitrary size and arbitrary maximum degree. Periodically exploring a graph means visiting every node infinitely often. We are interested in the length of the *period*, i.e., the maximum number of edge traversals between two consecutive visits of any node by the automaton in the same configuration (i.e., same position and same state). Budach [4] proved that no finite automaton can explore all graphs if the local orientation is given by an adversary. Given this impossibility result, we adress the following problem:

Problem. What is the mimimum function $\pi(n)$ such that there exist an algorithm for setting the local orientation, and a finite automaton using it, such that the automaton explores all graphs of size n within the period at most $\pi(n)$?

A trivial upper bound on the period is $2m$, where m is the number of edges of the explored graph. One can indeed set the local orientation such that a right-hand-on-the-wall walk defined by $f(s, i, d) = (s, (i \bmod d) + 1)$ induces an eulerian cycle of the graph, where all edges are traversed twice, once in each direction. Dobrev et al. [10] presented a port-numbering algorithm, and an automaton using it, achieving a period of at most $10n$ for graphs of size n. Hence $\pi(n) \leq 10n$. The main advantage of their approach is that their automaton is ultimately simple: it is oblivious (i.e. it uses only one state). Using an oblivious automaton naturally solves the problem of setting the initial state. However, the good performance of the automaton in [10] relies on the fact that the agent must start the exploration by the edge with port number 1.

In this paper we prove that $\pi(n) \leq 4n - 2$. Our automaton is not oblivious but has only three states. Moreover, it performs periodic exploration independently from its starting position and initial state. Our port-numbering algorithm is based on a spanning tree of the graph and can be easily implemented in a distributed environment, and extended to dynamic networks.

1.1   Related Work

Exploration of unknown environments have been extensively studied in the literature (cf. [19, 21]). The environment can be modeled using geometry as a plan with obstacles or as a graph. In the latter case, moves are restricted to the edges of the graph. The graph setting can be further specified in two different ways. In [3, 8, 13, 17] the robot explores strongly connected directed graphs and it can move only in the head-to-tail direction of an edge, not vice-versa. In [4, 9, 11, 12, 15, 20, 23] the explored graph is undirected and the robot can traverse edges in both directions. Again two different assumptions are used in the literature: it is either assumed that nodes of the graph have unique labels which the robot can recognize (as in, e.g., [8, 12, 20]), or it is assumed that nodes are anonymous (as in, e.g., [3, 4, 11, 23]). We are concerned with the latter context.

It is often assumed that the robot has an unlimited amount of memory to perform his task. In this paper, we are interested in robots using very little memory. More precisely we want the robots to have only a constant number of memory bits. A very natural model in this case is the finite automaton. Budach [4] proved that no finite automaton can explore all graphs. Rollik [23] proved that even a finite team of finite automata cannot explore all planar cubic graphs. This result is improved in [6], in which the authors introduced an even more powerful machine, called the JAG, for Jumping Automaton for Graphs. A JAG is a finite team of finite automata that can permanently cooperate and that can use "teleportation" to move from their current location to the location of any other automaton. Cook and Rackoff [6] proved that no JAG can explore all graphs. It was proved later in [18] that an automaton requires at least n states to explore all graphs of size n. Reingold [22] proved a very challenging result stating that $SL = L$ by providing a log-space algorithm solving the USTCON problem. A consequence of his work is the existence of a robot with $O(\log n)$ bits performing exploration in n-node graphs, matching the lower bound of $\Omega(\log n)$ bits in [18].

Several papers investigated graph exploration in which nodes of the graph are provided with a whiteboard (as in, e.g., [1, 7, 17]). A whiteboard is a memory where the automaton can read, write and erase information. Initially, all whiteboards are empty. In this setting, exploration requires at least m edge traversals, where m is the number of edges in the graph, because any unexplored edge may lead to an unexplored node. It is proved in [5] that there is an algorithm coloring the nodes using only three colors, and a finite automaton using this coloring which can explore all graphs. The traversal is of length approximately $20m$. Other assumptions are used in the literature to improve the performances of algorithms (see, e.g., [14, 16]).

In this paper we restrict attention to fully anonymous graphs: nodes are not labeled and not colored, no whiteboard is provided, and the automaton is not allowed to use any marker on nodes or edges. Having in mind the impossibility result of Budach [4], the only freedom is the setting of the local orientation. This method is used by Dobrev et al. [10]. As stated before, the authors presented a port-numbering algorithm, and an oblivious automaton using it, achieving a period of at most $10n$ for graphs of size n.

1.2   Our Results

Our main result is the design of a very simple algorithm for setting the local orientation of any graph and the design of a 3-state automaton performing periodic exploration using the local orientation computed by the algorithm. The periodic traversal of the agent is of length at most $4n - 2$, where n is the number of vertices of G. Hence $\pi(n) \leq 4n - 2$. Moreover, the good performances of the exploration do not depend on the initial state and starting position of the automaton.

Our port-numbering algorithm is based on computing a spanning tree of the graph and constructing the local orientation from this spanning tree. We prove that our labeling scheme can be easily transformed in a distributed algorithm or used in a dynamic environment, answering open problems stated in [10].

2 The Port-Numbering and the Corresponding Automaton

We first describe our algorithm computing the local orientation of the edges. This algorithm is mainly based on coding a spanning tree of the graph by choosing the small port numbers for the edges of the spanning tree.

Next, we will present a 3-state Mealy automaton that explores the constructed spanning tree (plus some additionnal edges) in a DFS manner.

We will conclude this section by proving the correctness of our algorithm and of the corresponding automaton.

2.1 Local Orientation Algorithm

Let $G = (V, E)$ be a graph. Let us consider an arbitrary spanning tree T of G. Let $F \subseteq E$ be the set of edges of T. For any node $v \in V$, let F_v be the set of edges in F that are incident to v.

Definition 1. *A local orientation of the edges of the graph G is* compatible with *a spanning tree $T = (V, F)$ if and only if:*

- *for any edge $e \in E$, at least one of its two port numbers is 1 if and only if $e \in F$;*
- *for any node $v \in V$, the edges in F_v have their port numbers from 1 to $|F_v|$.*

We say that a local orientation of G is tree-oriented *if there exists a spanning tree T of G such that the local orientation is compatible with T.*

Our algorithm, called SMALL-PORTS, constructs local orientations that are tree-oriented. To fix attention, the algorithm uses the following local orientation.

Algorithm SMALL-PORTS

1. Pick a rooted spanning tree T of G. Let r be its root.
2. For any node $v \neq r$, assign port number 1 to the edge of T leading toward the root. At r, assign 1 to an arbitrary edge in F_r.
3. For any node v of G, assign arbitrarily port numbers from 2 to $|F_v|$ to the remaining edges of F_v, if any.

4. Finally, assign arbitrarily port numbers from $|F_v| + 1$ to d_v (the degree of v) to the edges that have not yet assigned port numbers, if any.

Clearly this local orientation is compatible with T.

Remark 1. SMALL-PORTS is very simple since it only requires the computation of a spanning tree to set the local orientation. Moreover, many applications use a spanning tree as underlying structure and in this case, SMALL-PORTS gets the spanning tree for free. The performance and simplicity of SMALL-PORTS has to be compared with the ones of the algorithm presented in [10]. SMALL-PORTS performs in time $O(m)$ whereas the algorithm in [10] performs in time $O(n^3)$.

Remark 2. Consider a graph G and a tree-oriented local orientation of G. There is a unique spanning tree T such that this local orientation is compatible with T. Namely, T is the tree composed of the $n - 1$ edges of G that have at least one of their port numbers equal to 1. Moreover there exist exactly two possible roots for T such that the local orientation can be obtained by running Algorithm SMALL-PORTS with this rooted spanning tree. These two roots are the two extremities of the unique edge with both port numbers equal to 1.

2.2 Description of the Exploring Automaton

Our exploring automaton, called $\mathcal{A}$, has three states: N (for *Normal*), T (for *Test*), and B (for *Backtrack*). The transition function f of the automaton is defined as follows. Here d denotes the degree of the current node, and i the incoming port number. (Recall that the second parameter outputed by the transition function is the output port number.)

$$f(N, i, d) = \begin{cases} (N, 1) & \text{if } i = d \\ (T, i + 1) & \text{if } i \neq d \end{cases}$$

$$f(T, i, d) = \begin{cases} (N, 1) & \text{if } i = 1 \text{ and } d = 1 \\ (T, i + 1) & \text{if } i = 1 \text{ and } d \neq 1 \\ (B, i) & \text{if } i \neq 1 \end{cases}$$

$$f(B, i, d) = (N, 1)$$

Intuitively, the automaton traverses an edge in state N when it knows that the edge is in the spanning tree, in state T when it does not know yet, and in state B when it knows that the edge does not belong to the spanning tree.

2.3 Correctness

Theorem 1. *Let G be a graph of size n, with a tree-oriented local orientation. Start the automaton $\mathcal{A}$ in an arbitrary state at any arbitrary position in the graph. After at most two steps, the automaton enters a closed walk P and explores it forever. Moreover, P is of length at most $4n - 2$ and contains all the nodes of G.*

Proof. Let G be an arbitrary graph and let n be its number of nodes. Assume that the local orientation is compatible with some spanning tree T. We first study the periodic behavior of the automaton, and then the initial transient regime.

Let v be an arbitrary node of G, and let e be its incident edge with port number 1. The removal of e in T results in two connected components (subtree). Let T' be the component containing v. Finally, let n' be the number of nodes of T'.

Claim. If the automaton $\mathcal{A}$ enters v through port 1 in a state different from B, then it eventually leaves v through port 1 in state N. Moreover, between these two events, it explores all nodes of T' in at most $4n' - 2$ edge traversals, and does not leave any node not in T' through port 1 during those traversals.

We prove this by induction on the height h of T' rooted in v, i.e., the eccentricity of v in T'. The case $h = 0$ corresponds to v leaf of T. If v is also a leaf in G, then the automaton immediately leaves v through port 1 in state N and the claim is proved. Therefore we assume that $\deg(v) > 1$. By hypothesis of the claim, the automaton enters v in state T or N. In both cases, it switches to state T, and traverses the edge e' of port number 2. v is a leaf of T and since e is in T, e' is not. Thus the port number of e' at the other extremity is not equal to 1. Hence the automaton comes back to v in state B, and finally leaves v through port 1 after $4 \cdot 1 - 2 = 2$ edge traversals, which proves the basis of the induction.

Let us now consider the case $h > 0$. Let d be the degree of v. We have $d \neq 1$ because v is incident to e and $\mathrm{depth}(T') > 0$. For $i \geq 2$, let v_i be the node at the other extremity of the edge e_i with port number i at v. If e_i is in T, then let T_i be the connected component of $T' \setminus \{e_i\}$ containing v_i. Finally, let p be the largest port number of an edge in T incident to v. We have $p \geq 2$ because v is not a leaf in T. By hypothesis of the claim, the automaton enters v in state T or N. In both cases, it switches to state T, and traverses the edge e_2 of port number 2. Assume that the automaton leaves v through port i in state T, with $2 \leq i \leq p$. It reaches node v_i. By induction hypothesis on h, the automaton eventually comes back from v_i to v through port i, in state N, after at most $4n_i - 2$ edge traversals. (Note that during these traversals, the automaton may have visited nodes outside T_i but it never left these nodes through port 1.) If $i \neq d$, then the automaton leaves v through port $i + 1$ in state T. Hence, the automaton successively explores the subtrees T_i.

If $p = d$, then the automaton eventually leaves v through port 1 in state N after finishing the exploration of T_p. If $p < d$, then the automaton takes the edge e_{p+1} in state T. Since e_{p+1} is not in the tree T, the port number of e_{p+1} at the other extremity is not equal to 1. Thus the automaton comes back to v in state B and finally leaves v through port 1 in state N. In both cases, it remains to bound the number of edge traversals. The automaton traversed $\sum_{i=2}^{p}(4n_i - 2) = 4(n'-1) - 2(p-1)$ edges during the exploration of the subtrees T_i. It also traversed twice each edge e_i, $2 \leq i \leq p$. Finally there are possibly two additional edge traversals, in the case $p \neq d$. To summarize, the number of edge traversals is at most $4(n' - 1) - 2(p - 1) + 2(p - 1) + 2 = 4n' - 2$. This concludes the proof of the claim.

We now use the previous claim to exhibit the closed path traversed periodically by the automaton. There is a unique edge $e = \{v, v'\}$ with both its port numbers equal to 1. Assume that the automaton is at position (e, v) in state N. Applying the claim, the automaton explores the subtree of $T \setminus \{e\}$ rooted in v, comes back to v and goes at position (e, v') in state N. Applying again the claim, the automaton explores the subtree of $T \setminus \{e\}$ rooted in v', comes back to v' and goes at position (e, v) in state N. Therefore the automaton traverses a closed walk P of length at most $4n - 2$ visiting all nodes of T, and thus of G.

It remains to prove that the automaton enters P after at most two edge traversals. The automaton starts in an arbitrary state at an arbitrary position. By definition of the transition function of the automaton, there are three cases:

- Case 1: the automaton leaves the node through port 1 in state N. This implies that the automaton immediately enters the closed walk P.
- Case 2: the automaton leaves the node in state B. The next edge traversal is then along the edge with port number 1, in state N. This implies that the automaton enters P during the second traversal.
- Case 3: the automaton leaves the node v by edge e of port number i, with $i \geq 2$, in state T. Assume that either e is in T or e is the edge with the smallest port number that is not in T. In this case the edge traversal is in the closed walk. If it is not the case, then the port number j at the other extremity u of e is not equal to 1 because e is not in T. Hence the automaton switches to state B at u, and comes back to v by e. Then, it leaves v through port 1 in state N. This latter edge traversal is in P.

Finally, in all cases, the automaton enters the closed walk after at most two edge traversals. □

3   Additional Properties

In the previous section, we presented a simple algorithm, using a spanning tree of the graph, to set the local orientation of the edges, and a 3-state automaton that performs periodic exploration in time at most $4n$ using this orientation, where n is the number of nodes of the explored graph. We prove that thanks to the robustness and simplicity of our approach, it is possible to use our algorithm in a distributed environment, and in dynamic networks.

3.1   The Distributed Variant

The distributed construction of a tree spanning an anonymous graph may be impossible if the graph has symmetry. However this task is possible if a single node initiates it. In our setting, we use the automaton to break the symmetry between nodes. The starting position of the automaton is used as the distinguished node, that becomes the root of the spanning tree. This node will wake up all the other nodes of the network by flooding. A node distinct from the root chooses his parent as the node from which it received the wakeup message (ties are broken arbitrarily). Finally, the technique described in Section 2.1 is used to set up the local orientation, based on the constructed spanning tree.

More precisely, the distributed variant of our algorithm, called DISTRIBUTED-SMALL-PORTS, proceeds as follows. At the beginning, only the node hosting the automaton is awake. This node is the root r of the future spanning tree. It starts the process by sending a "Hello" message to all its neighbors. A node v, except the root, is said to be awake when it has received at least one message. An awake node v chooses as parent the sender of the first message it has received. Ties are broken arbitrarily. Finally v sends a "Parent" message to the neighbor choosen as its parent and a "Hello" message to all its other neighbors.

When a node u has received a message from all its neighbors, it chooses the local orientation as follows. Let p be the number of "Parent" messages node u has received.

- If u is the root, then it assigns arbitrarily port numbers from 1 to p to the p edges leading to the senders of "Parent" messages. It assigns the remaining port numbers, if any, to the remaining edges arbitrarily.
- If u is not the root, then it assigns port number 1 to the neighbor that was choosen as its parent. Then it assigns arbitrarily port numbers from 2 to $p + 1$ to the p edges leading to the senders of a "Parent" message, if any. Finally it assigns the remaining port numbers, if any, to the remaining edges arbitrarily.

Theorem 2. *Algorithm* DISTRIBUTED-SMALL-PORTS *constructs a spanning tree of the graph and sets a local orientation compatible with it, using $2m$ messages.*

3.2   Exploration of Dynamic Networks

As proved in Theorem 1, the automaton periodically explores any graph in at most $4n$ steps, whatever the starting position and the initial state are, provided that the local orientation is compatible with some spanning tree of the graph. Therefore the automaton can be used in dynamic networks under the unique constraint that the local orientation of the network remains tree-oriented after every change.

We consider changes of the graph that keep it connected. A change of a graph can be decomposed in a sequence of the following basic changes.

- Addition of a new edge between two existing nodes;
- Addition of a new node, connected by a new edge to an existing node;
- Removal of an edge, without disconnecting the graph;
- Removal of a degree-1 node and of its unique incident edge.

Theorem 3. *In the case of a removal of an edge belonging to the spanning tree of a n-node graph G, $\Theta(n)$ modifications in the local orientation are necessary and sufficient to maintain it tree-oriented. Our algorithm updating the local orientation performs in time $O(m)$ in this case, where m is the number of edges. In all other cases, the update of the local orientation can be done in constant time, and thus with a constant number of modifications.*

Proof. Local orientations are updated as follows:

- *Addition of an edge.* This edge is not placed in the spanning tree. Let u and v be the two extremities of the new edge e and let d_u and d_v be their new respective degree. We set d_u, respectively d_v, as the port number of edge e at u, respectively v.

- *Addition of a leaf.* The new edge e connecting the new node u to node v of the existing graph is necessarily in the spanning tree. Let d be the degree of v and let p be the largest port number at v corresponding to an edge in the spanning tree, before modification. If $p = d$, then the port number of edge e at v is $d + 1$. Otherwise $(p \neq d)$, the edge with port number $p + 1$ has now the port number $d + 1$ and edge e has the port number $p + 1$ at v. Edge e is assigned port number 1 at u.

- *Removal of an edge.* If the removed edge e does not belong to the spanning tree T, then let u and v be its two extremities. We describe the modifications of the local orientation in node u. The modifications in v are done similarly. Let i be the port number of e at u. Let d be the degree of u before the removal of e. Finally, let e' be the edge incident to u with port number d. If $e = e'$ (i.e., $i = d$), then no port number is modified at u. If $e \neq e'$, then we set i as the new port number of e' at u.

  If edge e belongs to the spanning tree T, then T without edge e is not connected. Since we assume that the graph remains connected, there exists an edge e' in the new graph connecting the two parts of T. This edge e' is added to the tree. Some port numbers have to be changed so that the local orientation become compatible with the resulting spanning tree. We claim that only a constant number of port numbers have to be modified at each node. At the extremities of e and e', the set of tree-edges incident to it changes. However, at most two edges are concerned. Apart from this, the only modifications to do concern the choice of the incident edge with port number 1. A switch between two port numbers is sufficient. Therefore, at most a constant number of port numbers are modified at each node.

- *Removal of a leaf.* Let v be the node connected to the removed leaf u. Let i be the port number at v of the edge leading to u. Let p be the largest port number at v of an edge in T. Let d be the degree of v before the removal of e. Finally, let e', resp. e'', be the edge incident to v with port number p, resp. d. Since edge e is in the tree T, we have $i \leq p \leq d$. We modify the port number of e', resp. e'', if and only if $i \neq p$, resp. $p \neq d$. If $i \neq p$, then we set i as the new port number of e'. If $p \neq d$, then we set p as the new port number of e''.

In all cases, the other port numbers in the graph remain inchanged.

It may not be possible to avoid a linear number of modifications in the case of the removal of a tree-edge. For example, consider a cycle C of odd length $2n+1$. To simplify the description, let us give names from 1 to $2n+1$ to the nodes. For any node $i \leq n$, resp. $i > n$, 1 is the port number of the edge leading to node $i+1$, resp. $i-1$. Thus the local orientation is compatible with the path starting at node 1 and ending at node $2n+1$. Now assume that the edge $\{n, n+1\}$ is removed. The port numbers at node n and $n+1$ are set to 1 since they are now leaves. All edges are necessarily in the spanning tree but the edge $\{2n+1, 0\}$ has both its port numbers equal to 2. The local orientation is not tree-oriented. In fact, in a tree-oriented orientation, exactly one edge e must have both its port numbers equal to 1. Moreover, for any node v, excluding the extremities of e, the edge with port number 1 must point toward e, i.e., the edge must be in the path from v to the closer extremity of e. Hence, the local orientation has to be modified in at least n nodes to obtain a tree-oriented local orientation.   □

4   Further Investigations

In this paper, we proved the upper bound $4n - 2$ on the minimal period $\pi(n)$ for periodic graph exploration by a finite automaton. Our algorithm uses an arbitrary spanning tree to set the local orientations. The automaton explores this spanning tree plus at least one additional edge per node. It seems difficult to avoid these additional edge traversals. Hence $4n - O(1)$ may be optimal for tree-based approach. We conjecture that this bound cannot be improved even with other techniques.

Conjecture. $\pi(n) = 4n - O(1)$.

Since graphs are anonymous, using an extensive amount of memory does not help much. Therefore finding the minimal period for machine with unbounded memory may be very challenging.

Open problem. What is the mimimum period $\psi(n)$ such that there exists an algorithm setting the local orientations and a robot with unlimited memory such that the automaton explores any graph of size n within the period $\psi(n)$?

Finally, it remains open if the period $10n$ proved in [10] can be improved if the robot is restricted to be oblivious.

References

1. Y. Afek and E. Gafni. Distributed Algorithms for Unidirectional Networks. SIAM J. Computing 23(6):1152-1178, 1994.
2. S. Albers and M. R. Henzinger. Exploring unknown environments. SIAM J. Computing 29:1164-1188, 2000.
3. M. Bender, A. Fernandez, D. Ron, A. Sahai and S. Vadhan. The power of a pebble: Exploring and mapping directed graphs. Information and Computation 176(1):1-21, 2002.
4. L. Budach. Automata and labyrinths. Math. Nachrichten, pages 195-282, 1978.
5. R. Cohen, P. Fraigniaud, D. Ilcinkas, A. Korman and D. Peleg. Label-Guided Graph Exploration by a Finite Automaton. In 32nd Int. Colloq. on Automata, Languages & Prog. (ICALP), LNCS 3580, pages 335-346, 2005.
6. S. Cook and C. Rackoff. Space lower bounds for maze threadability on restricted machines. SIAM J. on Computing 9(3):636–652, 1980.

7. S. Das, P. Flocchini, A. Nayak, and N. Santoro. Distributed Exploration of an Unknown Graph. In 12th Colloquium on Structural Information and Communication Complexity (SIROCCO), LNCS 3499, pages 99-114, 2005.

8. X. Deng and C. H. Papadimitriou. Exploring an unknown graph. J. Graph Theory 32(3):265-297, 1999.

9. K. Diks, P. Fraigniaud, E. Kranakis, and A. Pelc. Tree Exploration with Little Memory. J. Algorithms 51(1):38-63, 2004.

10. S. Dobrev, J. Jansson, K. Sadakane, and W.-K. Sung. Finding Short Right-Hand-on-the-Wall Walks in Graphs. In 12th Colloquium on Structural Information and Communication Complexity (SIROCCO), LNCS 3499, pages 127-139, 2005.

11. G. Dudek, M. Jenkin, E. Milios, and D. Wilkes. Robotic Exploration as Graph Construction. IEEE Transaction on Robotics and Automation 7(6):859-865, 1991.

12. C. Duncan, S. Kobourov, and V. Kumar. Optimal constrained graph exploration. In 12th Ann. ACM-SIAM Symp. on Discrete Algorithms (SODA), pages 807-814, 2001.

13. R. Fleischer and G. Trippen. Exploring an unknown graph efficiently. In 13th Annual International Symposium on Algorithms (ESA), LNCS 3669, pages 11-22, 2005.

14. P. Flocchini, B. Mans, and N. Santoro. Sense of direction in distributed computing. Theoretical Computer Science 291(1):29-53, 2003.

15. P. Fraigniaud, L. Gasieniec, D. Kowalski, and A. Pelc. Collective Tree Exploration. In 6th Latin American Theoretical Informatics (LATIN), LNCS 2976, pages 141-151, 2004.

16. P. Fraigniaud, C. Gavoille, and B. Mans. Interval routing schemes allow broadcasting with linear message-complexity. Distributed Computing 14(4):217-229, 2001.

17. P. Fraigniaud, and D. Ilcinkas. Digraphs Exploration with Little Memory. In 21st Symposium on Theoretical Aspects of Computer Science (STACS), LNCS 1996, pages 246-257, 2004.

18. P. Fraigniaud, D. Ilcinkas, G. Peer, A. Pelc, and D. Peleg. Graph Exploration by a Finite Automaton. In 29th International Symposium on Mathematical Foundations of Computer Science (MFCS), LNCS 3153, pages 451-462, 2004.

19. A. Hemmerling. Labyrinth Problems: Labyrinth-Searching Abilities of Automata. Volume 114 of Teubner-Texte zur Mathematik. B. G. Teubner Verlagsgesellschaft, Leipzig, 1989.

20. P. Panaite and A. Pelc. Exploring unknown undirected graphs. J. Algorithms 33(2):281-295, 1999.

21. N. Rao, S. Kareti, W. Shi, and S. Iyengar. Robot navigation in unknown terrains: Introductory survey of non-heuristic algorithms. Tech. Report ORNL/TM-12410, Oak Ridge National Lab., 1993.

22. O. Reingold. Undirected ST-Connectivity in Log-Space. In 37th ACM Symp. on Theory of Computing (STOC), pages 376-385, 2005.

23. H. Rollik. Automaten in planaren Graphen. Acta Informatica 13:287-298, 1980 (also in LNCS 67, pages 266-275, 1979).

Distributed Chasing of Network Intruders

Lélia Blin[1], Pierre Fraigniaud[2], Nicolas Nisse[2], and Sandrine Vial[1]

[1] IBISC, University of Evry, 91000 Evry, France
[2] LRI, CNRS and Université Paris-Sud, 91405 Orsay, France

Abstract. This paper addresses the graph searching problem in a distributed setting. We describe a distributed protocol that enables searchers with logarithmic size memory to clear any network, in a fully decentralized manner. The search strategy for the network in which the searchers are launched is computed online by the searchers themselves *without knowing the topology of the network in advance*. It performs in an asynchronous environment, i.e., it implements the necessary synchronization mechanism in a decentralized manner. In every network, our protocol performs a connected strategy using at most $k + 1$ searchers, where k is the minimum number of searchers required to clear the network in a monotone connected way, computed in the centralized and synchronous setting.

1 Introduction

Graph searching [18] is one of the most popular tool for analyzing the chase for a powerful and hostile agent, by a set of software agents in a network. Roughly speaking, graph searching involves an *intruder* and a set of *searchers*, all moving from node to node along the links of a network. The intruder is powerful in the sense that it is supposed to move arbitrarily fast, and to be permanently aware of the positions of the searchers. However, the intruder cannot cross a node or an edge occupied by a searcher without being caught. Conversely, the searchers are unaware of the position of the intruder. They are aiming at surrounding the intruder in the network. The intruder is caught by the searchers when a searcher enters the node it occupies. For instance, one searcher can catch an intruder in a path (by moving from one extremity of the path to the other extremity), while two searchers are required to catch an intruder in a cycle (starting from the same node, the two searchers move in opposite directions). In addition to network security, graph searching has several other practical motivations, such as rescuing speleologists in caves [6] or decontaminating a set of polluted pipes [19]. It has also several applications to the Graph Minor theory as it provides a dynamic approach to the analysis of static graph parameters such as treewidth and pathwidth [4].

1.1 The Problem

The main question addressed by graph searching is: given a graph G, what is the *search number* of G? That is, what is the minimum number of searchers, $s(G)$,

P. Flocchini and L. Gąsieniec (Eds.): SIROCCO 2006, LNCS 4056, pp. 70–84, 2006.

required to *clear* the graph G, i.e., to capture the intruder? This question is motivated by, e.g., the need for consuming the minimum amount of computing resources of the network at any time, while clearing it. The decision problem corresponding to computing the search number $s(G)$ of a graph G is NP-hard [18], and NP-completeness follows from [5, 16]. Computing the search number is however polynomial for trees [17, 18], and the corresponding *search strategy* can be computed in linear time [20]. In fact, the search number of a graph is known to be roughly equal to the pathwidth, pw, of the graph, and therefore the search number of an n-node graph can be approximated in polynomial time, up to multiplicative factor $O(\log n \sqrt{\log tw})$ where tw denotes the treewidth of the graph (see [7], and use the fact that $pw/tw \leq O(\log n)$).

The graph searching problem has given rise to a vast literature, and several variants of the problem have been considered (see, e.g., [14, 15]). Nevertheless, from a distributed systems point of view, the existing solutions for the graph searching problem (cf., e.g., [17, 18, 20]) suffer from a serious drawback: they are mostly centralized. In particular, (1) the search strategy for every network is computed based on the knowledge of the entire topology of the network, and (2) the moves of the searchers are controlled by a centralized mechanism that decides at every step which searcher has to move, and what movement it has to perform. These two facts limit the applicability of the solutions. Indeed, as far as networking or speleology is concerned, the topology of the network is often unknown, or its map unprecise. The topology can even evolve with time (either slowly as for, e.g., Internet, or rapidly as for, e.g., P2P networks). Moreover, the mobile entities involved in the search strategy can hardly be controlled by a central mechanism dictating their actions. All these constraints make centralized algorithms inappropriate for many instances of the graph searching problem.

This paper addresses the graph searching problem in a *distributed* setting, that is the searchers must compute their own search strategy for the network in which they are currently running. This distributed computation must not require knowing the topology of the network in advance, and the searchers must act in absence of any global synchronization mechanism, hence they must be able to perform in a fully asynchronous environment. Distributed strategies have been proposed for specific topologies only, such as trees [2], hypercubes [9], and rings and tori [8]. In this paper, we address the problem in arbitrary topologies.

1.2   The Model

The searchers are modeled by autonomous mobile computing entities with distinct IDs. More precisely, they are labeled from 1 to the current number k of searchers in the network (if a new searcher has to join the team, it will take number $k+1$). Otherwise searchers are all identical, and run the same program. The network and the searchers are asynchronous in the sense that every action of a searcher takes a finite but unpredictable amount of time. Moreover, motivated by the fact that the intruder models a potentially hostile agent that can, e.g., corrupt the node memories, the search strategy must perform independently from any local information stored at nodes a priori, and even independently from the node IDs. We thus consider *anonymous* networks, i.e., networks in which nodes do not have labels, or these labels are not accessible to the searchers. The $\deg(u)$ edges incident to any node u are labeled from 1 to $\deg(u)$, so that the searchers can distinguish the different edges incident to a node. These labels are called *port numbers*. Every node of the network has a whiteboard in which searchers can read, erase, and write symbols. (A whiteboard is modeling a specific zone of the local node memory that is reserved for the purpose of exchanging information between software agents). At every node, the local whiteboard is assumed to be accessible by the searchers in fair mutual exclusion. Since the content of the whiteboard at every node accessible by the intruder is corruptible, it is the role of the searchers to protect information stored at nodes' whiteboards.

The decisions taken by a searcher at a node (moving via port number p, writing the word w on the whiteboard, etc.) is local and depends only on (1) the current state of the searcher, and (2) the content of the node's whiteboard (plus possibly (3) the incoming port number, if the searcher just entered the node).

The powerful intruder is assumed to be aware of the edge-labeled network topology, and thus it does not need the whiteboards to navigate. In fact, as mentioned before, when the intruder enters a node that is not occupied by a searcher, then it can modify or even remove the content of the local whiteboard.

All searchers start from the same node u_0, called the *entrance* of the network, or the *homebase* of the searchers. This node u_0 is also a *source* of searchers, in the sense that if the current team of searchers realize that they are not numerous enough for clearing the network, then they can ask for a new searcher, that will appear at the source. Initially, one searcher spontaneously appears at the source. The size of the team will increase until it becomes large enough to clear the network. Basically, the searchers are aiming at expanding a cleared zone around their homebase u_0, that is at expanding a *connected* sub-network of the network G, containing u_0, until the whole network is clear. In particular, as the entrance u_0 of the network is a critical node, it has to be permanently protected from the intruder in the sense that the intruder must never be able to access it.

Among all search strategies, *monotone* ones play an important role. A monotone strategy insures that, once an edge has been cleared, it will always remain clear. Monotone strategies guaranty a polynomial number of moves: exactly one move for clearing every edge, plus few moves required by the searchers to set up their positions before clearing the next edge. In the connected setting, the corresponding graph searching parameter is called *monotone connected search number* starting at u_0 (cf., [2, 3, 13]), and is denoted by $\mathtt{mcs}(G, u_0)$.

1.3   Our Results

We describe a distributed protocol, called $\mathtt{dist\_search}$, that enables the searchers to clear any asynchronous network in a fully decentralized manner, i.e., the search strategy is computed online by the searchers themselves, after being launched in the network without any information about its topology. To the best of our knowledge, this is the first distributed protocol that addresses the graph searching problem in its whole generality, i.e., for arbitrary network topologies.

The distributed search strategy self-computed by the searchers in an asynchronous environment uses a number of searchers very close to the optimal. Indeed, we prove that the number of searchers involved in the strategy computed by our protocol in a network G is equal to 1 plus the minimum number of searchers required to clear G by a monotone connected search strategy starting at u_0, i.e., is equal to $\mathtt{mcs}(G, u_0) + 1$. It is known [13] that $\mathtt{mcs}(G, u_0) \leq \mathtt{s}(G) \lceil \log n \rceil$. Hence our protocol is optimal up to a logarithmic factor.

Our protocol is space-efficient from many respects. In particular, it requires only $O(\log k)$ bits of memory for each of the k searchers involved in the search. This amount of memory is independent from the size n of the network. Moreover, the amount of information stored at every whiteboard never exceeds $O(m \log n)$ bits, where m is the number of edges of the network.

To obtain our results, we had to address several problems. First, since the network is a priori unknown to the searchers, they have to explore it. However, this exploration cannot be achieved easily because of the potential corruption of the whiteboards by the intruder. Our protocol insures that exploration and searching are performed somehow simultaneously, and that the whiteboards of cleared nodes remain permanently protected unless there is no need to protect the stored information anymore. Second, as the searchers asynchronously spread out in the network, they become rapidly unaware of their relative positions. Our protocol synchronizes the searchers in a non trivial manner so that an action by a searcher is not ruined by the action of another searcher. Finally, to obtain space-efficient solutions, our protocol takes advantage from the accesses to the whiteboards, to store and read information useful to the searchers: it maintains a stack at every whiteboard, and every searcher at a node has access only to the top of a stack stored locally on the current node's whiteboard, and to few other variables also stored on the whiteboard.

2   Main Result and Sketch of the Protocol

The following theorem summarizes the main characteristics of $\mathtt{dist\_search}$.

Theorem 1. *For any connected, asynchronous, and anonymous network G, and any $u_0 \in V(G)$, $\mathtt{dist\_search}$ enables capturing an intruder in G using searchers starting from the homebase u_0, and initially unaware of G. The main characteristics of $\mathtt{dist\_search}$ are the following: (1) $\mathtt{dist\_search}$ uses at most $k = \mathtt{mcs}(G, u_0) + 1$ searchers if $\mathtt{mcs}(G, u_0) > 1$, and $k = 1$ searcher if $\mathtt{mcs}(G, u_0) = 1$; (2) Every searcher involved in the search strategy computed by $\mathtt{dist\_search}$ uses $O(\log k)$ bits of memory; (3) During the execution of $\mathtt{dist\_search}$, at most $O(m \log n)$ bits of information are stored at every whiteboard.*

Note that the theorem above implies that for networks searchable by a monotone connected search strategy using a constant number of searchers, the protocol $\mathtt{dist\_search}$ can be implemented using finite state automata.

Let us briefly sketch Protocol `dist_search` and its proof. Given a connected network G, and $X \subseteq E(G)$, we denote by $\delta(X)$ the nodes in $V(G)$ that are incident to an edge in X and an edge in $E(G) \setminus X$. Given $k \geq 1$, we call k-configuration any set $X \subseteq E(G)$ such that $|\delta(X)| \leq k$. The k-*configuration di-graph* $\mathcal{C}_k$ of G is defined as follows. $V(\mathcal{C}_k)$ is the set of all possible k-configurations. There is an arc from X to X' in $\mathcal{C}_k$ if the configuration X' can be reached from X by one step of a monotone connected search strategy using at most k searchers (a *step* of a monotone connected search strategy starting at node u_0 is the action consisting in moving a searcher along an edge, all searchers being initially at u_0). The objective of Protocol `dist_search` is essentially to try, for successive $k = 1, 2, \ldots$, whether the configuration graph $\mathcal{C}_k$ can be traversed from $\emptyset$ to $E(G)$ under the constraint that the searchers starts at u_0. If yes, then `dist_search` completes after having captured the intruder using $\leq k$ searchers. Otherwise, `dist_search` tries with $k + 1$ searchers. Note that this approach is similar to the (centralized) parametrized algorithms of the literature (cf., e.g., [1, 10, 11]). However, the difficulty of our approach is to discover whether the configuration digraph $\mathcal{C}_k$ can be traversed from $\emptyset$ to $E(G)$ in a *decentralized* manner.

For a fixed k, the objective of `dist_search` is to organize the movements of the searchers so that they perform a DFS of $\mathcal{C}_k$ (again, ignoring the topology of G, and in an asynchronous environment). This objective is achieved according to an order specified by a *virtual* stack in which are stored information related to the moves of the searchers. Roughly, Protocol `dist_search` constructs all possible states for the virtual stack, according to a lexicographic order on the states of the stack. The difficulty of the protocol is to distribute the virtual stack on the whiteboards so that when a searcher visits a node, it finds on the whiteboard enough information for computing the next step of the search strategy that it should perform. Since the intruder can corrupt the whiteboards, withdrawals from previously visited nodes must be scheduled so that to make sure that no information will be lost. Note here that, albeit the search strategy eventually computed by the searchers is monotone (in the sense that the contents of all the whiteboards describe a monotone search strategy when the protocol completes), failing search strategies investigated before (according to the lexicographic order on the states of the virtual stack) lead to withdrawals, and therefore to recontamination. If all strategies with k searchers have failed, then the searchers terminate at the homebase, call a new searcher, and restart searching the network with $k+1$ searchers.

The additional searcher used by `dist_search`, compared to $\text{mcs}(G, u_0)$, is required for avoiding deadlocks. It is also used to schedule the moves of the other searchers and to transmit few information between the searchers. It could be replaced by simple communication facilities. For instance, if the searchers would have the ability to send to and read from a mailbox available at the homebase, this additional searcher could be avoided. In particular, in the Internet, each searcher would just have to keep in its memory the IP address of the homebase.

The proof of correctness of Protocol `dist_search` is twofold. First, we prove the correctness of an algorithm, denoted by $\mathcal{A}$, that uses a centralized stack for traversing the configuration digraph $\mathcal{C}_k$. The second part of the proof consists in proving a one-to-one correspondence between every execution of dist_search using a virtual (i.e., decentralized) stack, and every execution of $\mathcal{A}$ using a centralized stack.

3   Search Strategy Using a Centralized Stack

In this section, we describe the algorithm $\mathcal{A}$ enabling a team of searchers launched in an unknown network to capture an intruder hidden in this network. Algorithm $\mathcal{A}$ is not fully distributed because it uses a centralized stack whose top is accessible from every node by every searchers.

3.1   Description of Algorithm $\mathcal{A}$

Algorithm $\mathcal{A}$ uses the notion of *extended moves*, that are triples (a_i, a_j, p) where a_i and a_j denote searchers, and p is a port number.

Definition 1. *An extended move (a_i, a_j, p) corresponds to the following: (1) searcher a_i joins searcher a_j, and (2) the searcher with the smallest ID among a_i and a_j leaves the node now occupied by the two searchers via port p. (Note that $i = j$ is allowed, in which case a_i leaves the node it occupies by port p).*

The central stack stores extended moves and thus describes a sequence of operations performed by the searchers. More precisely, reading the stack bottom-up defines a sequence of operations that describes a partial execution of a search strategy.

Definition 2. *For a fix parameter $k \geq 1$, a state of the virtual stack is* valid *if there exists a monotone connected search strategy using at most k searchers whose partial execution is described by this state.*

By some abuse of terminology, we sometime say that a stack Q is valid, meaning that the current state S of the stack Q is valid. Given a valid state S of a stack Q, we denote by X_S the configuration induced by S, that is X_S is the set of clear edges after the execution of the extended moves in S.

   The principle of Algorithm $\mathcal{A}$ is to try, for each $k = 1, 2, \ldots$, every possible monotone connected search strategy using k searchers, until one reaches a situation in which either the whole network is clear, or all search strategies have been exhausted. In the latter case, Algorithm $\mathcal{A}$ proceeds with $k+1$ searchers by calling for a new searcher at the homebase u_0. From now on, we assume that k is fixed. The k searchers are denoted by $a_1, \ldots, a_k$, where the ID of a_i is simply its index i. Algorithm $\mathcal{A}$ is described in Figure 1. It returns a boolean *possible*. If *possible* is true then clearing the network with k searchers is possible, in which case the stack Q returned by Algorithm $\mathcal{A}$ is valid, and contains a monotone connected search strategy clearing G with k searchers.

   In Algorithm $\mathcal{A}$, the stack Q is initially empty, and only a_1 is placed at u_0. the other searchers $a_2, \ldots, a_k$ are *available*. In addition to the centralized stack Q,

Algorithm $\mathcal{A}$ uses a global variable *state* that takes two possible values CLEAR or BACKTRACK whose meaning will appear clear later on. Finally, Algorithm $\mathcal{A}$ uses a boolean variable *decided* that is false until either a monotone connected search strategy using k searchers clearing the network is discovered, or all possible monotone connected search strategies using k searchers have been considered. Hence the main while-loop of Algorithm $\mathcal{A}$ is based on the value of *decided* (cf. Figure 1). This main while-loop mainly contains two blocks of instructions. These blocks are executed depending on the value of *state* (CLEAR or BACKTRACK).

Case CLEAR corresponds to a situation in which Algorithm $\mathcal{A}$ has just cleared an edge, i.e., the last execution of the main while-loop has resulted in pushing some extended move in Q. Case BACKTRACK corresponds to a situation when the last execution of main while-loop has resulted in popping the stack Q, i.e., has resulted in the recontamination of an edge.

Let us first focus on the case *state* $=$ CLEAR. Algorithm $\mathcal{A}$ focuses on specific extended moves, only those that do not imply recontamination (this is because $\mathcal{A}$ eventually computes a monotone strategy). More formally, let us consider a valid state S of the stack Q, i.e., S is a sequence of extended moves denoted by $M_1|\ldots|M_r$. Pushing an extended move M in Q results in a new state, denoted by $S|M$. We say that a extended move M is *valid according to Q* if $S' = S|M$ is a valid state. Note that $\mathcal{A}$ does not maintain the set X of clear edges and the set of available searchers. Indeed, given a valid state S of the stack Q, one can easily construct X_S by executing the partial search strategy described by S. A searcher is then *available* if either it stands at a node not in $\delta(X_S)$ or it stands at a node also occupied by a searcher of lower index. There is therefore a simple characterization of a valid extended move M according to a valid state S of Q: If $S = \emptyset$, then M is valid if and only if either u_0 is a 1-degree node and $M = (a_1, a_1, 1)$, or $k > 1$ and $M = (a_2, a_1, 1)$. If $S \neq \emptyset$, $M = (a_i, a_j, p)$ is valid according to Q if and only if either $i = j$, a_i stands at a node $u \in \delta(X_S)$, and p is the only contaminated port of node u, or $i \neq j$, a_i is available, a_j stands at a node $u \in \delta(X_S)$, and p is a contaminated port of node u.

The first instruction of the case *state* $=$ CLEAR consists in checking whether there exists a valid extended move according to Q. The key issue is to choose which extended move to apply, among all possible valid extended moves. For this choice, the extended moves are ordered in lexicographic order.

Definition 3. *Let $M = (a_i, a_j, p)$ and $M' = (a_{i'}, a_{j'}, p')$ be two extended moves. We define $M \prec M'$ if and only if either $(i < i')$, or $(i = i'$, and $j < j')$, or $(i = i'$, $j = j'$, and $p < p')$.*

If there is an extended move that is valid according to Q then Algorithm $\mathcal{A}$ chooses the one that has minimum lexicographic order among all extended moves that are valid according to Q. If there is no extended moves that are valid according to Q, then $\mathcal{A}$ switches to the state BACKTRACK. For this purpose, the last move in Q is popped out, and stored in the global variable M_{last}. If fact, if $Q = \emptyset$, then backtracking is not possible, and $\mathcal{A}$ decides that k searchers are not sufficient to clear the network.

```
Input: k ≥ 1 searchers a₁, a₂, ···, aₖ and a node u₀ of a graph G.
Output: a boolean possible, and a stack Q of extended moves.
begin
    Q ← ∅;
    state ← CLEAR;
    decided ← false;
    while not decided do
        if all searchers are available then
            decided ← true;
            possible ← true;
        else
            /* case state = CLEAR */
            if state = CLEAR then
                if there exists a valid extended move according to Q then
                    (aᵢ, aⱼ, p) ← minimum valid extended move according to Q;
                    push(aᵢ, aⱼ, p);
                else
                    if Q ≠ ∅ then
                        M_last ← pop();
                        state ← BACKTRACK;
                    else
                        decided ← true;
                        possible ← false;
            /* case state = BACKTRACK */
            else
                Let M_last = (aᵢ, aⱼ, p);
                if there exists a valid extended move according to Q larger than (aᵢ, aⱼ, p) then
                    (a'ᵢ, a'ⱼ, p') ← min valid extended move according to Q larger than (aᵢ, aⱼ, p);
                    push(a'ᵢ, a'ⱼ, p');
                    state ← CLEAR;
                else
                    if Q ≠ ∅ then M_last ← pop();
                    else
                        decided ← true;
                        possible ← false;
            endif
        endif
    endwhile
    return(possible, Q);
end.
```

Fig. 1. The Algorithm $\mathcal{A}$

Let us now focus on the case $state$ = BACKTRACK. $\mathcal{A}$ considers the move M_{last}. If there is an extend move $M \succ M_{last}$ that is valid according to the stack, then $\mathcal{A}$ performs the smallest such move by pushing M in the stack, and going back to state CLEAR. Otherwise $\mathcal{A}$ carries on backtracking by popping out the last extended move from the stack.

3.2 Property of Algorithm $\mathcal{A}$

Lemma 1. *Algorithm $\mathcal{A}$ completes for $k = \mathrm{mcs}(G, u_0)$, and then the stack Q describes a monotone connected search strategy for G starting at u_0 and using k searchers.*

Sketch of proof. First we prove that, after any execution of the *while*-loop, the state of the stack is valid. The main tools for the proof in then an ordering of the states of the stack. We order them the same way we ordered extended moves. Precisely, given $S = M_1 | \cdots | M_r$ and $S' = M'_1 | \cdots | M'_{r'}$, two states of the stack Q, $S \prec S'$ if and only if there exists $i \leq \min\{r, r'\}$ such that $M_i \prec M_i'$ and, for any $j < i$, $M_j = M_j'$. Also, let us say that a valid sequence of extended moves is *complete* if the corresponding search strategy clears the whole network. Consider $S = M_1 | \ldots | M_r$ a sequence of extended moves corresponding to a partial execution of a search strategy using at most k searchers. We prove that either there exists a complete sequence S' of extended moves with $S' \prec S$, or Algorithm $\mathcal{A}$ eventually computes state S of the stack. Based on these preliminary results, we prove that if $\text{mcs}(G, u_0) > k$ then Algorithm $\mathcal{A}$ returns $(false, \emptyset)$ for k. Conversely, we prove that if $\text{mcs}(G, u_0) = k$, and if S is the smallest complete sequence of valid extended moves corresponding to a monotone connected search strategy in G starting from u_0, then Algorithm $\mathcal{A}$ returns $(true, Q)$ for k, where Q is in state S. As a direct consequence of these results, we get that Algorithm $\mathcal{A}$ computes a minimal monotone connected search strategy starting from u_0 in G. □

4   Fully Distributed Search Strategy

In this section, we describe the main features of protocol `dist_search`. In this description, we assume that searchers are able to communicate by exchanging messages of size $O(\log k)$ bits where k is the number of searchers currently involved in the search. With this facility, we will show that `dist_search` captures the intruder with $\text{mcs}(G, u_0)$ searchers. Using an additional searcher for implementing communications between the $\text{mcs}(G, u_0)$ other searchers, `dist_search` captures the intruder with $\text{mcs}(G, u_0) + 1$ searchers in total. Assuming that the searchers can communicate by exchanging messages is only for the purpose of simplifying the presentation. Moreover, for the sake of simplicity, we assume that two searchers on the same node can "see" each other. Obviously, this can be implemented with the whiteboards, but would unnecessarily complicate the presentation. First, we describe the data structure used by `dist_search`.

4.1   Data Structure of `dist_search`

Every searcher has a state variable that can take $k+2$ different values where k is the current number of searchers. These $k+2$ states are: CLEAR, BACKTRACK, and (HELP, j), for $j = 1, \ldots, k$. Initially, all searchers are in state CLEAR. During the execution of the protocol, (1) a searcher is in state CLEAR if it has just cleared an edge; (2) a searcher is in state BACKTRACK if it has just backtracked through an edge that it has previously cleared; and (3) a searcher is in state (HELP, j) if it is aiming at joining the searcher j to help him clearing the network (i.e., one of them will guard a node, while the other will clear an edge incident to this node).

The messages that searchers can exchange are of four types: start, move, help and sorry. (1) start is an initialization message that is only used to start Protocol `dist_search` (only agent a_1 receives this message, at the very beginning of the protocol execution). (2) If a searcher i receives a message (move, j) from some searcher a_j, then it is the turn of searcher a_i to proceed. (As it should appear clear later, the searchers schedule themselves so that exactly one searcher performs an action at a time). (3) If a searcher a_i receives a message $(\texttt{help}, j)$ from some searcher a_j, then a_j is currently just arriving at the same node as a_i to help a_i. (Note that a_i and a_j could use the whiteboard to communicate, and this type of messages is just used for a purpose of unification with the other message types). (4) If a searcher a_i had received a message $(\texttt{move}, j)$ or $(\texttt{help}, j)$ from some searcher a_j and, after having possibly performed several actions, it turns out that these actions are useless, then a_i sends a message $(\texttt{sorry}, i)$ back to searcher a_j.

The whiteboard of every node contains a local stack, and two vectors $\texttt{direc-}$ $\texttt{tion}[]$ and $\texttt{cleared\_port}[]$. The protocol insures that, after the node has been visited by a searcher, $\texttt{direction}[0]$ indicates the port number to take for reaching the homebase, and, for $i > 0$, $\texttt{direction}[i]$ is the port number of the edge that searcher a_i has used to leave the current node the last time it was at this node. At node v, for any $1 \leq p \leq \deg(v)$, $\texttt{cleared\_port}[p] = 1$ if and only if the edge corresponding to the port number p is clear.

When a searcher at a node v decides to perform any action, it saves a *trace* of this action in the local stack. A trace is a triple (X, a, x) where X is a symbol, a is a searcher's ID, and x is either a port number, or a searcher's ID, depending on symbol X. More precisely: (1) (CC, i, p) means that p is the only contaminated (C) port, and searcher a_i decided to clear (C) the edge that corresponds to p; (2) (CJ, i, p) means that some searcher joined (J) a_i at this node, and a_i decided to clear (C) the edge that corresponds to p; (3) (JJ, i, j) means that searcher a_i decided to join (J) the searcher a_j; (4) (RT, i, j) means that searcher a_i received (R) a message from searcher a_j; (5) (ST, i, j) means that searcher a_i decided to send (S) a message to searcher a_j; (6) (AC, i, p) means that searcher a_i arrived (A) at v by port p after clearing (C) the corresponding edge; (7) (AH, i, p) means that searcher a_i arrived (A) at v by port p in order to join another (H) searcher.

4.2   The Protocol Dist_Search

The protocol $\texttt{dist\_search}$ organizes the movements of the searchers, and the messages exchanged between the searchers, in a specific order. Based on a lexicographic order of the searchers' actions, $\texttt{dist\_search}$ orders them in order to always execute the smallest action that can be performed. The principle of $\texttt{dist\_search}$ is to try every possible monotone connected search strategy using k searchers, until either the whole graph is clear, or no searcher can move without implying recontamination. In the latter case, the searcher that made the last move backtracks, and $\texttt{dist\_search}$ tries the next action according to the lexicographic order on the actions.

The termination of $\texttt{dist\_search}$ is insured as follows. The graph is cleared at time t if and only if all searchers are occupying clear nodes at this time, i.e., nodes whose all incident edges are clear. This configuration is identified by the searchers because searcher a_1 tries to help all the other searchers, from a_2 to a_k, but none of them needed help. Conversely, the searchers identify that k searchers are not sufficient to clear the graph when they are all occupying the homebase, and try to pop the local stack that is empty. In this case, a_1 calls for a new

Program of searcher i at node v.	/* Searcher i arrives at node v by port p */

begin
 /* Searcher i receives a message */
 Case:
     message = start
       $decide()$;
     message = (move, j)
       $push(RT, i, j)$;
       $decide()$;
     message = (help, j)
       $push(RT, i, j)$;
       $p \leftarrow$ smallest contaminated port;
       $clear\_edge(CJ, i, p)$
     message = (sorry, j)
       $back()$;
	Case:
	     state = CLEAR
	       if no other searcher is at v **then**
	         erase whiteboard;
	       direction[0] $\leftarrow p$;
	       cleared_port[p] $\leftarrow 1$;
	       $push(AC, i, p)$;
	       if $i \neq 1$ **then**
	         $push(ST, i, 1)$;
	         send message (move, i) to 1;
	       else $decide()$;
	     state = (HELP, j)
	       $push(AH, i, p)$;
	       $join(j)$;
	     state = BACKTRACK
	       $back()$;
	 end

Fig. 2. Skeleton of Protocol dist_search

searcher, and the $k + 1$ searchers are ready to try again capturing the intruder from the homebase.

A skeleton of the protocol dist_search is given in Figures 2-3. More precisely, Figure 2 describe the global behavior of a searchers, using subroutines described in Figure 3. A searcher reacts to either the reception of a message (cf. left part of Figure 2), or to its arrival at a node (cf. right part of Figure 2). The message type start is uniquely for the purpose of the initialization: initially, searcher a_1 receives a message start (and hence calls procedure $decide()$).

If searcher a_i receives a message (move, j), then, by definition of such a message, it simply means that it is the turn of a_i to proceed. Therefore, a_i writes on the whiteboard of the node where it is currently standing that received a message from searcher a_j giving it turn to proceed. For this purpose, a_i pushes (RT, i, j) in the local stack. The nature of the next actions of a_i depends on the result of procedure $decide()$. Let us list all other cases depending on the message received by a_i. If a_i receives a message (help, j) then it means that a_j has just arrived at the same node as a_i to help him. Thus, a_i pushes (RT, i, j) in the local stack, and clears the edge with the smallest port number p among all contaminated edges incident to the node where a_i is standing. This action is performed by calling procedure $clear\_edge(CJ, i, p)$. Finally, if a_i receives a message (sorry, j), then it means that a_i had sent a message (move, i) or a message (help, i) to a_j but a_j could not do anything, or all actions a_j attempted lead to backtracking. Therefore, a_i calls procedure $back()$ to figure out which searcher it can help next.

The action of searcher a_i arriving at some node v by port p depends on its local state. In state (HELP, j), a_i aims at joining a_j to help him clearing the network. Hence a_i pushes (AH, i, p) in the local stack to indicate that it arrived here by port p in order to join another searcher, and then calls procedure $join()$ to figure out what to do next in order to join a_j. Procedure $join()$ uses indications on whiteboards. Recall that if a_j was at a node, the whiteboard contains in direction[j] the port number through which a_j left that node.

```
clear_edge(action X, ID i, port p)          next_searcher(searcher_ID i)
/* X ∈ {CC; CJ} */                          begin
begin                                           j ← i + 1;
    push(X, i, p);                              if i is not smallest searcher at v then
    cleared_port[p] ← 1;                            while (j is at node v) and (j ≤ k) do
    state ← CLEAR;                                      j ← j + 1;
    move(p);                                    if j ≤ k then
end                                                 push(ST, i, j);
                                                    send (move, i) to j;
                                            else
move(port_number p)                                 back()
begin                                       end
    direction[i] ← p;
    leave current vertex by port number p;
end
```

Fig. 3. Procedures `clear_edge`, `next_searcher` and `move`

Agent a_i returns to the homebase using `direction[0]` until it passes through a node where `direction[j]` is set, in which case a_i starts following this direction to eventually find a_j. In state BACKTRACK, a_i simply calls procedure $back()$ to carry on its backtracking. The case where a_i arrive at a node v in state CLEAR is more evolved. If there is no other searcher at v then a_i erases the whiteboard since it was accessible to the intruder, and thus its content is meaningless (when a searchers arases a whiteboard, it reset all local variables to 0, and the stack to ∅). Then a_i sets `direction[0]` to p to indicate that it arrived here via port p, and sets `cleared_port[p]` to 1 to indicate that the edge of port p is clear. a_i then pushes (AC, i, p) in the local stack at v to indicate that indeed a_i arrived at v by port p after clearing the corresponding edge. At this point, the behavior of a_i depends on whether $i = 1$ or not. While a_1 simply calls $decide()$ to figure out what to do next, a_i for $i > 1$ proposes to a_1 to proceed next. For this purpose, a_i sends a message (`move`, i) to a_1. Of course, to keep trace of this action, a_i pushes $(ST, i, 1)$ in the local stack.

Remark. Note that the actions are ordered. For instance, if several incident edges can be cleared then the cleared one is with the smallest port number. Similarly, after clearing an edge, a_i proposes to the smallest searcher a_1 to proceed next. Protocol `dist_search` always tries to perform the smallest action. This is in particular the role of procedure $next\_searcher(i)$ described on the right side of Figure 3. This procedure aims at determining which searcher a_j proceeds next. In the case where a_i is the searcher with smallest index occupying the node, $j = i + 1$. Otherwise, i.e., a_i is not the searcher with smallest index occupying the node, j is the smallest index $> i$ such that a_j is not occupying the same node as a_i. Once j is found, a_i offers to a_j to proceed next, by sending it a message (`move`, i). As always, a trace of this action is kept at the current node by pushing (ST, i, j) in the local stack. If there is no a_j with $j > i$ occupying a node different from the one occupied by a_i, then a_i calls $back()$ for the purpose of backtracking.

The procedures $clear\_edge()$ and $move()$ described in the left side of Figure 3 execute clearing an edge, and traversing an edge, respectively. (Of course, clearing an edge requires traversing it). Procedures $decide()$, $back()$, and $join()$ are avoided due to lack of space.

5   Sketch of Proof of `Dist_Search`

First, one can check that at any step of `dist_search` there is only one operation performed, on only one of the stacks distributed over all nodes of the network. Indeed, only the searcher who has just received a message can perform an action, and in particular modify a stack. Thus we can define a *virtual stack*, $Q_{virtual}$, where we push or pop all the moves performed by the searchers, instead of pushing or popping them in and out of the distributed stacks.

Precisely, a *move* is a pair $(a_i \rightarrow a_j, p)$ to be interpreted as follows. If $i \neq j$, then $(a_i \rightarrow a_j, p)$ means that a_i leaves its current node by port p with the objective of joining a_j. The move $(a_i \rightarrow a_i, p)$ means that a_i leaves its current node by port p, for clearing the corresponding edge. Clearly, an extended move corresponds to a sequence of moves. From the interpretation above, the extended move (a_i, a_i, p) is identical to the move $(a_i \rightarrow a_i, p)$, and if $i \neq j$ then the extended move (a_i, a_j, p) is identical to the sequence of moves

$$(a_i \rightarrow a_j, p_1), (a_i \rightarrow a_j, p_2), \ldots, (a_i \rightarrow a_j, p_\ell), (\min\{a_i, a_j\} \rightarrow \min\{a_i, a_j\}, p)$$

where $p_1, \ldots, p_\ell$ is a sequence of port numbers corresponding to a clear path from the node occupied by a_i to the node occupied by a_j when the extended move (a_i, a_j, p) is considered.

$Q_{virtual}$ is updated in the following way. At every execution of the Procedure *move()*, we push or pop a move in $Q_{virtual}$ as follows. If a_i applies *move(p)* during the execution of Procedure *clear_edge(X, i, p)*, then the move $(a_i \rightarrow a_i, p)$ is pushed in $Q_{virtual}$. If a_i applies *move(p)* during the execution of Procedure *join(j)*, then the move $(a_i \rightarrow a_j, p)$ is pushed in $Q_{virtual}$. Finally, if a searcher applies *move(p)* during the execution of Procedure *back()*, then $Q_{virtual}$ is popped.

With this definition of $Q_{virtual}$, we show that the stack Q of the centralized algorithm $\mathcal{A}$, and the virtual stack $Q_{virtual}$ are equivalent in the following way. Let $Q = M_1 | \cdots | M_r$ be a sequence of extended moves (possibly empty). $Q_{virtual}$ is *strongly equivalent* to Q if, for any $1 \leq j \leq r$, there exists a sequence of moves S_j equivalent to M_j such that $Q_{virtual} = S_1 | \cdots | S_r$. $Q_{virtual}$ is *weakly equivalent* to Q if for any $1 \leq j \leq r$, there exists a sequence of moves S_j equivalent to M_j such that $Q_{virtual} = S_1 | \cdots | S_r | S_{r+1}$ where $S_{r+1} = (a_i \rightarrow a_{i'}, p_1), (a_i \rightarrow a_{i'}, p_2), \ldots, (a_i \rightarrow a_{i'}, p_\ell)$ where $p_1, \cdots, p_\ell$ is a sequence of port numbers corresponding to a path from a searcher a_i to a searcher $a_{i'}$, in the cleared part of the graph corresponding to the configuration associated to Q in state $M_1 | \cdots | M_r$.

Two strongly equivalent stacks correspond to exactly the same strategy (i.e., at the end of both strategies, the set of cleared edges, and the positions of the searchers are the same). If Q and $Q_{virtual}$ are weakly equivalent, then the strategy associated to $Q_{virtual}$ consists in performing the strategy associated to Q and then to move some searcher to the node occupied by some other searcher (via a path in the cleared part of the graph, and without recontamination).

The proof of `dist_search` proceeds by considering the algorithm step by step, where a *step* is a moment of the execution where an edge is either cleared or recontaminated. That is, a step of `dist_search` denotes a step of its execution when a move of type $(a_i \rightarrow a_i, p)$ is pushed in or popped out $Q_{virtual}$.

Formally, we prove that, for any $t \geq 0$, the virtual stack $Q_{virtual}$ after step t of `dist_search` is equivalent to the stack Q constructed by $\mathcal{A}$. In other words, we prove that, at any step $t \geq 0$, both algorithms construct the same partial strategy, that is the cleared subgraph and the positions of the searchers that guard the border of this cleared subgraph are the same for both strategies. Simultaneously, we prove that for any step, when an extended move is popped out in $\mathcal{A}$, all the traces of the equivalent sequence of moves in `dist_search` are removed from the distributed whiteboards.

Our proof is by induction on number of steps. Let us assume that the centralized stack Q and the virtual stack $Q_{virtual}$ are equivalent up to step t. We consider the next step. The difficulty of the proof is in the number of different cases to consider. There are actually exactly fourteen cases to consider, grouped in two groups:

- Group A: Q and $Q_{virtual}$ just cleared an edge e. The first case is if the graph is entirely clear. Otherwise there are 3 cases: (1) a searcher can clear a new edge alone, or (2) a searcher can join another searcher and one of them can clear a new edge, or (3) no other edge can be cleared and the clearing of e has to be canceled. These cases have to be combined with 3 other cases depending on the way e has been cleared. Thus Group A yields 7 cases in total.
- Group B: Q and $Q_{virtual}$ just cancelled the clearing of an edge. Then, either another edge e can be cleared, or no other edge can be cleared (and the last cleared edge, say e', has to be canceled). In the former case, there are 3 subcases depending on the type of move that has been popped out the stack (canceling corresponding to popping out the stack). In the latter case, there are 4 subcases depending on the way e' had been cleared. Thus Group B yields 7 additional cases.

The proof of correctness consists in a careful analysis of each of these 14 cases. Finally, every agent uses at most $O(\log k)$ bits of memory to store the label of another agent in state (HELP, j). The whiteboard size is $O(m \log n)$ by a careful analysis of the protocol.

Acknowledgments. The first and fourth authors received additional supports from the project "ALGOL" of the ACI Masses de Données, and from the project "ROM-EO" of the RNRT program. The second and third authors received additional supports from the project "PairAPair" of the ACI Masses de Données, from the project "Fragile" of the ACI Sécurité Informatique, and from the project "Grand Large" of INRIA.

References

1. S. Arnborg, D. Corneil, and A. Proskurowski. Complexity of finding embeddings in a k-tree. SIAM J. Alg. Disc. Meth. 8(2):277-284, 1987.
2. L. Barrière, P. Flocchini, P. Fraigniaud, and N. Santoro. Capture of an intruder by mobile agents. In 14th ACM Symp. on Parallel Algorithms and Architectures (SPAA), pages 200-209, 2002.

3. L. Barrière, P. Fraigniaud, N. Santoro, and D. M. Thilikos. Searching is not jump-ing. In 29th Workshop on Graph Theoretic Concepts in Computer Science (WG), Springer-Verlag, LNCS 2880, pages 34–45, 2003.

4. D. Bienstock, Graph searching, path-width, tree-width and related problems (a survey), DIMACS Ser. in Discrete Mathematics and Theoretical Computer Science, 5 (1991), pp. 33–49.

5. D. Bienstock and P. Seymour. Monotonicity in graph searching. Journal of Algo-rithms 12:239–245, 1991.

6. R. Breisch.   An intuitive approach to speleotopology.   *Southwestern Cavers* VI(5):72–78, 1967.

7. U. Feige, M. Hajiaghayi, and J. Lee.   Improved approximation algorithms for minimum-weight vertex separators. In 37th ACM Symposium on Theory of Com-puting (STOC), 2005.

8. P. Flocchini, F.L. Luccio, and L. Song. Decontamination of chordal rings and tori. Proc. of 8th Workshop on Advances in Parallel and Distributed Computational Models (APDCM), 2006.

9. P. Flocchini, M. J. Huang, F.L. Luccio. Contiguous search in the hypercube for capturing an intruder. Proc. of 18th IEEE Int. Parallel and Distributed Processing Symposium (IPDPS), 2005.

10. F. Fomin, P. Fraigniaud and N. Nisse. Nondeterministic Graph Searching: From Pathwidth to Treewidth. In 30th International Symposium on Mathematical Foun-dations of Computer Science (MFCS), LNCS 3618, pages 364-375, Springer, 2005.

11. F. V. Fomin, D. Kratsch, and I. Todinca. Exact algorithms for treewidth and min-imum fill-in. In 31st Int. Colloquium on Automata, Languages and Programming (ICALP 2004), LNCS vol. 3142, Springer, pp. 568–580, 2004.

12. P. Fraigniaud and D. Ilcinkas.   Directed Graphs Exploration with Little Mem-ory. Proc. 21st Symposium on Theoretical Aspects of Computer Science (STACS), LNCS 2296, pages 246-257, 2004.

13. P. Fraigniaud and N. Nisse. Connected Treewidth and Connected Graph Searching. In 7th Latin American Theoretical Informatics, LNCS 3887, pages 470-490, 2005.

14. L. Kirousis, C. Papadimitriou. Interval graphs and searching. Discrete Math. 55, pages 181-184, 1985.

15. L. Kirousis, C. Papadimitriou.   Searching and Pebbling. Theoretical Computer Science 47, pages 205-218, 1986.

16. A. Lapaugh.   Recontamination does not help to search a graph. Journal of the ACM 40(2):224–245, 1993.

17. F. S. Makedon and I. H. Sudborough, On minimizing width in linear layouts, Discrete Appl. Math., 23:243–265, 1989.

18. N. Megiddo, S. Hakimi, M. Garey, D. Johnson and C. Papadimitriou. The com-plexity of searching a graph. Journal of the ACM 35(1):18–44, 1988.

19. T. Parsons. Pursuit-evasion in a graph. *Theory and Applications of Graphs*, Lecture Notes in Mathematics, Springer-Verlag, pages 426–441, 1976.

20. K. Skodinis Computing optimal linear layout of trees in linear time. In 8th Eu-ropean Symp. on Algorithms (ESA), Springer, LNCS 1879, pages 403-414, 2000. (Also, to appear in SIAM Journal on Computing).

21. B. Yang, D. Dyer, and B. Alspach. Sweeping Graphs with Large Clique Number. In 15th Annual International Symposium on Algorithms and Computation (ISAAC), pages 908-920, 2004.

Election in the Qualitative World

Jérémie Chalopin

LaBRI, Université Bordeaux 1
351 cours de la Libération
33405 Talence, France
chalopin@labri.fr

Abstract. In [3], Barrière et al. consider a *qualitative* model of distributed computing, where the labels of the entities are distinct but mutually incomparable. They study the leader election problem in a distributed mobile environment and they wonder whether there exists an algorithm such that for each distributed mobile environment, it either states that the problem cannot be solved in this environment, or it successfully elects a leader. In this paper, we give a positive answer to this question. We also give a characterization of the distributed mobile environments where the election problem can be solved.

1 Introduction

Consider an intercontinental highway network linking different cities in different countries. In each city, the directions to the other cities are written in the language that is locally spoken. Consider now a set of different drivers coming from different countries. Initially, each driver starts in his town and all the drivers want to meet at a single place. The only mean they have to communicate is to leave messages in each city they reach, but each driver can only speak his mother tongue: he can see that another driver left some message, but he cannot understand it. Moreover, each driver can consistently distinguish the different directions in each city, but the drivers cannot agree on an alphabetical order on these directions: a French driver would not be able to figure out how to order Chinese words in the Chinese way, for example. We wonder whether there exists a procedure that enables them to meet at a single point in a finite time.

In distributed computing, the links incident to each process are usually labelled by distinct numbers in order to allow each process (or each mobile agent) to consistently distinguish its neighbours; this labelling is usually called a ports-numbering. In fact, these numbers allow not only to distinguish the links, but also to order them. Many distributed algorithms assume also that all the processes can be unambiguously identified, and therefore the processes are given numbers. Again, one can see that this allows to order the different processes according to their labels. This usual setting is a *quantitative* model, since each label can be seen as a number.

Nevertheless, as in the example presented above, one may be able to distinguish labels without being able to order them. In this paper, we consider

P. Flocchini and L. Gąsieniec (Eds.): SIROCCO 2006, LNCS 4056, pp. 85–99, 2006.

distributed mobile environments where mobile agents are scattered all over a network. All the agents have distinct *colors* (their labels), which are mutually incomparable: each agent can just check whether two colors are equal or not. The links incident to each vertex are also given distinct incomparable colors. This model is close to the one introduced by Barrière et al. in [3]; it is *qualitative*, in the sense that there is no a priori order between the labels. As in [3], we study the impact of the lack of a total order on the set of labels in a distributed mobile environment. In this way, we investigate the leader election problem, that is a classical problem to highlight the differences between various models of distributed computing.

In usual models, there is always an implicit order over the set of labels, since for each agent, each information is just a sequence of bits. Nevertheless, consider an algorithm designed to be executed by mobile agents over a network. If the agents have been implemented by different companies, and if the specifications of the algorithm do not specify how the integers must be represented, some agents can for instance store numbers with most significant bit first whereas other agents store numbers with least significant bit first; in this case, the agents would not agree on the meaning of the sequence 01101. Moreover, it is always interesting to deal with algorithms that need less specifications, since they are generally more robust, and easier to implement in different models of distributed computing.

The Model. In this paper, an agent is an entity which executes an algorithm: it can move from place to place (with some data and its algorithm) through communication links, it can make local computations on a place (a place provides tools for local computations: data, memories and process) and leave messages on a place. In our model, the environment is represented by a simple undirected connected graph $G = (V(G), E(G))$ and a set $\mathcal{E}$ of mobile agents is scattered over G. Communications between agents is achieved through writing messages on *whiteboards*, where agents can read, write, and erase messages. There is one whiteboard on each vertex of G, and access to a whiteboard is in mutual exclusion. Initially, all the whiteboards are empty. Let $p : \mathcal{E} \to V(G)$ be the injection describing the initial placement of the agents in G. The vertex $p(r)$ is called the *homebase* of the agent $r \in \mathcal{E}$. We will denote such a distributed mobile environment by $(G, \mathcal{E}, p)$.

We consider a set of colors C and a function $color : \mathcal{E} \to C$ that associates to each agent a unique color. There is no a priori order on the set of colors: each agent can give its own order on the set of colors, but the agents do not agree on a particular order. Each agent can understand a message it has written, but it cannot understand a message written by another agent, it can just know the color of the message. We also suppose that initially, the homebases are marked: they contain a marker that enables each agent to know that a place is a homebase and to detect the color of this homebase. In each place, the incident links are labelled by different colors that enable each agent to consistently distinguish the neighbours of the place: for each vertex u, there exists an injective function δ_u that associates a color from a new set C' (i.e., $C' \cap C = \emptyset$) to each edge incident to u. The set $\delta = \{\delta_u : u \in V(G)\}$ constitutes the *ports-labelling* of G. Thanks to this labelling δ, each agent can make a distinction between the incident edges of each vertex. Such a distributed colored mobile environment will be denoted by $(G, \delta, \mathcal{E}, p, color)$.

The agents are asynchronous, in the sense that every action they perform (computing, moving, etc.) takes a finite but otherwise unpredictable amount of time. Moreover, we suppose that an agent has not an initial knowledge of the network topology, neither of its size nor of the number of agents in the system. The actions an agent a located at a node v can perform depends on the current state of a, the current state of the whiteboard at v, and the color of the port through which a entered v. According to these informations, a can decide to write a message on the whiteboard of v, to leave v (through a port whose color may result from some computation), or to stay at v (for example, to wait that another agent leaves a message on the whiteboard).

This model is more restrictive than the one presented in [3], since in the model of Barrière et al. the agents cannot agree on an order on the set of colors, but they fully understand the symbols written by the other agents. However, the necessary condition presented in our model is the same as the one presented in [3]: the results presented in this paper remain true in the model of [3].

The Election Problem. The election problem is one of the paradigms of the theory of distributed computing. In the distributed mobile setting, the aim of a leader election algorithm is to distinguish one agent among the others. All the agents execute the same protocol, i.e., the only initial difference between two agents is their colors. At the end of the execution of the algorithm, there is exactly one agent in the state *elected*, whereas all the other agents enter the state *non-elected*. Moreover, it is supposed that once an agent enters in the state *elected* or *non-elected*, it remains in such a state until the end of the computation. Another important problem in this setting is the rendez-vous problem. The aim of a rendez-vous algorithm is to reach a configuration where all the mobile agents gather in the same vertex of the graph. These two problems are equivalent, since once an agent has been elected, if all the agents agree on the label *elected*, all the agents can gather in the homebase of the elected agent. Reversely, once all the agents have gathered in some place, the first agent that writes on the whiteboard of this place is elected, whereas all the others become non-elected. There exists a large variety of results for these problems in the mobile agent setting assuming different properties of the environment [2, 4, 5, 10, 11]. The election problem has also been extensively studied in the distributed setting, and particularly in anonymous networks, where the processes do not have distinct labels [1, 6, 9, 13].

Consider a graph G and a set of agents $\mathcal{E}$ scattered over the network according to a function p. We say that we can solve the election problem on $(G, \mathcal{E}, p)$ if the problem can be solved on $(G, \delta, \mathcal{E}, p, color)$ for all ports-labellings δ and all agent-coloring functions *color*. This implies that an election algorithm in the distributed mobile environment $(G, \mathcal{E}, p)$ must not use some particularity of the ports-labelling or make any assumption on the set of colors (for example, if one know that there is always a red agent, one can design an algorithm that elects the red agent). Note that, as for anonymous networks in the distributed setting [9, 13], the protocols must not depend on the ports-labelling. Indeed, the role of the ports-labelling is just to enable an agent to make a distinction between the different neighbours of a vertex.

As in [3], we say that an algorithm $\mathcal{A}$ is an *effective* election algorithm if for each distributed mobile environment $(G, \mathcal{E}, p)$, each ports-labelling δ and each coloring function *color*, for all the executions of $\mathcal{A}$ on $(G, \delta, \mathcal{E}, p, color)$, either all the agents detect that the election problem cannot be solved in $(G, \mathcal{E}, p)$, or the agents successfully elect one of them. In particular, note that such an algorithm does not need any initial knowledge about the topology, the size, the diameter of the network or about the number of agents.

Main Results. In this work, we give a characterization (Theorem 1) of distributed mobile environments, where the election problem can be solved.

In [3], Barrière et al. wonder whether there exists an effective algorithm for the qualitative world. The algorithm we describe gives a positive answer to this question (Theorem 2).

To obtain a necessary condition (Proposition 2), we use *well-balanced* automorphisms that have been introduced by Bougé in [8].

Then, we show that this necessary condition is also sufficient: we use some links between fibrations and automorphisms presented in [7] to describe an effective algorithm in Section 4.2 that solves the election problem when the necessary condition is satisfied.

2 Preliminaries

Labelled Digraphs. In the following, we will consider directed graphs (digraphs) with multiple arcs and self-loops. A digraph $D = (V(D), A(D), s_D, t_D)$ is defined by a set $V(D)$ of vertices, a set $A(D)$ of arcs and by two maps s_D and t_D that assign to each arc two elements of $V(D)$: a source and a target (in general, the subscripts will be omitted); if a is an arc, the arc a is said to be going out of $s(a)$ and coming into $t(a)$. We say that $s(a)$ is a predecessor of $t(a)$ and that $t(a)$ is a successor of $s(a)$. A digraph D is strongly connected if for all vertices $u, v \in V(D)$, there exists a sequence of arcs $a_1, a_2, \ldots a_p$ such that $s(a_1) = u, \forall i \in [1, p-1], t(a_i) = s(a_{i+1})$ and $t(a_p) = v$. In the following, we will only consider strongly connected digraphs. A *symmetric* digraph D is a digraph endowed with a symmetry, that is, an involution $Sym : A(D) \to A(D)$ such that for every $a \in A(D), s(a) = t(Sym(a))$.

A digraph homomorphism γ between the digraph D and the digraph D' is a mapping $\gamma \colon V(D) \cup A(D) \to V(D') \cup A(D')$ such that if u, v are vertices of D and a is an arc such that $u = s(a)$ and $v = t(a)$ then $\gamma(u) = s(\gamma(a))$ and $\gamma(v) = t(\gamma(a))$. We say that γ is an isomorphism if γ is bijective and γ^{-1} is a homomorphism, too.

Throughout the paper we will consider digraphs where the vertices and the arcs are labelled with labels from a recursive label set L. A digraph G labelled over L will be denoted by (D, λ), where $\lambda \colon V(D) \cup A(D) \to L$ is the labelling function. The digraph D is called the underlying digraph and the mapping λ is a labelling of D. A mapping $\gamma : V(D) \cup A(D) \to V(D') \cup A(D')$ is a homomorphism from (D, λ) to (D', λ') if γ is a digraph homomorphism from D to D' which preserves the labelling, i.e., such that $\lambda'(\gamma(x)) = \lambda(x)$ for every $x \in V(D) \cup A(D)$. Labelled digraphs will be designated by bold letters like $\mathbf{D}, \mathbf{G}, \ldots$ If $\mathbf{D}$ is a labelled digraph, then D denotes the underlying digraph.

Let $G = (V(G), E(G))$ be a connected simple graph. The symmetric strongly connected digraph associated to G and denoted by $Dir(G)$ is (V, A) defined by: there is an arc a_1 from v_1 to v_2 and an arc a_2 from v_2 to v_1 in A if $\{v_1, v_2\} \in E(G)$ and $Sym(a_1) = a_2$. Note that this digraph does not contain multiple arcs or self-loops. Given a mobile environment $(G, \mathcal{E}, p)$, we define the labelling function χ_p of the vertices by $\chi_p(v) = 1$ if there exists an agent a such that $p(a) = v$, and $\chi_p(v) = 0$ otherwise. A distributed mobile environment $(G, \mathcal{E}, p)$ can therefore be represented by the labelled digraph $(Dir(G), \chi_p)$.

For any set S, $|S|$ denotes the cardinality of S. For any integer q, we denote by $[1, q]$ the set of integers $\{1, 2, \ldots, q\}$.

Fibrations and Coverings. The notions of fibrations and coverings are fundamental in this work; definitions and main properties are presented in [7].

A *fibration* between the digraphs D and D' is a homomorphism φ from D to D' such that for each arc a' of $A(D')$ and for each vertex v of $V(D)$ such that $\varphi(v) = v' = t(a')$ there exists a unique arc a in $A(D)$ such that $t(a) = v$ and $\varphi(a) = a'$. The arc a is called the lifting of a' at v, D is called the total digraph and D' the base of φ. We shall also say that D is fibred (over D'). The fibre over a vertex v' (resp. an arc a') of D' is the set $\varphi^{-1}(v')$ of vertices of D (resp. the set $\varphi^{-1}(a')$ of arcs of D).

An *opfibration* between the digraphs D and D' is a homomorphism φ from D to D' such that for each arc a' of $A(D')$ and for each vertex v of $V(D)$ such that $\varphi(v) = v' = s(a')$ there exists a unique arc a in $A(D)$ such that $s(a) = v$ and $\varphi(a) = a'$.

A *covering projection* is a fibration that is also an opfibration. If a covering projection $\varphi : D \to D'$ exists, D is said to be a *covering* of D' via φ. A symmetric digraph D is a *symmetric covering* of a symmetric digraph D' via a homomorphism φ if D is a covering of D' via φ such that $\forall a \in A(D), \varphi(Sym(a)) = Sym(\varphi(a))$. A digraph D is *symmetric-covering-minimal* if there does not exist any digraph D' not isomorphic to D such that D is a symmetric covering of D'.

Given two strongly connected digraphs D and D', an interesting property satisfied by any covering projection φ from D to D' is that there exists $q \in \mathbb{N}$ such that $\forall x' \in V(D') \cup A(D'), |\varphi^{-1}(x')| = q$.

The notions of fibrations and coverings extend to labelled digraphs in a natural way: the homomorphisms must preserve the labelling. Examples of fibrations and coverings are given in Figure 1.

Fibrations, Coverings and Automorphisms. We now describe some properties of the relations that exist between fibrations and the automorphisms of a digraph. These results are described and proved in [7].

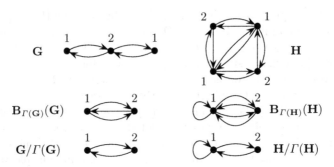

Fig. 1. The digraph **G** is fibred over $\mathbf{B}_{\Gamma(\mathbf{G})}(\mathbf{G})$ via the homomorphism $\varphi_{\mathbf{G}}$ that maps each vertex of **G** labelled i to the unique vertex labelled i of $\mathbf{B}_{\Gamma(\mathbf{G})}(\mathbf{G})$. The digraph **H** is a covering of $\mathbf{B}_{\Gamma(\mathbf{H})}(\mathbf{H})$ via the homomorphism $\varphi_{\mathbf{H}}$ defined in the same way. The digraph $\mathbf{G}/\Gamma(\mathbf{G})$ (resp. $\mathbf{H}/\Gamma(\mathbf{H})$) is the digraph whose vertices and arcs correspond to equivalence classes of vertices and arcs of **G** (resp. **H**) under the action of $\Gamma(\mathbf{G})$ (resp. $\Gamma(\mathbf{H})$).

An automorphism σ of a digraph **G** is an isomorphism from the digraph **G** onto itself. Consider a subgroup Γ of the group $\Gamma(\mathbf{G}) = Aut(\mathbf{G})$ of the automorphisms of a digraph $\mathbf{G} = (G, \lambda)$; we will denote by Id the identity automorphism of **G**. The action of this group on **G** induces an equivalence relation over the vertices and the arcs of **G**: for each $x, x' \in V(G) \cup A(G)$, $x \sim_\Gamma x'$ if there exists $\sigma \in \Gamma$ such that $\sigma(x) = x'$. The equivalence class of x is called the orbit of x and is denoted by $[x]_\Gamma$. Recall that an automorphism of (G, λ) must preserve the labelling, and therefore for all elements $x_1, x_2 \in [x]_\Gamma$, $\lambda(x_1) = \lambda(x_2)$. If $\Gamma = \Gamma(\mathbf{G})$, we will note $x \sim x'$ (resp. $[x]$) for $x \sim_\Gamma x'$ (resp. $[x]_\Gamma$).

Remark 1. For all vertices $v, v' \in V(G)$, if $v \sim_\Gamma v'$, then there is a bijection between the incoming arcs of v and the incoming arcs of v'.

We will now describe two kinds of constructions. The first one allows us to build a digraph $\mathbf{B}_\Gamma(\mathbf{G})$ from a digraph **G** such that **G** is fibred over $\mathbf{B}_\Gamma(\mathbf{G})$. The second one allows to build the quotient-graph $\mathbf{G}/\Gamma$. Examples are presented in Figure 1 where $\Gamma = \Gamma(\mathbf{G})$.

From the relation $\sim_\Gamma$, we construct the directed graph $\mathbf{B}_\Gamma(\mathbf{G})$ defined as follows: $V(\mathbf{B}_\Gamma(\mathbf{G}))$ is the set of the equivalence classes of $V(G)$ under the action of Γ and there are as many arcs from $[v]_\Gamma$ to $[w]_\Gamma$ as each vertex in $[w]_\Gamma$ has predecessors in $[v]_\Gamma$. Due to Remark 1, this does not depend on the choice of the element of $[w]_\Gamma$. We define the labelling ν of $\mathbf{B}_\Gamma(\mathbf{G})$ by $\nu([v]_\Gamma) = \lambda(v)$ for each $v \in V(G)$. We label the arcs from $[v]$ to $[w]$ with the labels of the arcs from the elements of $[v]$ to w in **G**. By Remark 1, there exists a fibration φ from **G** to $(\mathbf{B}_\Gamma(\mathbf{G}), \nu)$.

We consider also the quotient-graph $\mathbf{G}/\Gamma$ whose vertices and arcs are the equivalence classes of the vertices and the arcs of G under the action of Γ and whose labelling μ is defined by $\mu([x]_\Gamma) = \lambda(x)$ for each $x \in V(G) \cup A(G)$. There exists a natural surjective homomorphism from $(\mathbf{B}_\Gamma(\mathbf{G}), \nu)$ to $\mathbf{G}/\Gamma$ which is the identity on the vertices and which maps an arc a to $[a]_\Gamma$ (a can be seen as an arc of G).

We say that a subgroup Γ of $\Gamma(\mathbf{G})$ acts *freely* on $\mathbf{G}$ if for each $x, y \in V(G) \cup A(G)$, there is at most one $\sigma \in \Gamma$ such that $\sigma(x) = y$. Equivalently, Γ acts freely on $\mathbf{G}$ if and only if for each $\sigma \in \Gamma \setminus \{Id\}$, σ has no fixpoint.

In the following, we will use a particular class of automorphisms: the class of *well-balanced* automorphisms. These automorphisms have been introduced by Bougé in [8] to study the importance of the guards in CSP through the symmetric election problem. In [12], Palamidessi uses also well-balanced automorphisms to study the same problem in order to give a hierarchy between different subsets of the π-calculus. An automorphism σ of a digraph $\mathbf{G}$ is *well-balanced* if there exists an integer q such that for each vertex or arc x of $\mathbf{G}$, $|\{\sigma^k(x) \mid k \in \mathbb{N}\}| = q$. Equivalently, σ is well-balanced if and only if the subgroup Γ_σ generated by σ acts freely on $\mathbf{G}$.

The group Γ contains only well-balanced automorphisms if and only if Γ acts freely on $\mathbf{G}$. Thanks to this equivalence and the results of Boldi and Vigna [7], we have the following property.

Proposition 1. *Given any strongly connected digraph $\mathbf{G}$, the quotient projection $\Gamma : \mathbf{G} \to \mathbf{G}/\Gamma$ is a covering projection if and only if for each $\sigma \in \Gamma$, σ is well-balanced.*

3   Impossibility Result

The following proposition gives a necessary condition that the distributed mobile environment $(G, \mathcal{E}, p)$ must verify if there exists an election algorithm for $(G, \mathcal{E}, p)$. This necessary condition is equivalent to the one presented in [3].

Proposition 2. *Consider a graph G and an initial placement of the agents p. If there exists a non-trivial well-balanced automorphism σ of the digraph $\mathbf{G}' = (Dir(G), \chi_p)$, then there is no election algorithm over the graph G with the initial placement of the agents p.*

Using known results in distributed computing [6, 9, 13], we can show that in the anonymous setting, i.e., when the agents can understand each other but do not have distinct labels, there exists an election algorithm for an environment $(G, \mathcal{E}, p)$ if and only if the labelled digraph $\mathbf{G}' = (Dir(G), \chi_p)$ is symmetric-covering-minimal. Moreover, from Proposition 1, we know that if the symmetric digraph $\mathbf{G}' = (Dir(G), \chi_p)$ admits a non-trivial well-balanced automorphism σ, then $\mathbf{G}'$ is a symmetric covering of $\mathbf{G}'/\Gamma_\sigma$ that is not isomorphic to $\mathbf{G}'$.

Consequently, an interesting corollary of Proposition 2 is that if the election problem cannot be solved on $(G, \mathcal{E}, p)$ in the qualitative setting, then it cannot be solved on $(G, \mathcal{E}, p)$ in the anonymous setting. On the other hand, we will show in the following that this necessary condition is also sufficient. Note that there exist symmetric digraphs that are not symmetric-covering-minimal and that does not admit any non-trivial well-balanced automorphism. It means that one can solve the election problem in strictly more environments in the qualitative setting than in the anonymous one.

4   An Effective Election Algorithm

4.1   How to Order the Equivalence Classes?

We use the same ideas as Barrière et al. [3] to define a total order between
the different equivalent classes. The idea is to construct an ordering on the
unlabelled digraphs of size n; we extend it to digraphs labelled by elements of a
totally ordered set.

Consider a labelled digraph $\mathbf{G} = (G, \lambda)$ without multiple arcs where λ is a
labelling function from $V(G) \cup A(G)$ to a totally ordered set L with a minimal
element $\perp$. We suppose that $\forall x \in V(G) \cup A(G), \lambda(x) \in L \setminus \{\perp\}$.

Let $n = |V(G)|$ and consider an enumeration function num of the vertices
(i.e., num is a one-to-one mapping from $V(G)$ onto $[1, n]$). We say that num
is an *increasing* enumeration of the vertices if for all vertices $v, v' \in V(G)$,
if $num(v) \leq num(v')$, then $\lambda(v) \leq_L \lambda(v')$. Given an increasing enumeration
num, we define the adjacency matrix M_{num} as follows: for all vertices v, v',
$M_{num}[num(v), num(v')] = \ell$, if there is an arc from v to v' labelled by ℓ, and
$M_{num}[num(v), num(v')] = \perp$ otherwise. To this matrix, we associate the word
$w(M_{num})$ obtained by the concatenation of the n rows of M_{num}.

To each vertex $v \in V(G)$ (resp. arc $a \in A(G)$), we choose num such that
$(num(v), w(M_{num}))$ (resp. $(num(s(a)), num(t(a)), w(M_{num}))$) is minimum for
the lexicographic order and associate this value, denoted by $\pi(v)$ (resp. $\pi(a)$), to
v (resp. a). Note that there exists an automorphism σ of $\mathbf{G}$ such that $\sigma(x) = x'$
if and only if $\pi(x) = \pi(x')$. Consequently, this induces a total ordering of the
equivalence classes of vertices and arcs: we will write $[x'] \prec [x]$ if $\pi(x)$ is greater
than $\pi(x')$ in the lexicographic order.

Remark 2. In the following, we will show that all the agents agree on a total
order of the classes and all the agents use the same order. Actually, as it was
already explained in [3], even if the agents cannot agree on an a priori order over
the set of colors, they can agree on an order on the different classes, provided that
all the agents have the same representation of the graph (up to isomorphism).

In fact, we suppose that each agent has its own totally ordered set isomorphic
to $(\mathbb{N}, \leq)$ and each agent can use its own way to compute its order: the algorithm
does not make any assumption on the way the order is implemented by each
agent.

4.2   An Election Algorithm

In this subsection, we describe our effective election algorithm. In a first phase
all the agents reconstruct the digraph $(Dir(G), \chi_p)$ and check that the election
problem can be solved on $(G, \mathcal{E}, p)$. Then, using its knowledge of the graph, each
agent constructs the equivalence classes induced by $\Gamma((Dir(G), \chi_p))$. During successive rounds, using the order between the different classes defined above, some
agents become passive and get the label *non-elected*, whereas the active agents
mark some vertices and some arcs of the digraph to obtain a new labelling μ of
the digraph on which all the active agents agree. At the end of the computation,

the automorphism group of $(Dir(G), \mu)$ consists only of the identity and each vertex has a unique label. At this point, there is exactly one active agent that is elected. A high level description of the algorithm is presented in Algorithm 1.

Algorithm 1. The Election Algorithm

Every agent builds a map of the graph;
Synchronization;
if *there exists a non-trivial well-balanced automorphism of* $\mathbf{G'} = (Dir(G), \chi_p)$
then
  | Every agent knows that it is impossible to solve the election problem;
else
  Every agent marks as many vertices as possible;
  Synchronization;
  /* Initially, all the agents are active */
  repeat
    The active agents compute the classes of all the vertices and all the arcs;
    The active agents give different numbers to different classes of vertices and arcs;
    if *All the active agents are not in the same class* **then**
      Select active agents;
      The passive agents take the label *non-elected*;
      The active agents mark the homebases of the passive agents;
      Synchronization;
    else if $\mathbf{G'}$ *is not a covering of* $\mathbf{B}_{\Gamma(\mathbf{G'})}(\mathbf{G'})$ **then**
      The active agents mark a class of vertices;
      Synchronization;
    else if $\mathbf{G'}$ *is not a covering of* $\mathbf{G'}/\Gamma(\mathbf{G'})$ **then**
      The active agents mark a class of arcs;
      Synchronization;
    else
      /* In this case, there is exactly one active agent */
      The active agent takes the label *elected*;
  until *An agent is elected*;

A Synchronization Procedure. In the algorithm we describe below, we distinguish different rounds. An important point is that an active agent does not enter in a new round if another active agent has not finished the previous one. To be able to avoid this kind of situation, we synchronize the active agents.

Each agent can consistently distinguish its homebase; therefore, we can construct an algorithm such that no agent needs to write anything on its homebase. Moreover, we suppose that each agent has already built its own map of the graph and does not need to write anything on any whiteboard in order to perform a traversal of the graph.

In the following, the active agents will do some traversals of the network and they will store the colors of the marks that appear on each vertex to construct what we will call a *colored map* of the network. The marks that appear in a colored map of an active agent will correspond to marks that have been put by other active agents during the round (but it will not necessary contain all the marks the active agents should put during this round). In the algorithm described below, each active agent can know from a colored map if any other active agent has marked all the vertices it should have marked during the round. Furthermore, in each round of the algorithm, each agent will mark at least one vertex (which is not its homebase).

In the synchronization procedure described below, some active agents will have to wait on some particular vertices for other agents to put (resp. remove) some marks. Each time an agent arrives on a place where it has to wait for a mark to be put (resp. removed), it can immediately continue to execute the procedure if this mark is present (resp. not present).

To synchronize the agents, we proceed as follows. During each round, each active agent r executes the following instructions.

(1) The agent r marks some vertices (but not its homebase) according to the computation rules of the round.
(2) The agent r does a traversal of the network and stores all the colors of the marks that appear on each vertex to construct a colored map of the network.
(3) If there exists another active agent r' that has not finished Step (1) (the agent r can detect it from the colored map it has of the network), then the agent r goes to the homebase of r' and waits until the agent r' puts a mark on its homebase. Then the agent r does a traversal of the network and stores all the colors of the marks that appear on each vertex in order to update its colored map of the network.
(4) The agent r puts a mark on its homebase.
(5) The agent r does a traversal of the network and each time it arrives on the homebase of another agent r', it waits until the agent r' marks its homebase.
(6) The agent r does a traversal of the network and it removes the marks it puts during Step (1), but not the mark on its homebase.
(7) The agent r does a traversal of the network. Each time it arrives on a vertex that has been marked by another active agent r' during this round (but that is not the homebase of r'), it waits until the agent r' removes its mark.
(8) The agent r removes the mark it puts on its homebase.
(9) The agent r does a traversal of the network. Each time it arrives on the homebase of an active agent r', it waits until the agent r' removes its mark on its homebase.

We can note that the synchronization procedure enables also to erase all the marks that have been put on the vertices during the round, i.e., when one agent has finished Step (9) of a round, then all the marks that have been left by the active agents during this round have been erased. The following proposition ensures that the procedure is indeed a synchronization procedure.

Proposition 3. *Each time an agent starts executing Step (1) of the $i + 1$th round, then each active agent knows what vertices have been marked by the other active agents during Step (1) of the ith round and all the marks that have been put during the ith round have been removed. Moreover, the synchronization procedure avoids any deadlock.*

Initialization. During the first phase of the algorithm, each agent reconstructs the graph with the position and the colors of the different homebases. Using the whiteboards, each agent performs a depth first traversal of the graph.

Since each agent can distinguish all the homebases, we suppose that during this traversal, the agents do not write anything on the whiteboard of any of the homebases. Once an agent has reconstructed the whole graph, it performs a traversal of the network using the information it has stored to erase what it has written on the whiteboards. At this point, each agent puts a mark on the homebase of another agent.

During this first phase, no agent has written anything on its homebase. Furthermore, an agent has finished to perform this phase if and only if it has marked the homebase of another agent and this can be checked from a colored map of the network. Moreover, at the end of this phase, each agent has reconstructed a map of the network and it knows the position of all the homebases. We can therefore use the synchronization procedure defined above at this point.

If the digraph $(Dir(G), \chi_p)$ admits a well-balanced automorphism σ different from Id, then each agent detects it and declares that the election problem is unsolvable in this environment. We will now suppose that $(Dir(G), \chi_p)$ does not admit such an automorphism.

Once the graph is known by all the agents, each agent tries to mark as many vertices of the networks as possible. It does a traversal of the network and each time it arrives on a vertex that is not a homebase, it performs one of the two following actions. Either the whiteboard is blank and it puts a mark with its color on the whiteboard, or there is already a mark on the whiteboard and it stores the color of the mark. Once an agent has finished this traversal, it puts a mark on the homebase of another agent. Again we use the synchronization procedure at this point. Then each agent is aware of the different vertices marked by the other agents during this round.

At the end of this phase, each agent reconstructs a graph where all the vertices are colored (they belong to the agent that has this color) and it knows the position and the color of the homebases of all the other agents.

How can the agents increase their territory? During the different phases of the algorithm, some agents become passive whereas the others continue to execute the protocol in order to elect one of them. In our algorithm, in order to break the symmetry between the agents, all the vertices must belong to one active agent, and all the active agents must agree on which agent a vertex belongs to. During the initialization, each vertex is marked by one agent and we say that it belongs to this agent. Once an agent becomes passive, the vertices that were belonging to this agent must be given to another agent.

Once a selection between agents is done, the agents that become passive take the label *non-elected* and become passive until the end of the algorithm, whereas the others try to mark the homebases of these agents that have just become passive. Each active agent knows what are the colors of the other active agents. From its representation of the graph, each active agent can reach the homebases of the passive agents.

The first agent that reaches such a homebase during this round puts a mark with its color on the vertex. The other agents (there is already a mark on the homebase when they reach it) store the color of the agent that owns this vertex (i.e., the color of the mark). Again, at the end of its traversal of the graph, each agent puts a mark on the homebase of another active agent. Therefore, each active agent can detect from a colored map if another active agent has finished this phase. Then the active agents apply the synchronization procedure.

If an agent has marked the homebase of a passive agent, then all the vertices that were belonging to this passive agent belong now to this active agent. For each vertex of the graph, all the active agents agree on the color of the agent that owns this vertex.

How to refine the labelling μ? During the execution of the algorithm, the agents mark vertices and arcs to break the symmetry that may exist in the network. In this way, at each round, numbers will be associated to some vertices and arcs and we will obtain a labelling of the graph μ. Initially, all the homebases have the label 1 whereas all the other vertices have the label 0 and all the arcs are labelled 0.

At the beginning of each round, from its representation $(Dir(G), \mu)$ of the graph, each agent computes the value $\pi(v)$ (resp. $\pi(a)$) for each vertex $v \in V(Dir(G))$ (resp. for each arc $a \in A(Dir(G))$). We say that two agents are equivalent if their homebases are in the same equivalence class, and we use the order $\prec$ on the homebases of the agents to order the classes of agents.

Since all the agents agree on the order to compare the equivalence classes, we can use the following procedure. If there exist two vertices v, v' such that $\mu(v) = \mu(v')$ and $\pi(v) \neq \pi(v')$, then let m be the lowest number such that there exist v, v' with $\mu(v) = \mu(v') = m$ and $[v] \prec [v']$. Suppose that there exist exactly j classes $\{[v_i] | i \in [1, j]\}$ such that $\mu(v_i) = m$ and $[v_1] \prec [v_2] \prec \cdots \prec [v_j]$. For each vertex $v \in [v_i]$ with $i < j$, we define $\mu'(v) = q + i$, where q is the greatest label that appears on a vertex in (G, μ). The labels of the other vertices are not changed.

We apply the same method to arcs using the order we have on the classes of arcs, i.e., the lexicographic order over the $\pi(a)$. Thanks to this procedure, two arcs that are not in the same class are given distinct labels. We repeat this procedure, until all the vertices (resp. all the arcs) that have the same label are in the same equivalence class.

If some active agents do not own the same number of vertices in a given class. We consider now a configuration such that two vertices (resp. arcs) in different classes have different numbers. Consider a class of agents $[r]$ and a class of vertices $[v]$. We define $NotBalanced([r], [v])$ to be false if all the agents of $[r]$ own the same number of vertices in $[v]$, and true otherwise. If there exist $[r], [v]$ such that $NotBalanced([r], [v])$ is true, then we apply the following technique to split some class of vertices.

Consider the minimum class $[r]$ of agents, according to $\prec$, such that there exists a class $[v]$ of vertices satisfying $NotBalanced([r], [v])$. Consider the minimum class of vertices $[v]$ such that $NotBalanced([r], [v])$ is true. In this case, we give different numbers to the homebases of the agents that do not own the same number of vertices in $[v]$. We subdivide the class $[r]$ into a partition $R_1, \ldots, R_j$ such that the agents in R_i own strictly more vertices in $[v]$ than the agents in $R_{i'}$ when $i < i'$. Using the same technique as before, we give different numbers to the homebases of the agents that are not in the same R_i and then obtain a new representation of the digraph $(Dir(G), \mu')$. Then, the agents try to refine again this new labelling.

How to split the arc classes thanks to the colors of their ends? We will say that an arc a belongs to an agent r, if r owns $s(a)$ and $t(a)$. Otherwise, the arc is such that $s(a)$ belongs to an agent r_1 and $t(a)$ to a distinct agent r_2. We will say that this arc is shared by r_1 and r_2. If there exists a class of arcs $[a]$ such that some arcs of $[a]$ belong to some agents, whereas the other arcs are shared by distinct agents, then we apply the following technique.

Consider a class of arcs a such that for each class $[a'] \prec [a]$, either $[a']$ contains only arcs that belong to some agents or $[a']$ contains only arcs shared by different agents. We suppose also that $[a]$ contains arcs that belong to some agents and arcs that are shared. All the arcs in $[a]$ that are shared by distinct agents are relabelled $q + 1$, where q is the greatest label that appears on an arc in $(Dir(G), \mu)$. Then, the agents try to refine again this new labelling.

If some active agents are in different classes. At this point, if the active agents are not in the same equivalence class, we are able to select some agents. Consider all the equivalence classes of active agents that contains a minimal number of agents. Among these classes, we select the class $[r]$ such that $\pi(v)$ is minimal, where v is the homebase of r. The agents that do not belong to this class take the label *non-elected* and become passive. The agents of the class $[r]$ remain active and try to increase their territory as explained above. Then they try to refine again the labelling μ.

If $\mathbf{G'} = (\mathbf{Dir(G)}, \mu)$ is not a covering of $\mathbf{B}_{\Gamma(\mathbf{G'})}(\mathbf{G'})$. There exist some configurations where it is impossible to select some agents just by using the representation the agents have of the graph, because there is too much symmetry in the graph. Nevertheless, we now explain how active agents can break these symmetries by marking some vertices or arcs.

All the active agents agree on the graph $\mathbf{G'} = (Dir(G), \mu)$. All these agents consider the automorphism group $\Gamma(\mathbf{G'})$ and construct the graphs $\mathbf{B}_{\Gamma(\mathbf{G'})}(\mathbf{G'})$ and $\mathbf{G'}/\Gamma(\mathbf{G'})$. We already know that $\mathbf{G'}$ is fibred over $\mathbf{B}_{\Gamma(\mathbf{G'})}(\mathbf{G'})$. If $\mathbf{G'}$ is not a covering of $\mathbf{B}_{\Gamma(\mathbf{G'})}(\mathbf{G'})$, it implies that there exist two classes of vertices $[v]$ and $[v']$ such that $|[v]| \neq |[v']|$. Let $[r]$ be the class of the homebases of the active agents.

Consider a class $[v]$ such that for each class $[v'] \prec [v]$, $|[v']| = |[r]|$ and $|[v]| \neq |[r]|$. We already know that each active agent owns the same number of vertices in $[v]$ and therefore $|[r]|$ divides $|[v]|$. Each active agent then marks a vertex it owns that is in $[v]$. An agent has finished this round if and only if it has marked exactly one vertex in $[v]$: it can be detected from a colored map of the graph. Then, the agents synchronize.

At the end of this round, all the agents give the number $q+1$ to the vertices that have just been marked, where q is the greatest label that appears on a vertex in $(Dir(G), \mu)$. Using this new labelling μ', the active agents try to refine the labelling μ', as explained above.

If $\mathbf{G'} = (\mathbf{Dir(G)}, \mu)$ is not a covering of $\mathbf{G'}/\mathbf{\Gamma(G')}$. We suppose now that $\mathbf{G'}$ is a covering of $\mathbf{B}_{\Gamma(\mathbf{G'})}(\mathbf{G'})$ but not of $\mathbf{G'}/\mathbf{\Gamma(G')}$. It means that all the equivalence classes of vertices have the same size s, but there exists an equivalence class of arcs $[a]$ such that $\|[a]\| > s$. Instead of marking vertices, we mark arcs in this round.

Each class $[a]$ of arcs of $\mathbf{G'}$ corresponds to exactly one arc in $\mathbf{G'}/\mathbf{\Gamma(G')}$. Consider the class of arcs $[a]$ such that for each class $[a'] \prec [a]$, $\|[a']\| = s$ but $\|[a]\| > s$. We already know that each active agent owns exactly one vertex in $[s(a)]$ and one vertex in $[t(a)]$. Since $\|[a]\| > s$ and since two arcs in the same class are either both owned by an agent or both shared by distinct agents, we know that each arc in $[a]$ is shared.

To select arcs from $[a]$, each agent r just chooses one arc a_r in $[a]$ such that $s(a_r)$ belongs to r and then puts a mark with its color on $t(a_r)$. An agent has finished this round if and only if it has marked exactly one vertex: it can be detected from a colored map of the graph. Then, the agents synchronize. Once an agent knows what vertices have been marked by the other agents, it knows what are the arcs that have been marked.

At the end of this round, all the agents give the number $q+1$ to the arcs that have just been marked, where q is the greatest label that appears on an arc in $(Dir(G), \mu)$. Using this new labelling μ', the active agents try to refine again the labelling μ', as explained above.

If $\mathbf{G'}$ is a covering of $\mathbf{G'}/\mathbf{\Gamma(G')}$. At this point, $\mathbf{G'} = (Dir(G), \mu)$ is a covering of $\mathbf{G'}/\mathbf{\Gamma(G')}$. From Proposition 1, it implies that $\Gamma(\mathbf{G'})$ contains only well-balanced automorphisms, and since we already know that there is no well-balanced isomorphism of $(Dir(G), \chi_p)$ different from Id, we have $\Gamma(\mathbf{G'}) = \{Id\}$. Consequently, there is exactly one active agent, since the set of active agents is an equivalence class of the relation induced by $\Gamma(\mathbf{G'})$ and this agent takes the label *elected*.

4.3   The Characterization

In Section 3, we have shown that if the graph $(Dir(G), \chi_p)$ admits a well-balanced automorphism, then it is impossible to solve the election problem on $(G, \mathcal{E}, p)$. The algorithm described in Section 4.2 is an algorithm that answers that it is impossible to solve the problem if the graph $(Dir(G), \chi_p)$ admits a well-balanced automorphism, and otherwise it successfully elects an agent: it is an effective algorithm. We have therefore proved the following theorems.

Theorem 1. *There exists an election algorithm for a distributed mobile environment $(G, \mathcal{E}, p)$ if and only if $(Dir(G), \chi_p)$ does not admit a non-trivial well-balanced automorphism.*

Theorem 2. *Algorithm 1 is an effective election algorithm in the qualitative world.*

The traditional complexity measures for mobile agents are the number of agents moves and the amount of time of a synchronous execution of the algorithm, where in each round, each active agent traverses an edge.

In a distributed mobile environment $(G, \mathcal{E}, p)$ with $|V(G)| = n$, $|E(G)| = m$ and $|\mathcal{E}| = k$, when executing Algorithm 1, the agents detect with $O(mk)$ moves in time $O(m)$ if the election problem can be solved; if it is possible, they successfully elects a leader with $O(mn \log k)$ moves in time $O(mn)$.

References

1. D. Angluin. Local and global properties in networks of processors. In *Proceedings of the 12th Symposium on Theory of Computing, STOC'80*, pages 82–93, 1980.
2. B. Awerbuch, M. Betke, R. Rivest, and M. Singh. Piecemeal graph exploration by a mobile robot (extended abstract). In *Proc. of the 8th annual conference on Computational Learning Theory, COLT'95*, pages 321–328. ACM Press, 1995.
3. L. Barrière, P. Flocchini, P. Fraigniaud, and N. Santoro. Can we elect if we cannot compare? In *Proc. of the 15th annual ACM Symposium on Parallel Algorithms and Architectures, SPAA'03*, pages 324–332. ACM Press, 2003.
4. L. Barrière, P. Flocchini, P. Fraigniaud, and N. Santoro. Rendezvous and election of mobile agents: Impact of sense of direction. *Theory of Computing Systems*, to appear.
5. M. Bender and D. Slonim. The power of team exploration: Two robots can learn unlabeled directed graphs. In *Proc. of the 35th annual Symposium on Foundations of Computer Science, FOCS'94*, pages 75–85, 1994.
6. P. Boldi, B. Codenotti, P. Gemmell, S. Shammah, J. Simon, and S. Vigna. Symmetry breaking in anonymous networks: Characterizations. In *Proc. 4th Israeli Symposium on Theory of Computing and Systems*, pages 16–26. IEEE Press, 1996.
7. P. Boldi and S. Vigna. Fibrations of graphs. *Discrete Math.*, 243:21–66, 2002.
8. L. Bougé. On the existence of symmetric algorithms to find leaders in networks of communicating sequential processes. *Acta Informatica*, 25(2):179–201, 1988.
9. J. Chalopin and Y. Métivier. A bridge between the asynchronous message passing model and local computations in graphs (*extended abstract*). In *Proc. of Mathematical Foundations of Computer Science, MFCS'05*, volume 3618 of *LNCS*, pages 212–223, 2005.
10. S. Das, P. Flocchini, A.Nayak, and N. Santoro. Distributed exploration of an unknown graph. In *Proc. of the 12th international colloquium on Structural Information and Communication Complexity, SIROCCO'05*, volume 3499 of *LNCS*, pages 99–114, 2005.
11. A. Dessmark, P. Fraigniaud, and A. Pelc. Deterministic rendezvous in graphs. In *Proc. of the 11th annual European Symposium on Algorithms, ESA'03*, volume 2832 of *LNCS*, pages 184–195, 2003.
12. C. Palamidessi. Comparing the expressive power of the synchronous and the asynchronous π-calculus. *Mathematical Structures in Computer Science*, 13(5):685–719, 2003.
13. M. Yamashita and T. Kameda. Computing on anonymous networks: Part i - characterizing the solvable cases. *IEEE Transactions on parallel and distributed systems*, 7(1):69–89, 1996.

Fast Deterministic Distributed Algorithms for Sparse Spanners

Bilel Derbel and Cyril Gavoille*

LaBRI, Université Bordeaux 1
351, Cours de la Libération,
33405 Talence, France
{derbel, gavoille}@labri.fr

Abstract. This paper concerns the efficient construction of sparse and low stretch spanners for unweighted arbitrary graphs with n nodes. All previous deterministic distributed algorithms, for constant stretch spanner of $o(n^2)$ edges, have a running time $\Omega(n^\epsilon)$ for some constant $\epsilon > 0$ depending on the stretch. Our deterministic distributed algorithms construct constant stretch spanners of $o(n^2)$ edges in $o(n^\epsilon)$ time for any constant $\epsilon > 0$.

More precisely, in the Linial's free model, we construct in $n^{O(1/\sqrt{\log n})}$ time, for every graph, a 5-spanner of $O(n^{3/2})$ edges. The result is extended to $O(k^{2.322})$-spanners with $O(n^{1+1/k})$ edges for every parameter $k \geqslant 1$. If the minimum degree of the graph is $\Omega(\sqrt{n})$, then, in the same time complexity, a 9-spanner with $O(n)$ edges can be constructed.

Keywords: Distributed algorithms, graph spanners, time complexity, Linial's free model, deterministic and randomized algorithms.

1 Introduction

This paper deals with deterministic distributed construction of sparse and low stretch graph spanners. Intuitively, spanners can be thought of as a generalization of the concept of a spanning tree. We look for a spanning subgraph such that the distance between any two nodes in the subgraph is bounded by some constant times the distance in the whole graph. More formally, H is a *k-spanner* of a graph G if H is a spanning subgraph of G, and if $d_H(u, v) \leqslant k \cdot d_G(u, v)$ for all nodes u, v of G, where $d_X(u, v)$ denotes the distance from u to v in the graph X. The smallest k for which H is a k-spanner is called the *stretch* of H, and the *size* of H is its number of edges. The quality of a spanner refers to the trade-off between the stretch and the size of the spanner.

The distributed model of computation we will be concerned with is the Linial's free model [26], also known as $\mathcal{LOCAL}$ model in [34]. In this model, communication is completely synchronous and reliable. At every time unit, each node may send or receive a message of unlimited size to or from all its neighbors, and can

* Supported by the project "PairAPair" of the ACI Masses de Données.

P. Flocchini and L. Gąsieniec (Eds.): SIROCCO 2006, LNCS 4056, pp. 100–114, 2006.
© Springer-Verlag Berlin Heidelberg 2006

locally compute any function. The model also assumes that each node is equipped with a unique identifier. Much as PRAM algorithms in parallel computing give a good indication of parallelism, the free model gives a good indication of the locality and distributed time.

From a theoretical point of view, we are interested in the *locality nature* of constructing graph spanners, i.e., what spanners can we compute assuming only some local knowledge? The locality of a distributed problem is often expressed in term of the time needed to resolve it. In fact, in the distributed setting, the best a node can do in $O(t)$ time units is to collect its t neighborhood. For instance, $\Theta(\log^* n)$ time are necessary and sufficient to compute a maximal independent set for trees, bounded degree graphs, or bounded growth graphs with n nodes [11, 21, 27, 22]. Results are known for other fundamental problems such as non-uniform coloring [2, 33], minimum spanning tree [16, 17, 29, 28, 35], small dominating set [25, 38], and maximal matching [23, 27].

Graph spanners are in the basis of various applications in distributed systems. For instance, Peleg and Ullman [36] establish the relationship between the quality of spanners, and the time and message complexity of network synchronizers (see also [1, 32]). Spanners are also implicitly used for the design of low stretch routing schemes with compact tables [12, 14, 37, 39, 41], and appear in many parallel and distributed algorithms for computing approximate shortest paths and for the design of compact data-structures, a.k.a. distance oracles [9, 20, 40, 42, 10].

1.1   Related Works

Sparse and low stretch spanners can be constructed from (d, c)-decomposition of Awerbuch and Peleg [6], that is a partition of the graph into clusters of diameter at most d such that the graph obtained by contracting each cluster can be properly c-colored. There are several deterministic algorithms for constructing (d, c)-decompositions [3, 4, 5, 33]. The resulting distributed algorithms provide $O(k)$-spanners of size $O(n^{1+1/k})$, for any integral parameter $k \geqslant 1$. However, these algorithms run in $\Omega(n^{1/k+\epsilon})$ time, where $\epsilon = \Omega(1/\sqrt{\log n})$, and provide a stretch $s \geqslant 4k - 3$.

Better stretch-size trade-offs exist but with an increasing time complexity. Recently, a deterministic distributed algorithm has been proposed for constructing a $(2k - 1)$-spanner of size $O(n^{1+1/k})$ in $O(n^{1-1/k})$ time [13]. In particular, 3-spanners of size $O(n^{3/2})$ can be deterministically constructed in $O(\sqrt{n})$ time.

Elkin et al. [15, 19, 18] develop a distributed algorithm for spanners such that the distance between two nodes in the spanner is at most $1 + \epsilon$ times the distance in the original graph plus β. The size is $O(\beta n^{1+\delta})$ whereas the time is $O(n^\delta)$, where $\beta = \beta(\delta, \epsilon)$ is independent of n but grows super-polynomially in δ^{-1} and ϵ^{-1}.

Randomized algorithms achieving better performances exist. Baswana et al. [8, 7] gave a randomized algorithm which computes an optimal $(2k - 1)$-spanner with expected size $O(n^{1+1/k})$ in $O(k)$ time. The latter stretch-size trade-off is optimal since, according to an Erdös Conjecture verified for $k = 1, 2, 3, 5$ [43], there are graphs with $\Omega(n^{1+1/k})$ edges and girth $2k + 2$ (the length of the smallest induced cycle), thus for which every s-spanner requires $\Omega(n^{1+1/k})$ edges if $s < 2k + 1$. However, as mentioned in [4], a randomized solution might not be acceptable in some cases, especially for distributed computing applications. In the case of graph spanners, deterministic algorithms that *guarantee* a high quality spanner are more than of a theoretical interest. Indeed, one cannot just run a randomized distributed algorithm several times to guarantee a good decomposition, since it is impossible to efficiently check the global quality of the spanner in the distributed model.

1.2   Results

We consider unweighted connected graphs with n nodes. All previous deterministic distributed algorithm for $O(1)$-spanner of size $o(n^2)$ have a running time $\Omega(n^\delta)$ for some constant $\delta > 0$ depending on the stretch. In this paper we construct constant stretch spanner of size $o(n^2)$ in $o(n^\epsilon)$ time for any constant $\epsilon > 0$.

More precisely, in the free model we construct in $n^{O(1/\sqrt{\log n})}$ time for every graph a 5-spanner of $O(n^{3/2})$ edges. The result is extended to larger stretch spanner of size $O(n^{1+1/k})$ for every $k \geqslant 1$. More precisely, for k power of two, the stretch is at most $k^{\log_2 5} < k^{2.322}$. For other values of k, we obtain stretches $s = s(k)$ which surprisingly depend on the positions of the first two leading 1's in the binary written of k (cf. the table below for the first values).

k	1	2	3	4	5
$s(k)$	1	5	9	25	33

We also show that if the minimum degree of the graph is $\Omega(\sqrt{n})$, then, in the same time complexity, a 9-spanner with $O(n)$ edges can be constructed.

The previous algorithms have simple randomized versions with improved performances. In particular, we can compute a 5-spanner of size $O(n \log^2 n)$ in $O(\log n)$ time if the minimum degree is $\Omega(\sqrt{n})$.

1.3   Outline of the Paper

The main idea to break the $O(n^\delta)$ time barrier is to abandon the optimality on the stretch-size trade-off. We show that constant stretch spanners can be constructed on the basis of a maximal independent set, i.e., a set of pairwise non-adjacent nodes, maximal for inclusion. This can be deterministically computed in $n^{O(1/\sqrt{\log n})}$ time [4, 33]. Therefore, the time complexity to construct our spanners is improved by a factor of $n^{1/k}$.

The generic algorithm is described in Section 2 and analyzed in Section 3, where a distributed implementation is presented.

We mainly reduce the problem to the computation of an *independent ρ-dominating set*, that is a set X of pairwise non-adjacent nodes such that every node of the graph is at distance at most ρ from X. Using the terminology of [34], a ρ-dominating set if nothing else than a (ρ, s)-ruling set for some $s > 1$. Actually, in order to optimize the stretch, the main algorithm combines two strategies in a way depending on the binary written of k.

In Section 4, we present the main results about 5- and 9-spanners. Observing that for $\rho = 1$ an independent ρ-dominating set is a maximal independent set, we conclude that our generic algorithm can be implemented to run in $n^{O(1/\sqrt{\log n})}$ time for $\rho = 1$. Several optimizations are then proposed including randomization and graphs of large minimum degree.

2   A Generic Algorithm

2.1   Definitions

Let us consider an unweighted connected graph $G = (V, E)$. Given an integer $t \geqslant 1$, the t-th power of G, denoted by G^t, is the graph obtained from G by adding an edge between any two nodes at distance at most t in G. For a set of nodes H, $G[H]$ denotes the subgraph of G induced by H. For $X, Y \subseteq V$, let $d_G(X, Y) = \min\{d_G(x, y) \mid x \in X \text{ and } y \in Y\}$.

We associate with each $v \in V$ a *region*, denoted by $R(v)$, that is a set of nodes containing v and inducing a connected subgraph of G. Given $C \subseteq V$, G_C denotes the graph whose node set is C, and there is an edge between u and v in C if $d_G(R(v), R(u)) = 1$. We denote by $R^+(v) = \{u \in V \mid d_G(u, R(v)) \leqslant 1\}$ and by $R_C^+(v) = \{u \in C \mid d_G(R(u), R(v)) \leqslant 1\}$.

The *eccentricity* of a node v in G is defined as $\max_{u \in V}\{d_G(u, v)\}$. For a node $v \in X$, we denote by $\mathrm{BFS}(v, X)$ a Breadth First Search spanning tree in X rooted at v. We define $\mathrm{IDS}(G, \rho)$ as any independent ρ-dominating set of G. Finally, we define the integer $\ell(x)$ as follows:

$$\ell(x) = \begin{cases} -1 & \text{if } x \leqslant 0, \\ \lfloor \log_2 x \rfloor & \text{otherwise.} \end{cases}$$

In the reminder of the paper we assume the free model of computation as defined in the introduction. We define the time complexity of a distributed algorithm to be the worst-case number of time units from the beginning of the algorithm to its termination.

2.2   Description of the Algorithm

The main idea of the algorithm is to find an efficient clustering of dense regions in the graph. A high level description of the algorithm, named SPANNER, is given in Fig. 1. Intuitively, i_0 represents the relative position of the first two leading 1's in the binary written of k.

The algorithm works in many phases, where new regions are formed at each phase. There are two types: the light regions (L) and the heavy regions (H). At a given phase, some of the heavy regions are selected and enlarged by including nodes from other neighboring regions. One important observation is that each new enlarged region is connected and the new constructed regions are mutually disjoint.

At each phase of the algorithm, one of the two strategies depicted in Fig. 2 and Fig. 3 applies. The main idea behind the two strategies is the same: choose

---

SPANNER

Input: a graph $G = (V, E)$ with $n = |V|$, and integers $\rho, k \geqslant 1$
Output: a spanner S

1. $i_0 := \ell(k) - \ell(k - 2^{\ell(k)})$; $C := V$; $r = 0$; $\forall v \in V$, $R(v) := \{v\}$, and $c(v) := v$
2. **for** $i := 1$ **to** $\ell(k) + 1$ **do: if** $i = i_0$ **then** STRATEGY 1 **else** STRATEGY 2

Fig. 1. The algorithm SPANNER

some well selected dense regions and merge them with the other ones in order to form new larger regions. The main difference is that the density of a region is computed in a different way. The stretch of the output spanner depends on the way the radius of the regions increases and on the total number of phases of the algorithm, depending on the volume of the regions. And, radius and volume increase very differently.

On one hand, in STRATEGY 1, a region is dense if its neighborhood is $n^{1/k}$ times greater than its size. Applying only STRATEGY 1 allows to obtain small stretch for small values of k. However, asymptotically, the stretch is exponential in k. On the other hand, in STRATEGY 2, a region is dense if the number of its neighboring regions is $n^{1/k}$ times greater than its size which provides an exponential growth of the size of a region. Applying only STRATEGY 2 allows to obtain asymptotically stretches polynomial in k.

The main idea of algorithm SPANNER is to switch from one strategy to an other at each phase in order to obtain the smallest possible stretch. A full analysis shows that, by alternating STRATEGY 1 and STRATEGY 2, the best stretch can be obtained by applying STRATEGY 1 only once at a well chosen phase i_0. Typically, $i_0 = p - q$ if $k = 2^p + 2^q$ with $p > q$.

We associate with each region $R(v)$ an active node, called *center*, and the set of centers forms C. Initially, each node is the center of the region formed by itself. Each phase $i \in \{1, \ldots, \ell(k) + 1\}$ can be decomposed in seven parts we briefly sketch.

---

1. $L := \left\{ v \in C, |R^+(v)| \leqslant n^{1/k} \cdot |R(v)| \right\}$ and $H := C \setminus L$;
2. $\forall (u, v) \in L \times V$ such that $\exists$ edge e between $R(u)$ and v, $S := S \cup \{e\}$
3. $X := \mathrm{IDS}(G^{2(r+1)}[H], \rho)$
4. $\forall z \in V$, if $d_G(z, X) \leqslant (2\rho + 1)r + 2\rho$, then set $c(z)$ to be its closest node of X, breaking ties with identities.
5. $\forall v \in X$, $R(v) := \{z \in V \mid c(z) = v\}$
6. $\forall v \in X$, $S := S \cup \mathrm{BFS}(v, R(v))$
7. $C := X$ and $r := (2\rho + 1)r + 2\rho$

Fig. 2. STRATEGY 1

---

1. $L := \left\{ v \in C, |R_C^+(v)| \leqslant n^{1/k} \cdot |R(v)| \right\}$ and $H := C \setminus L$
2. $\forall (u, v) \in L \times C$ such that $\exists$ edge e between $R(u)$ and $R(v)$, $S := S \cup \{e\}$
3. $X := \mathrm{IDS}((G_C)^2[H], \rho)$
4. $\forall u \in C$, if $d_{G_C}(u, X) \leqslant 2\rho$, then set $c(u)$ to be its closest node of X in G_C, breaking ties with identities.
5. $\forall v \in X$, $R(v) := \{R(u) \mid u \in C$ and $c(u) = v\}$
6. $\forall v \in X$, $S := S \cup \mathrm{BFS}(v, R(v))$
7. $C := X$ and $r := (4\rho + 1)r + 2\rho$

Fig. 3. STRATEGY 2

In Step 1, we compute the two sets H and L corresponding respectively to heavy and light regions. In Step 2, a light region is connected with some neighboring nodes. This step is crucial in the stretch bound analysis. If STRATEGY 1 is applied, then each light region is connected with each neighboring node in V, i.e., $\forall u \in L, R^+(u)$ is spanned. If STRATEGY 2 is applied, then each light region is connected with every neighboring region. Note that at the beginning of a given phase, every region is spanned by a BFS tree constructed in Step 6 of the previous phase.

The nodes H are then processed at the aim of constructing new regions with a set of new centers. The key point of our construction is to efficiently merge *all* the regions defined by the set H into *more dense, connected* and *disjoint* regions. In order to guarantee that the algorithm terminates quickly, the dense regions must grow enough. More precisely, if a dense region $R(v)$ is enlarged it must contain at least its neighborhood $R^+(v)$ when STRATEGY 1 is applied or its neighborhood in the graph G_C if STRATEGY 2 is applied. It is clear that two regions at distance one or two (in G or in G_C depending on the strategy 1 or 2) cannot grow simultaneously without overlapping. Thus, a difficulty is to elect in an efficient way the centers of regions that are allowed to grow in parallel.

In Step 3, we compute an independent ρ-dominating set X in the graph $G^{2(r+1)}[H]$ if STRATEGY 1 is applied (resp. $(G_C)^2[H]$ for STRATEGY 2), where r is a radius that grows at each phase. The set X defines the set of nodes allowed to grow in parallel.

In order to guarantee that nodes in non selected regions in Step 3 (the set $H \setminus X$) will be spanned by the output spanner, we must merge them with nodes in the selected regions. Thus, in Step 4, we define a coloring strategy allowing a correct merge process. In fact, in order to ensure that the new regions are disjoint, we let nodes choose their new region in a consistent manner, i.e., a node chooses to be in the region of the closest node in X breaking ties using identities. If STRATEGY 1 is applied then each node chooses by itself its new dominator, i.e., its new region. However, once a node u chooses its new dominator node v, and in order to ensure that the new formed regions are connected, we include all the nodes in the shortest path between u and v, even those in non dense region. If STRATEGY 2 is applied then, the center of each region chooses a new region and merge its whole region with the new chosen region.

In Step 5, the new regions are formed according to the coloring step. Note that as soon as the new region are formed, they are spanned in Step 6. Finally, in Step 7, the set C and the variable r are updated for the next phase.

3 Analysis of the Algorithm

For every phase i, we denote by H_i (resp. X_i and L_i) the set H (resp. X and L) computed during phase i, i.e., after Steps 1 and 3 of phase i. Similarly, we denote by $c_i(z)$ the color of z assigned during phase i, i.e., after Step 4 of phase i. We denote by C_i the set C at the beginning of phase i, and r_i denotes the value of r at the beginning of phase i. For a node $v \in C_i$, we denote by $R_i(v)$ the region of v at the beginning of phase i. In the following we need the four important properties.

Lemma 1. *At the beginning of phase i, every $v \in C_i$ is of eccentricity at most r in $G[R_i(v)]$.*

Lemma 2. *At the beginning of phase i, for every two nodes $u \neq v \in C_i$, $R_i(u) \cap R_i(v) = \varnothing$.*

Lemma 3. *At the beginning of phase $i \neq i_0$, if $|R_i(v)| \geqslant \mathcal{V}_i$ for every $v \in C_i$, then $|R_{i+1}(v)| \geqslant n^{1/k} \cdot \mathcal{V}_i^2$ for every $v \in C_{i+1}$.*

Lemma 4. *For every node $u \in V$, there exists a phase i and a node $v \in V$ such that:*

- *at the beginning of phase i, $v \in C_i$ and $u \in R_i(v)$; and*
- *v is in the set L_i computed in Step 1 of phase i.*

3.1 Stretch and Size Analysis

Lemma 5. *For any integer $k, \rho \geqslant 1$, the stretch s of the output spanner S of algorithm SPANNER verifies*

$$
s \leqslant \begin{cases} (4\rho+1)^{\ell(k)} & \text{if } k = 2^{\ell(k)}, \\ 2(2\rho+1)(4\rho+1)^{\ell(k)-1} + 4\rho(4\rho+1)^{\ell(k-2^{\ell(k)})} - 1 & \text{otherwise.} \end{cases}
$$

Proof. As a consequence of Lemma 4 and Step 6 of the algorithm, every node $u \in V$ is spanned by the output S of the algorithm, i.e., S is a spanner of G. From the initialization step of the algorithm, we have $r_1 = 0$. Let us denote by $i_1 = \ell(k)$ and $i_2 = \ell(k - 2^{\ell(k)})$, i.e., $i_0 = i_1 - i_2$. For every $1 < i \leqslant i_0$, we have $r_i = (4\rho+1)r_{i-1} + 2\rho$. Thus, $r_i = \frac{1}{2} \cdot \left((4\rho+1)^{i-1} - 1\right)$ for $1 \leqslant i \leqslant i_0$.

Let us consider an edge $(z, z') \in E$. Using Lemma 4, there exists a phase j (resp. j') and a node v (resp. v') such that $v \in C_j$ (resp. $v' \in C_{j'}$), $z \in R_j(v)$ (resp. $z' \in R_{j'}(v')$) and $v \in L_j$ (resp. $v' \in L_{j'}$). We take v (resp. v') to be the first dominator of z, i.e., node in C whose region contains z, (resp. z') that fall into set L. In fact, one can see that node z (or z') can be in a sparse region at some phase and switch to a dense region at the next phase, because either its sparse region has been merged with a neighboring dense one (if STRATEGY 2 is applied), or it is in the neighborhood of a dense region, or it is on a shortest path leading to a dense region (if STRATEGY 1 is applied). W.l.o.g., suppose that $j \leqslant j'$.

Case 1: $i_2 = -1$. Hence, $i_0 = \ell(k) + 1$ and $k = 2^{\ell(k)}$. By induction and using Lemma 3, at the beginning of phase i_0, the size of the region of any node in C_{i_0} is at least $n^{(2^{i_0-1}-1)/k} = n^{(k-1)/k}$. Note that we apply STRATEGY 1 at phase i_0. Thus, every node in C_{i_0} will be in L.

Subcase 1.1: Suppose that $j \leqslant j' < i_0$. Thus, using Step 6, a BFS tree spanning $R_j(v)$ is added to the output spanner at phase $j - 1$. In addition, one can easily show that there exists a node $v'' \in C_j$ such that $z' \in R_j(v'')$. Hence, a BFS tree spanning $R_j(v'')$ is added to the output spanner at phase $j - 1$. Using Step 2, there exists an edge $e \in S$ connecting $R_j(v)$ and $R_j(v'')$. Thus, $d_S(z, z') \leqslant 4r_j + 1 = 2(4\rho + 1)^{j-1} - 1$.

Subcase 1.2: Suppose that $j = j' = i_0$. Hence, STRATEGY 1 is applied at phase j. Thus, a BFS tree spanning $R_j(v)$ is added to the output spanner at phase $j - 1$. Using Step 2, $R_j^+(v)$ is also spanned. Thus, because $z' \in R_j^+(v)$, $d_S(z, z') \leqslant 2r_j + 1 = 2r_{i_0} + 1 \leqslant (4\rho + 1)^{i_0-1}$.

Finally, because $\rho > 0$, in both subcases, the stretch is bounded by $(4\rho + 1)^{\ell(k)}$.

Case 2: $i_2 \geqslant 0$. Hence, at the beginning of phase $i_0 + 1$, the radius of a region is at most $(2\rho + 1)r_{i_0} + 2\rho$. Thus,

$$r_{i_0+1} = \frac{1}{2}(2\rho + 1) \cdot \left((4\rho + 1)^{i_0-1} - 1\right) + 2\rho \tag{1}$$

Suppose $i_2 \neq 0$. For every $i_0 + 1 < i \leqslant \ell(k) + 1$, we have $r_i = (4\rho + 1)r_{i-1} + 2\rho$. Thus, by induction, for every $i_0 + 1 < i \leqslant \ell(k) + 1$,

$$r_i = (4\rho + 1)^{i-i_0-1} \cdot r_{i_0+1} + \frac{1}{2}((4\rho + 1)^{i-i_0-1} - 1)$$

In particular,

$$r_{\ell(k)+1} = (4\rho + 1)^{\ell(k)-i_0} \cdot r_{i_0+1} + \frac{1}{2}((4\rho + 1)^{\ell(k)-i_0} - 1)$$

Thus,

$$r_{\ell(k)+1} = (4\rho + 1)^{i_2} \cdot r_{i_0+1} + \frac{1}{2}((4\rho + 1)^{i_2} - 1) \tag{2}$$

Now suppose that $i_2 = 0$. Hence, $i_0 = \ell(k)$ and it is easy to see that Eq. 2 is still true.

Subcase 2.1: Suppose that $j \neq i_0$. Thus, it easy to show that there exists a node $v'' \in C_j$ such that $z' \in R_j(v'')$. Using Step 6, $R_j(v'')$ and $R_j(v)$ were spanned by a BFS tree at phase $j - 1$. In addition, because STRATEGY 2 is applied at phase j, an edge e connecting $R_j(v)$ and $R_j(v'')$ is added at phase j (Step 2). Thus, $d_S(z, z') \leqslant 4r_j + 1$.

Subcase 2.2: Suppose that $j = i_0$. Thus, because $R_j^+(v)$ is spanned, $d_S(z, z') \leqslant 2r_j + 1$.

At phase $\ell(k) + 1$, all active nodes will be in the set $L_{\ell(k)+1}$. Thus the stretch is bounded by $4r_{\ell(k)+1} + 1$. Using Eq. (1) and (2), we have:

$$
\begin{aligned}
4r_{\ell(k)+1} + 1 &= 4\left((4\rho + 1)^{i_2} \cdot r_{i_0+1} + \tfrac{1}{2}((4\rho + 1)^{i_2} - 1)\right) + 1 \\
&= 2(2\rho + 1)(4\rho + 1)^{i_1-1} + 4\rho(4\rho + 1)^{i_2} - 1 \qquad \square
\end{aligned}
$$

Lemma 6. *For any integer $k, \rho \geqslant 1$, the size of the output spanner S of algorithm* SPANNER *is $O(\log k \cdot n^{1+1/k})$.*

3.2   Distributed Implementation and Time Complexity

In the free model, distributed computation of some distributed procedure A on $G^t[H]$ can be easily simulated on G as follows, charging the overall time by a factor of t. Hereafter, we assume that each node $u \in G$ can determine if it belongs or not to H. Indeed, consider one communication step in A running on some node u of $G^t[H]$ followed by one local computation step. In G, an original message in A is sent from u with a counter initialized to $t - 1$ as an extra field. Now, each node $v \in G$, upon the reception of a message with some counter in its header: 1) decrements the counter; 2) stores this message if $v \in H$; and 3) forwards the incoming message with the updated counter to all its neighbors in G if the updated counter is non-null (if many messages are received during a round, then they are concatenated before being sent). After t communication rounds in G, every node $u \in H$ starts the local computation step of A on the base of all received messages during the last t communication rounds.

Similarly, given $C \subseteq V$, the computation of some distributed procedure A on G_C can be simulated on G as follows, charging the overall time by a factor $O(r)$ where r is an upper bound of the eccentricity of a node $v \in C$ in $G[R(v)]$. At each time procedure A requires for a node v of G_C to send a message to a neighbor, v broadcasts the message in $G[R(v)]$ (which is connected). The nodes at the frontier of $R(v)$, i.e., nodes having neighbors in different regions, broadcast also the message out their region. Symmetrically, upon the reception of messages from different regions, messages are concatenated and a convergecast is performed to v. The time overhead for one step of A is at most $2r + 1$.

Relying on the above discussions, running procedure A on $G^{2(r+1)}[H]$ or on $(G_C)^2[H]$ can be simulated on G within a factor of $O(r)$ on the time complexity.

Lemma 7. *For any integer $k, \rho \geqslant 1$,* SPANNER *can be implemented with a deterministic distributed algorithm in $O(\log k \cdot \rho^{\log k} \cdot \tau)$ time, where τ is the time complexity to compute an independent ρ-dominating set in a graph of at most n nodes.*

4   Applications to Low Stretch Spanners

4.1   Constant Stretch Spanners with Sub-quadratic Size

Let $\mathrm{MIS}(n)$ denote the time complexity for computing, by a deterministic distributed algorithm, a maximal independent set (MIS) in a graph with at most n

nodes. The fastest deterministic algorithm [33] shows that $\text{MIS}(n) \leqslant n^{O(1/\sqrt{\log n})}$. It is also known that $\text{MIS}(n) \geqslant \Omega(\sqrt{\log n / \log \log n})$ [23].

It is not difficult to check that a set X is an independent 1-dominating set if and only if X is a maximal independent set (cf. [34, pp. 259, Ex. 4]). Thus, using the fast distributed MIS algorithm as a subroutine in algorithm SPANNER, we obtain:

Theorem 1. *There is a deterministic distributed algorithm that given a graph G with n nodes and any fixed integer $k = 2^p$ with $p \geqslant 0$, constructs a $(k^{\log_2 5})$-spanner for G with $O(n^{1+1/k})$ edges in $O(\text{MIS}(n))$ time.*

Proof. Size and time are direct consequences of lemmas 3 and 7 fixing k and $\rho = 1$. Note that $\ell(k) = p = \log k$. Thus, using Lemma 5, the stretch of the output spanner is bounded by $5^{\log k} = k^{\log 5}$. □

Theorem 2. *There is a deterministic distributed algorithm that given a graph G with n nodes and any fixed integer $k = 2^p + 2^q - 1$ with $p \geqslant q > 0$, constructs a $(6 \cdot 5^{p-1} + 4 \cdot 5^{q-1} - 1)$-spanner for G with $O(n^{1+1/k})$ edges in $O(\text{MIS}(n))$ time.*

Proof. Size and time are direct consequences of lemmas 3 and 7 fixing k and $\rho = 1$.

If $p = q$, then $k = 2^{p+1} - 1 = \sum_{j=0}^{p} 2^j$. Hence, $\ell(k) = p$ and $\ell(k - 2^{\ell(k)}) = \ell(2^{p+1} - 1 - 2^p) = \ell(2^p - 1) = p - 1$. Thus, using Lemma 5 (second case), the stretch of the output spanner is bounded by $6 \cdot 5^p + 4 \cdot 5^{p-1} - 1$.

If $p \neq q$, then $k = 2^p + \sum_{j=0}^{q-1} 2^j$. Hence, $\ell(k) = p$. In addition, $\ell(k - 2^{\ell(k)}) = \ell(2^p + 2^q - 1 - 2^p) = \ell(2^q - 1) = q - 1$. Thus, using Lemma 5 (second case), the stretch of the output spanner is bounded by $6 \cdot 5^p + 4 \cdot 5^{q-1} - 1$. □

Corollary 1. *For every integer k such that $k = 2^p + 2^q - 1$, where $p \geqslant q \geqslant 0$, there is a deterministic distributed algorithm that given a graph G with n nodes, constructs a $s[k]$-spanner for G with $O(n^{1+1/k})$ edges in $O(\text{MIS}(n))$ time, where $s[k]$ is given by Table 1.*

Table 1. Stretch and Strategy examples for $k = 2^p + 2^q - 1$

(p,q)	$(0,0)$	$(1,0)$	$(1,1)$	$(2,0)$	$(2,1)$	$(2,2)$	$(3,0)$	$(3,1)$	$(3,2)$	$(3,3)$
k	1	2	3	4	5	7	8	9	11	15
$s[k]$	1	5	9	25	33	49	125	153	169	249
i_0	1	2	1	3	2	1	4	3	2	1

4.2 Graphs with Large Minimum Degree

It is known that sparser spanners exist whenever the minimum degree increases (cf. the concluding remark of [7]). In this paragraph, we show that graphs with minimum degree large enough enjoy an $O(1)$-spanner with only $O(n)$ edges, moreover computable with a fast deterministic distributed algorithm.

Let us first note that if a graph G has a ρ-dominating set X, then G has a $(4\rho+1)$-spanner with at most $n+|X|^2/2$ edges. Assuming we are given such a dominating set, the spanner can be constructed distributively in $O(\rho)$ time by first clustering the nodes of the graph around the nodes in the dominating set, and then by connecting every two neighboring clusters. The two endpoints either belong to the same cluster, and thus the endpoints are at distance at most 2ρ in the spanner, or belong to two adjacent clusters. In that case the endpoints are at distance at most $2\rho + 1 + 2\rho$ in the spanner using the selected inter-cluster edge of the spanner.

Proposition 1. *For every parameter $\rho \geqslant 1$, there exists a deterministic distributed algorithm that given a graph G with n nodes and a ρ-dominating set X, constructs a $(4\rho+1)$-spanner for G with at most $n+|X|^2/2$ edges in $O(\rho)$ time.*

This proposition can be combined with the observation that if G has minimum degree $\delta \geqslant \sqrt{n\log n}$, then G has a 1-dominating set X of size $O(\sqrt{n\log n})$. Indeed, this can be proved using the following greedy algorithm [30]: one starts with $X = \varnothing$ and with the set of all radius-1 balls, $\mathcal{B} = \{N[v] \mid v \in V\}$, where $N[v] = \{u \in V \mid d_G(u,v) \leqslant 1\}$. Then, while $\mathcal{B}$ is nonempty, one selects a node $x \in V$ for X that belongs to the maximum number of balls in the current set $\mathcal{B}$. The set $\mathcal{B}$ is updated by removing all balls containing x. The constructed set X is a 1-dominating set and it can be shown that $|X| \leqslant n(1+\ln n)/\min_{v \in V} |N[v]|$ which is at most $O(\sqrt{n\log n})$ if $\delta \geqslant \sqrt{n\log n}$. Thus, the problem is to efficiently compute such 1-dominating set.

Unfortunately, no deterministic distributed implementation of the greedy algorithm faster than $O(|X|)$ is known. A small ρ-dominating set can be computed much more efficiently in $O(\rho\log^* n)$ time by the algorithm of [25]. Unfortunately, its guaranteed size for X is only of $O(n/\rho)$. Finally, no algorithm is known to run in $o(\sqrt{n\log n})$ time for this problem.

However, using our algorithm, we obtain a 9-spanner with only $O(n)$ edges, moreover with a better time complexity.

Theorem 3. *There exists a deterministic distributed algorithm that given a graph G with n nodes and minimum degree $\delta \geqslant \sqrt{n}$, constructs a 9-spanner for G with at most $3n/2$ edges in $O(\mathrm{MIS}(n))$ time.*

Proof. The algorithm consists in two stages. First, we construct an MIS for G^2. Then, each node of the MIS constructs its region using the coloring technique of SPANNER. The spanner is obtained by considering the edges spanning the regions and the edges connecting every two adjacent regions (cf. Proposition 1).

The number of nodes belonging to the MIS, and thus the number of regions, is at most $n/\delta \leqslant \sqrt{n}$. Therefore, the number of edges of the spanner is at most $n + \sqrt{n}^2/2 = 3n/2$. The radius of a region is bounded by 2. Thus, the stretch is $2\cdot 2 + 1 + 2\cdot 2 = 9$. $\qquad\square$

4.3   Randomized Distributed Implementation Issues

In [31], Luby gives a simple and efficient randomized PRAM algorithm for computing an MIS in $O(\log n)$ expected time. Luby's algorithm can be turned to run in the free model, and we obtain a distributed algorithm for computing an independent 1-dominating set which terminates within $O(\log n)$ expected time. We remark that upon termination of the algorithm, the constructed 1-dominating set is always correct, the randomization is only on the running time, i.e., it is a *Las Vegas* algorithm.

The two randomized algorithms we present below guarantee the stretch and the size bounds for the constructed spanners, while the $O(k)$ time (*Monte Carlo*) randomized algorithms [8] do not give any guarantee on the spanner size. This is of course achieved at the price of increasing the stretch factor of the spanner.

Thus, we obtain the following randomized version of Theorems 1 and 2:

Theorem 4. *There is a (Las Vegas) randomized distributed algorithm that given a graph G with n nodes and any fixed integer $k = 2^p$ with $p \geqslant 0$, constructs a $(k^{\log_2 5})$-spanner for G with $O(n^{1+1/k})$ edges in $O(\log n)$ expected time.*

Theorem 5. *For every fixed integer $k \geqslant 3$, there is a (Las Vegas) randomized distributed algorithm that given a graph G with n nodes and any fixed integer $k = 2^p + 2^q - 1$ with $p \geqslant q > 0$, constructs a $(6 \cdot 5^{p-1} + 4 \cdot 5^{q-1} - 1)$-spanner for G with $O(n^{1+1/k})$ edges in $O(\log n)$ expected time.*

Recently, in [24], Khun et al. show that every packing problem can be approximated by a constant factor with high probability in $O(\log n)$ time in the free model. Therefore, the (*Monte Carlo*) algorithm of [24] implies a randomized constant approximation algorithm for the minimum 1-dominating set problem with $O(\log n)$ time. Thus, using Proposition 1, we obtain the following result (to be compared with Theorem 3 and [8]):

Theorem 6. *There exists a (Monte Carlo) randomized distributed algorithm that given a graph G with n nodes of minimum degree $\delta \geqslant \sqrt{n}$, constructs a 5-spanner for G in $O(\log n)$ time. The size is $O(n \log^2 n)$ edges with high probability. More generally, for a minimum degree δ graph, we obtain a 5-spanner with $O(n + (n \log n/\delta)^2)$ edges.*

Let us remark that, in Theorem 6, 5 is the best possible bound on the stretch if $\delta \geqslant w(n^{1/4} \log n)$. In fact, there exist graphs with minimum degree $c\sqrt{n}$ (for some constant $c > 0$) and girth 6 (the length of its smallest cycle). Thus, the deletion of any edge implies a stretch of at least 5 for its endpoints. Therefore, any spanner with size less than $\frac{1}{2}cn\sqrt{n}$ have stretch at least 5, and $O(n + (n \log n/\delta)^2) = o(n\sqrt{n})$ if $\delta \geqslant w(n^{1/4} \log n)$.

5   Conclusion

In this paper we have considered deterministic distributed algorithm to construct low stretch and sparse spanners of unweighted arbitrary graphs. In particular we have shown that 5-spanner with $O(n^{3/2})$ edges can be constructed in $n^{O(1/\sqrt{\log n})}$ time. Let us observe that $\log n < n^{1/\sqrt{\log n}}$ only for $n > 2^{4^2}$. In other words, deterministic distributed $n^{1/\sqrt{\log n}}$ time algorithms might be competitive[1] over randomized $\log n$ time algorithms for distributed system up to $n \leqslant 32656$ processors. We left open the two following problems:

1. Reduce the stretch from 5 to optimal stretch 3, without increasing the size of the spanner and the running time. More generally, is it possible, for every $k \geqslant 1$, to compute with a deterministic distributed algorithm a $(2k-1)$-spanners of size $O(n^{1+1/k})$ in $O(\mathrm{MIS}(n))$ time?

2. Reduce the time complexity to $o(\mathrm{MIS}(n))$, possibly with some small stretch and size increasings. More precisely, is it possible to compute with a deterministic distributed algorithm a constant stretch spanner with $o(n^2)$ edges in $o(\mathrm{MIS}(n))$ time? Using our approach, it suffices to show that there is a constant ρ for which an independent ρ-dominating set can be computed in $o(\mathrm{MIS}(n))$ time for every graph.

References

1. Baruch Awerbuch. Complexity of network synchronization. *Journal of the Association for Computing Machinery*, 32:804–823, 1985.

2. Baruch Awerbuch. Optimal distributed algorithms for minimum weight spanning tree, counting, leader election and related problems. In 19th *Annual ACM Symposium on Theory of Computing (STOC)*, pages 230–240. ACM Press, May 1987.

3. Baruch Awerbuch, Bonnie Berger, Lenore J. Cowen, and David Peleg. Near-linear cost sequential and distributed constructions of sparse neighborhood coverss. In 34th *Annual IEEE Symposium on Foundations of Computer Science (FOCS)*, pages 638–647. IEEE Computer Society Press, November 1993.

4. Baruch Awerbuch, Bonnie Berger, Lenore J. Cowen, and David Peleg. Fast distributed network decompositions and covers. *Journal of Parallel and Distributed Computing*, 39:105–114, 1996.

5. Baruch Awerbuch, Bonnie Berger, Lenore J. Cowen, and David Peleg. Near-linear time construction of sparse neighborhood covers. *SIAM Journal on Computing*, 28(1):263–277, February 1998.

6. Baruch Awerbuch and David Peleg. Sparse partitions. In 31th *Annual IEEE Symposium on Foundations of Computer Science (FOCS)*, pages 503–513. IEEE Computer Society Press, October 1990.

7. Surender Baswana, Telikepalli Kavitha, Kurt Mehlhorn, and Seth Pettie. New constructions of (α, β)-spanners and purely additive spanners. In 16th *Symposium on Discrete Algorithms (SODA)*, pages 672–681. ACM-SIAM, January 2005.

8. Surender Baswana and Sandeep Sen. A simple linear time algorithm for computing a $(2k-1)$-spanner of $O(n^{1+1/k})$ size in weighted graphs. In 30th *International Colloquium on Automata, Languages and Programming (ICALP)*, volume 2719 of Lecture Notes in Computer Science, pages 384–396. Springer, July 2003.

9. Surender Baswana and Sandeep Sen. Approximate distance oracles for unweighted graphs in $\tilde{O}(n^2)$ time. In 15th *Symposium on Discrete Algorithms (SODA)*, pages 271–280. ACM-SIAM, January 2004.

10. Edith Cohen. Fast algorithms for constructing t-spanners and paths with stretch t. *SIAM Journal on Computing*, 28(1):210–236, 1998.

---

[1] This obviously depends on the constants hidden in the O-notation.

11. Richard Cole and Uzi Vishkin. Deterministic coin tossing with applications to optimal parallel list ranking. *Information and Control*, 70(1):32–53, 1986.
12. Lenore J. Cowen. Compact routing with minimum stretch. *Journal of Algorithms*, 38(1):170–183, 2001.
13. Bilel Derbel, Mohamed Mosbah, and Akka Zemmari. Fast distributed graph partition and application. In 20th *IEEE International Parallel & Distributed Processing Symposium (IPDPS)*. IEEE Computer Society Press, April 2006.
14. Tamar Eilam, Cyril Gavoille, and David Peleg. Compact routing schemes with low stretch factor. *Journal of Algorithms*, 46:97–114, 2003.
15. Michael Elkin. Computing almost shortest paths. In 20th *Annual ACM Symposium on Principles of Distributed Computing (PODC)*, pages 53–62. ACM Press, 2001.
16. Michael Elkin. A faster distributed protocol for constructing a minimum spanning tree. In 15th *Symposium on Discrete Algorithms (SODA)*, pages 359–368. ACM-SIAM, January 2004.
17. Michael Elkin. Unconditional lower bounds on the time-approximation tradeoffs for the distributed minimum spanning tree problems. In 36th *Annual ACM Symposium on Theory of Computing (STOC)*, pages 331–340. ACM Press, May 2004.
18. Michael Elkin and David Peleg. $(1+\epsilon,\beta)$-spanner constructions for general graphs. *SIAM Journal on Computing*, 33(3):608–631, 2004.
19. Michael Elkin and Jian Zhang. Efficient algorithms for constructing $(1 + \epsilon,\beta)$-spanners in the distributed and streaming models. In 23rd *Annual ACM Symposium on Principles of Distributed Computing (PODC)*, pages 160–168. ACM Press, July 2004.
20. Cyril Gavoille, David Peleg, Stéphane Pérennès, and Ran Raz. Distance labeling in graphs. *Journal of Algorithms*, 53(1):85–112, 2004.
21. Andrew V. Goldberg, Serge A. Plotkin, and Gregory E. Shannon. Parallel symmetry-breaking in sparse graphs. *SIAM Journal on Discrete Mathematics*, 1(4):434–446, 1988.
22. Fabian Kuhn, Thomas Moscibroda, Tim Nieberg, and Roger Wattenhofer. Fast deterministic distributed maximal independent set computation on growth-bounded graphs. In 19th *International Symposium on Distributed Computing (DISC)*, volume Lecture Notes in Computer Science. Springer, September 2005.
23. Fabian Kuhn, Thomas Moscibroda, and Roger Wattenhofer. What cannot be computed locally! In 23rd *Annual ACM Symposium on Principles of Distributed Computing (PODC)*, pages 300–309. ACM Press, July 2004.
24. Fabian Kuhn, Thomas Moscibroda, and Roger Wattenhofer. The price of being near-sighted. In 17th *Symposium on Discrete Algorithms (SODA)*, pages 980–989. ACM-SIAM, January 2006.
25. Shay Kutten and David Peleg. Fast distributed construction of small k-dominating sets and applications. *Journal of Algorithms*, 28(1):40–66, 1998.
26. Nathan Linial. Distributive graph algorithms - Global solutions from local data. In 28th *Annual IEEE Symposium on Foundations of Computer Science (FOCS)*, pages 331–335. IEEE Computer Society Press, October 1987.
27. Nathan Linial. Locality in distributed graphs algorithms. *SIAM Journal on Computing*, 21(1):193–201, 1992.
28. Zvi Lotker, Boaz Patt-Shamir, Elan Pavlov, and David Peleg. Minimum-weight spanning tree construction in $O(\log \log n)$ communication rounds. *SIAM Journal on Discrete Mathematics*, 35(1):120–131, 2005.
29. Zvi Lotker, Boaz Patt-Shamir, and David Peleg. Distributed MST for constant diameter graphs. In 20th *Annual ACM Symposium on Principles of Distributed Computing (PODC)*, pages 63–71. ACM Press, 2001.

30. Laszlo Lovász. On the ratio of optimal integral and fractional covers. *Discrete Mathematics*, 13:383–390, 1975.
31. Michael Luby. A simple parallel algorithm for the maximal independent set problem. *SIAM Journal on Computing*, 15(4):1036–1053, November 1986.
32. Shlomo Moran and Sagi Snir. Simple and efficient network decomposition and synchronization. *Theoretical Computer Science*, 243(1-2):217–241, 2000.
33. Alessandro Panconesi and Aravind Srinivasan. On the complexity of distributed network decomposition. *Journal of Algorithms*, 20(2):356–374, 1996.
34. David Peleg. *Distributed Computing: A Locality-Sensitive Approach*. SIAM Monographs on Discrete Mathematics and Applications, 2000.
35. David Peleg and Vitaly Rubinovich. A near-tight lower bound on the time complexity of distributed minimum-weight spanning tree construction. *SIAM Journal on Computing*, 30(5):1427–1442, 2000.
36. David Peleg and Jeffrey D. Ullman. An optimal synchornizer for the hypercube. *SIAM Journal on Computing*, 18(4):740–747, 1989.
37. David Peleg and Eli Upfal. A trade-off between space and efficiency for routing tables. *Journal of the ACM*, 36(3):510–530, July 1989.
38. Lucia Draque Penso and C. Barbosa Valmir. A distributed algorithm to find k-dominating sets. *Discrete Applied Mathematics*, 141(1-3):243–253, May 2004.
39. Liam Roditty, Mikkel Thorup, and Uri Zwick. Roundtrip spanners and roundtrip routing in directed graphs. In 13th *Symposium on Discrete Algorithms (SODA)*, pages 844–851. ACM-SIAM, January 2002.
40. Liam Roditty, Mikkel Thorup, and Uri Zwick. Deterministic constructions of approximate distance oracles and spanners. In 32nd *International Colloquium on Automata, Languages and Programming (ICALP)*, volume Lecture Notes in Computer Science, 2005.
41. Mikkel Thorup and Uri Zwick. Compact routing schemes. In 13th *Annual ACM Symposium on Parallel Algorithms and Architectures (SPAA)*, pages 1–10. ACM Press, July 2001.
42. Mikkel Thorup and Uri Zwick. Approximate distance oracles. *Journal of the ACM*, 52(1):1–24, January 2005.
43. Rephael Wenger. Extremal graphs with no C^4's, C^6's, or C^{10}'s. *Journal of Combinatorial Theory, Series B*, 52(1):113–116, 1991.

Efficient Distributed Weighted Matchings on Trees[*]

Jaap-Henk Hoepman[1], Shay Kutten[2], and Zvi Lotker[3]

[1] Institute for Computing and Information Sciences
Radboud University Nijmegen, Nijmegen, the Netherlands
jhh@cs.ru.nl
[2] Faculty of Industrial Engineering and Management
Technion, Haifa, Israel
kutten@ie.technion.ac.il
[3] CWI, Amsterdam, the Netherlands
Z.Lotker@cwi.nl

Abstract. In this paper, we study distributed algorithms to compute a weighted matching that have constant (or at least sub-logarithmic) running time and that achieve approximation ratio $2 + \epsilon$ or better. In fact we present two such synchronous algorithms, that work on arbitrary weighted trees.

The first algorithm is a randomised distributed algorithm that computes a weighted matching of an arbitrary weighted tree, that approximates the maximum weighted matching by a factor $2 + \epsilon$. The running time is $O(1)$. The second algorithm is deterministic, and approximates the maximum weighted matching by a factor $2 + \epsilon$, but has running time $O(\log^* |V|)$. Our algorithms can also be used to compute maximum unweighted matchings on regular and almost regular graphs within a constant approximation.

1 Introduction

A *matching* $M(G)$ of a graph $G = (V, E)$ is any subgraph of G where no two edges are incident to the same vertex. A matching is *maximal* if no other edge from G can be added to the matching without violating this requirement. Let $w(e)$ be the weight of an edge $e \in E$ of G, where $w(e) > 0$. Define the weight $w(G)$ of a graph G to be the sum of the weights of all its edges. Then a *maximum weighted matching* $M^*(G)$ of G is a matching whose weight is the maximum among all matchings of G. We say that an algorithm achieves *approximation ratio* α if for all graphs G, the matching it returns has weight at least $\frac{1}{\alpha} w(M^*(G))$, i.e., $\frac{1}{\alpha}$ of the weight of the maximum weighted matching of that graph.

For sequential algorithms, the problem is well studied. For unweighted graphs, Micali and Vazirani [MV80] present an $O(\sqrt{|V|}|E|)$ time algorithm that computes a maximum matching. For weighted graphs Gabow [Gab90] gives an $O(|V||E|+$

---

[*] Id: random-matchings.tex,v 1.18 2006/04/20 08:28:26 jhh Exp .

P. Flocchini and L. Gąsieniec (Eds.): SIROCCO 2006, LNCS 4056, pp. 115–129, 2006.
© Springer-Verlag Berlin Heidelberg 2006

$|V|^2 \log |V|)$ time algorithm, computing the maximum weighted matching. Both return an exact solution, and not approximations.

Surprisingly, few distributed algorithms to compute (an approximation of) the maximum weighted matching of the communication graph are known. For unweighted graphs, there are deterministic distributed algorithms computing a *maximal* matching in trees [KS00], and bipartite and general graphs [CHS02]. A randomised algorithm for the general case also exists: Israeli and Itai [II86] compute a maximal matching (i.e., no approximation) in running time $O(\log |V|)$.

For weighted graphs, Uehara *et al.* [UC00] present a constant time distributed algorithm with approximation ratio $O(\Delta)$ (where Δ is the maximum degree of the graph). Recently, Wattenhofer *et al.* [WW04] presented a randomised distributed algorithm to compute a weighted matching $M(G)$ with approximation ratio 5 and running time $O(\log^2 |V|)$ for general graphs, and approximation ratio 4 and $O(1)$ running time for trees.

Hoepman [Hoe04] presents an $O(|E|)$ time[1] deterministic distributed algorithm that computes a weighted matching for general graphs with approximation ratio 2. This algorithm is based on sequential algorithms by Preis [Pre99] and Avis [Avi83], and does not require collecting all information in one node (which increases the message complexity).

In this paper, we study distributed algorithms to compute a weighted matching that have constant (or at least sub-logarithmic) running time and that achieve approximation ratio $2 + \epsilon$ or better. In fact, we present two such algorithms for arbitrary weighted trees, thus improving the previous algorithm of Wattenhofer *et al.* [WW04].

The first algorithm — presented in Sect. 3 — is randomised, and achieves approximation ratio $2 + \epsilon$ in running time $O(1)$. An interesting feature of this algorithm is that the quality of the approximation depends on the number of rounds the algorithm is allowed to run. The second algorithm — presented in Sect. 4 — is deterministic, and achieves approximation ratio $2 + \epsilon$ in running time $O(\log^* |V|)$. We start by introducing our computation model and some notation in Sect. 2, show how our algorithms can also be applied to achieve constant approximations to the maximum (unweighted) matchings for regular and almost regular graphs in Sect. 5, and finish with some pointers to further research in Sect. 6.

2   Model and Notation

Consider a distributed system with n nodes, whose communication graph is $G = (V, E)$. In this paper, G is a tree (denoted T) or a regular graph. Nodes can exchange point-to-point messages with their neighbours over the edges E in the graph. Each edge e has a weight $w(e)$, known to both endpoints of that edge. The system is synchronous and operates in rounds of message exchanges. We measure time complexity of our algorithms as the number of rounds needed to perform

---

[1] Careful analysis shows that the time complexity is actually at most $O(\operatorname{diam} G)$, the diameter of the graph.

the computation. We note that our algorithms also work in the asynchronous setting, after some minor modifications.

We write G for general graphs, T for trees, P for paths and S for segments (that are pieces of a path). The number of edges in G is $|G|$, and $w(G)$ is the weight of G, i.e., the sum of the weights of all edges of G. $M(G)$ is a weighted matching of graph G, and $M^*(G)$ is the maximum weighted matching of graph G.

Let X be a random variable. We write $\mathbf{E}\left[X \mid Q(X)\right]$ for the conditional expectation over X given that $Q(X)$ holds. By definition

$$\mathbf{E}\left[X \mid Q(X)\right] = \sum_{x:Q(x)} x\mathbf{Pr}\left[X = x \mid Q(X)\right] \qquad (1)$$

and for disjoint Q_i that together span the whole range of X we have

$$\mathbf{E}\left[X\right] = \sum_{i} \mathbf{E}\left[X \mid Q_i(X)\right]\mathbf{Pr}\left[Q_i(X)\right] . \qquad (2)$$

Our protocols are described in plain English, and not in any formal protocol notation, because they are quite straightforward.

3   Randomised Case: Constant Running Time

The protocol runs in four phases, and is parameterised by a real-valued constant p between 0 and 1 and an integer constant k greater than 1.

First, given input tree T, a set of paths $P(T)$ is generated by letting nodes select their heaviest incident edge as a potential member of a path. In the second phase, each path is cut into short segments by randomly removing edges from the path, each with probability p. Subsequently, each segment is tested to see whether its length is shorter than k. In the fourth phase, for these short segments an optimal matching is computed in time $O(k)$, while for the remaining longer segments a constant-time randomised algorithm computes a 2 approximation of the optimal matching for this segment. Combining all matchings, and compensating for the loss of dropped edges when cutting paths into segments, this gives an $O(k)$ algorithm to compute a matching, for which we prove an approximation ratio of $2 + \epsilon$ for arbitrary $\epsilon > 0$. Because k is a constant, the running time is $O(1)$.

3.1   Computing the Paths

We use the same procedure to construct a set of paths $P(T)$ from a given tree T as presented by Wattenhofer *et al.* [WW04]. That is, a node u requests the addition of its heaviest incident edge (u, v) to this set of paths from its neighbour v. Such a requested edge is only added to the set of paths if either u also receives a request from v (for the same edge), or if v sent a grant to u for its request. Nodes request exactly one edge each. Nodes only grant at most one request, being the heaviest request coming in over an edge it didn't request itself (assuming unique weights,

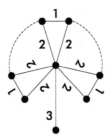

Fig. 1. Counterexample for non-tree graphs

or breaking ties). All remaining edges (either not granted or not requested) are dropped and will not be members of any path. Because at most 2 edges per node remain, a node is a part of at most one path, and the procedure yields a set of disjoint paths. Because nodes select the heaviest incident edge, these paths cannot be cycles (we assume unique edge weights).

The following lemma shows that we do not loose too much weight constructing paths this way, provided that the graph we start with is a tree.

Lemma 3.1 ([WW04]). *For trees T and $P(T)$ computed as described above,*

$$w(P(T)) \geq w(M^*(T)) .$$

In contrast, the following counterexample shows that for non-trees the difference in weight between the paths constructed for that graph and its maximum weight matching can be unbounded. Consider the graph in Fig. 1. One node connects to the central node over an edge with weight 3. This node requests this edge as a path member, and the central node grants that request. Furthermore, $2n$ nodes are connected to the central node through an edge with weight 2, while these nodes are connected pairwise through edges with weight 1. These $2n$ nodes each request the edge with weight 2 as a potential path member, but all these requests are rejected by the central node. Hence, the path consists of just one edge, and has total weight $w(P(G)) = 3$. However, the maximum weighted matching consists of the edge with weight 3, as well as all n edges with weight 1. Hence $w(M^*(G)) = 3 + n$.

3.2   Cutting the Paths into Segments

In the next phase, each path P is cut into segments $S(P)$ as follows. Every vertex sends, for each of its edges on the path, a cut request over this edge with probability $\sqrt{p}$. An edge is cut if both endpoints sent a cut request to each other. The other edges remain. Each connected component forms a segment S in $S(P)$. We see that the probability for each edge to be cut is exactly p. The expected number of removed edges is $p|P|$, and the expected weight of all edges removed together is $p \cdot w(P)$.

Lemma 3.2. *Let $S(P)$ be a random variable corresponding to the segments computed for a path P by the random process described above. Then*

$$E\left[\sum_{S \in S(P)} w(S)\right] = (1 - p)w(P) .$$

Proof. Only edges in some $S \in S(P)$ contribute to the weight. Let P consist of edges e_i, and define random variables X_{e_i} such that $X_{e_i} = 1$ iff e_i is a member of some segment (i.e., not cut), and 0 otherwise. Then $\mathbf{Pr}\left[X_{e_i} = 1\right] = (1 - p)$ and

$$
\begin{aligned}
E\left[\sum_{S \in S(P)} w(S)\right] &= \mathbf{E}\left[\sum_{e_i \in P} w(e_i)X_{e_i}\right] \\
&= \sum_{e_i \in P} \mathbf{E}\left[w(e_i)X_{e_i}\right] \\
&= \sum_{e_i \in P} w(e_i)\mathbf{Pr}\left[X_{e_i} = 1\right] = (1 - p)\sum_{e_i \in P} w(e_i) .
\end{aligned}
$$

This proves the lemma. □

3.3   Estimating the Size of the Segment

After cutting the path into segments, each vertex determines the size of the segment it is a member of. Or, to be more precise, it determines whether the segment is smaller or larger than k edges. It does so in the following manner. Vertices at the edge of a segment start the computation, by sending a *distance* message with value 1 along the only incident edge that belongs to a segment (and recording 0 as the distance from the other end). Nodes that receive a distance message record the distance coming in over that edge, and add one before forwarding it over the other segment edge. Forwarding stops if a segment endpoint is reached, or when the distance in the message equals k. If nodes do not record a distance for both edges, the total length of the segment is larger than k. Otherwise, the sum of both distances equals the length of the segment. In either case, all nodes know whether the length of the segment is $\leq k$ or $> k$. The running time is at most k.

3.4   Computing Matchings on the Segments

Segments compute their matching depending on their sizes, as determined in the previous phase. If a node finds it is not a member of a segment (or rather, it is a member of a segment of size 0) it does nothing. Otherwise, it cooperates with all other nodes in the same segment as described below.

Consider a segment $S \in S(P)$. If $|S| \leq k$, we compute a good matching $M(S)$ for S in time k by computing two matchings M and M' by adding edges in S alternately to M or M', and selecting the matching with maximum weight. Such a computation could be initiated by both endpoints of the segment. We note the running time is $O(k)$.

In the following lemma we bound the weight of the resulting matching from below[2].

Lemma 3.3. *For a matching $M(S)$ of S (with $|S| \leq k$) computed as described above,*

$$\boldsymbol{E}[w(M(S))] \geq \frac{1}{2}w(S) .$$

Proof. By construction we find two matchings M and M' with $S = M + M'$, and therefore $w(S) = w(M) + w(M')$. As we select the heaviest matching, the lemma follows. □

If $|S| > k$, we compute a matching $M(S)$ for S in one round by letting vertices vote for the incident edge that should be added to the matching with equal probability (segment endpoints vote for their only edge with $p = 1$). An edge is only added if both its endpoints vote for it.

Lemma 3.4. *For a matching $M(S)$ of S (with $|S| > k$) computed as described above,*

$$\boldsymbol{E}[w(M(S))] \geq \frac{1}{4}w(S) .$$

Proof. As each edge has at most 2 incident edges on the segment, the probability that an edge is added is at least $\frac{1}{4}$. Define random variables X_{e_i} such that $X_{e_i} = 1$ if the i-th edge e_i of S is in the matching. Then, similar to the proof of lemma 3.2,

$$\mathbf{E}[w(M(S))] = \sum_{e_i \in S} w(e_i)\mathbf{Pr}[X_{e_i} = 1] .$$

As $\mathbf{Pr}[X_{e_i} = 1] \geq \frac{1}{4}$, the lemma follows. □

3.5   Merging the Results

The final matching $M(T)$ for tree T is obtained by merging all matchings computed for all segments in $S(P)$ for each path P in $P(T)$. We conclude our analysis by estimating the weight of the resulting matching.

First we look at a single, but arbitrary, path P. In what follows, let $S(P)$ be a random variable corresponding to the segments computed for P by the random process described in section 3.2. Let S be a random variable ranging over all members of $S(P)$.

For such a path P, define $C(P)$ to be the cycle obtained by merging the two endpoints of P into a single node. Given a segmentation of P, define s_b and s_e to be the first and last segment of P, respectively, where s_b starts at the endpoint and s_e ends at the endpoint (and where either segment may be empty if the

---

[2] Even though the process is deterministic, we state the bound in terms of an expectation, because that is more useful further on.

first and/or last edge of P happened to be cut). A segmentation of P induces a segmentation on the cycle $C(P)$, by taking all segments, and merging s_b and s_e into a single segment $\bar{s}_c$. Let $S(C(P))$ be a random variable corresponding to the segments computed for $C(P)$. Let $\bar{S}$ be a random variable ranging over all members of $S(C(P))$.

For segments computed on this cycle, we have the following proposition.

Proposition 3.5. *For all $i \leq |P|$,*

$$\boldsymbol{Pr}\left[|\bar{S}| = i\right] \leq (1 - p)^i .$$

Proof. Clearly, to have a segment of length i, we need i uncut edges. This happens with probability $(1 - p)^i$. If $i = |P|$, then this is the exact probability (the segment happens to be the whole cycle), otherwise we need at least 1 ($i = |P|-1$) or 2 cut edges, that each lower the probability with a factor p.   □

We also need the following uniformity property on the distribution of the weights over the segments computed for the cycle.

Proposition 3.6. *For any $k \geq 0$,*

$$\boldsymbol{E}\left[w(\bar{S}) \mid |\bar{S}| = k\right] = k\frac{\boldsymbol{E}\left[w(\bar{S})\right]}{\boldsymbol{E}\left[|\bar{S}|\right]}$$

Proof. Let σ be a random variable, ranging over the single edges in a segment $\bar{S}$. Let $\sigma_1, \sigma_2, \ldots$ be the edges in $\bar{S}$. Then

$$\boldsymbol{E}\left[w(\bar{S}) \mid |\bar{S}| = k\right] = \{\text{Using the fact that } \bar{S} \text{ consists of } k \text{ edges } \sigma_i.\}$$

$$\boldsymbol{E}\left[\sum_{i=1}^{k} w(\sigma_i) \mid |\bar{S}| = k\right]$$

$$= \{\text{Independent of length of } \bar{S} \text{ now.}\}$$

$$\boldsymbol{E}\left[\sum_{i=1}^{k} w(\sigma_i)\right]$$

$$= \{\text{By symmetry of } C(p)$$
$$\text{all edges appear the same number of times. }\}$$
$$k\boldsymbol{E}\left[w(\sigma)\right]$$

By Eq. 2 we have

$$\boldsymbol{E}\left[w(\bar{S})\right] = \sum_i \boldsymbol{E}\left[w(\bar{S}) \mid |\bar{S}| = i\right] \boldsymbol{Pr}\left[|\bar{S}| = i\right]$$

$$= \{\text{By the above.}\}$$

$$\sum_i i\boldsymbol{E}\left[w(\sigma)\right] \boldsymbol{Pr}\left[|\bar{S}| = i\right]$$

$$= \boldsymbol{E}\left[w(\sigma)\right] \boldsymbol{E}\left[|\bar{S}|\right]$$

Combining both equations proves the proposition.   □

Next, we bound the expected weight of a matching of an arbitrary segment $\bar{S}$ from $S(C(P))$ in terms of the expected weight of $\bar{S}$ itself.

Lemma 3.7. *For a matching $M(\bar{S})$ of $\bar{S}$ computed as described above,*

$$\boldsymbol{E}\left[w(M(\bar{S}))\right] \geq \frac{1}{2}\left(1 - \frac{(1+kp)(1-p)^{k+1}}{2p^2}\right)\boldsymbol{E}\left[w(\bar{S})\right] .$$

Proof. First observe that $\mathbf{E}\left[w(M(\bar{S}))\right]$ depends on two random processes: the selection of a segment $\bar{s}$ from $S(C(P))$, and the random variable C denoting the coin sequence thrown by the randomised algorithm that computes $M(\bar{s})$. Let $M(c, \bar{s})$ denote the (deterministic) result of $M(\bar{s})$ when the coins thrown are fixed to sequence c. Then

$\mathbf{E}\left[w(M(\bar{S}))\right] =\{\bar{S}$ and C are independent$\}$

$$\sum_{\bar{s}}(\sum_{c} M(c, \bar{s})\mathbf{Pr}\left[C = c\right])\mathbf{Pr}\left[\bar{S} = \bar{s}\right]$$

$\geq\{$Split according to $|\bar{s}|$ and using Lemma 3.3 and 3.4$\}$

$$= \sum_{\bar{s}:|\bar{s}|\leq k} \frac{1}{2}w(\bar{s})\mathbf{Pr}\left[\bar{S} = \bar{s}\right] + \sum_{\bar{s}:|\bar{s}|>k} \frac{1}{4}w(\bar{s})\mathbf{Pr}\left[\bar{S} = \bar{s}\right]$$

$=\{$Rearranging sums and definition of $\mathbf{E}\left[w(\bar{S})\right]$. $\}$

$$\frac{1}{2}\mathbf{E}\left[w(\bar{S})\right] - \sum_{\bar{s}:|\bar{s}|>k} \frac{1}{4}w(\bar{s})\mathbf{Pr}\left[\bar{S} = \bar{s}\right]$$

$=\{$Using Eq. 1 and $\mathbf{Pr}\left[\bar{S} = \bar{s} \mid |\bar{S}|>k\right]\mathbf{Pr}\left[|\bar{S}|>k\right]=\mathbf{Pr}\left[\bar{S} = \bar{s}\right]\}$

$$\frac{1}{2}\mathbf{E}\left[w(\bar{S})\right] - \frac{1}{4}\mathbf{E}\left[w(\bar{S}) \mid |\bar{S}| > k\right]\mathbf{Pr}\left[|\bar{S}| > k\right]$$

$=\{$Using Prop. 3.6$\}$

$$\frac{1}{2}\mathbf{E}\left[w(\bar{S})\right] - \frac{1}{4}\mathbf{E}\left[w(\bar{S})\right] \sum_{i>k} i\frac{\mathbf{Pr}\left[|\bar{S}| = i\right]}{\mathbf{E}\left[|\bar{S}|\right]}$$

$\geq\{$Using Prop. 3.5 and $\mathbf{E}\left[|\bar{S}|\right] \geq 1\}$

$$\left(\frac{1}{2} - \frac{1}{4}\sum_{i>k} i(1-p)^i\right)\mathbf{E}\left[w(\bar{S})\right]$$

$=\{$Rearranging sums and computing geometric series.$\}$

$$= \left(\frac{1}{2} - \frac{1}{4}\left(\frac{1-p}{p^2} - \frac{k(1-p)^{k+2} - (k+1)(1-p)^{k+1} + (1-p)}{(-p)^2}\right)\right)$$
$$\mathbf{E}\left[w(\bar{S})\right]$$

$$= \left(\frac{1}{2} - \frac{1}{4}(1+kp)\frac{(1-p)^{k+1}}{p^2}\right)\mathbf{E}\left[w(\bar{S})\right]$$

This proves the lemma.   □

We also need the following two propositions.

Proposition 3.8.

$$\boldsymbol{E}\left[w(\bar{S}\right] \geq \boldsymbol{E}\left[w(S)\right] .$$

Proof. Any random segmentation for $S(P)$ induces a segmentation of $S(C(P))$, with s_e and s_b merged into $\bar{s}_c$ where $w(\bar{s}_c) = w(s_b) + w(s_e)$.   □

Proposition 3.9.

$$\boldsymbol{E}\left[\sum_{S \in S(P)} w(M(S))\right] \geq \boldsymbol{E}\left[\sum_{\bar{S} \in S(C(P))} w(M(\bar{S}))\right]$$

Proof. In what follows, let $S(P)$ be a random variable corresponding to the segments computed for a path P by the random process described in section 3.2. Let $S(C(P))$ be the corresponding set of segments for the cycle $C(P)$.

For all $S \in S(P)$ unequal to the end segments s_b and s_e, the corresponding segment $\bar{S}$ in $S(C(P))$ is the same, and hence $w(M(S)) = w(M(\bar{S}))$. It remains to show that $\boldsymbol{E}\left[w(M(s_b)) + w(M(s_e))\right] \geq \boldsymbol{E}\left[w(M(\bar{s}_c))\right]$. Split the matching $M(\bar{s}_c)$ into two parts, m_b (for $\bar{s}_b$) and m_e (for $\bar{s}_e$), that cover s_b and s_e respectively. Let $i \in \{e, b\}$. We show $\boldsymbol{E}\left[w(M(s_i))\right] \geq \boldsymbol{E}\left[w(m_i)\right]$. There are two cases.

$|s_i| \leq k$: In this case (see proof Lemma 3.3), the matching computed for s_i is optimal. As matching m_i on $\bar{s}_i$ is also a matching for s_i, the statement follows.

$|s_i| > k$: Then $|\bar{s}| > k$ as well, and the matching $M(\bar{s}_c)$ is computed using the probabilistic method for long segments (cf. Lemma 3.4). Consider the random process that selects edges for inclusion in $M(\bar{s}_c)$ and hence m_i. The 'end-edge' of $\bar{s}_i$ (the splitting edge at which the cycle is cut into the path P) has probability $1/2$ to be included in $M(s_i)$ but only probability $1/4$ to be included in $m_i = M(\bar{s}_i)$. Hence the expected weight of the matching m_i is lower than $M(s_i)$.

This completes the proof.   □

We now bound the weight of the matching computed for P as a whole.

Theorem 3.10. *For any path P, and matching $M(P)$ computed as above,*

$$\boldsymbol{E}[w(M(P))] \geq \frac{1}{2}\left(1 - \frac{(1+kp)(1-p)^{k+1}}{2p^2}\right)(1-p)w(P) .$$

Proof. In what follows, let $S(P)$ be a random variable corresponding to the segments computed for a path P by the random process described in section 3.2.

We have

$$\mathbf{E}\left[w(M(P))\right] = \mathbf{E}\left[w\left(\bigcup_{S \in S(P)} M(S)\right)\right] = \mathbf{E}\left[\sum_{S \in S(P)} w(M(S))\right]$$

$$\geq \{\text{By Prop. 3.9}\}$$

$$\mathbf{E}\left[\sum_{\bar{S} \in S(C(P))} w(M(\bar{S}))\right] = \sum_{\bar{S} \in S(C(P))} \mathbf{E}\left[w(M(\bar{S}))\right]$$

By Lemma 3.7 this is greater than or equal to

$$\frac{1}{2}\left(1 - \frac{(1+kp)(1-p)^{k+1}}{2p^2}\right) \sum_{\bar{S} \in S(C(P))} \mathbf{E}\left[w(\bar{S})\right]$$

which is bounded from below through Prop 3.8

$$\frac{1}{2}\left(1 - \frac{(1+kp)(1-p)^{k+1}}{2p^2}\right) \sum_{S \in S(P)} \mathbf{E}\left[w(S)\right]$$

which is further bounded from below using Lemma 3.2 by

$$\frac{1}{2}\left(1 - \frac{(1+kp)(1-p)^{k+1}}{2p^2}\right)(1-p)w(P) .$$

This completes the proof.   □

We finish by combining all matchings for all paths.

Corollary 3.11. *For any tree T, and matching $M(T)$ computed as above, we have*

$$2\mathbf{E}\left[w(M(T))\right] \geq \left(1 - \frac{(1+kp)(1-p)^{k+1}}{2p^2}\right)(1-p)M^*(T) .$$

Hence, the approximation ratio of the algorithm is $2 + \epsilon$ for arbitrary $\epsilon > 0$.

Proof. Follows from Theorem 3.10 and Lemma 3.1 and the fact that $M(T) = \cup_{P \in P(T)} M(P)$.

To see that the approximation ratio is equivalent to $2 + \epsilon$, set $\epsilon = \frac{p}{1-p}$ and observe that $1 - \frac{(1+kp)(1-p)^{k+1}}{2p^2}$ tends to 1 for k going to infinity.   □

4   Deterministic Case: $O(\log^* n)$ Running Time

Now consider a model where every node v has a unique identity $ID(v)$. Again, fix a constant k. In [KP98] a deterministic distributed algorithm is presented to partition a tree (or a forest) into clusters of diameter $O(k)$, each containing at least $k+1$ nodes. This constructions is done in $O(k \log^* n)$ time. More precisely:

Lemma 4.1. *[KP98] The collection P_{out} output by Algorithm DOM-Partition(k) is a partition (of the input tree T). Furthermore, if T is of size $n \geq k+1$, then every cluster C in P_{out} has the following properties:*

- *(a) $|C| \geq k+1$.*
- *(b) $Radius(C) \leq 5k+2$.*

Moreover, Algorithm DOM-Partition(k) requires time $O(k \log^* n)$.

This algorithm uses, as a subroutine, the algorithm of [GPS87].

Using this algorithm, it is easy to modify the randomised algorithm of the previous sections to become a deterministic algorithm (with a slightly higher time complexity of $O(\log^* n)$) as follows. Replace the second phase of the probabilistic algorithm (the randomised cutting of the paths into segments) by Algorithm DOM-Partition(k).

This change would have been enough, had the weights of the edges been equal. We need to take some care here, because DOM-Partition(k) may have deleted heavy edges. The nodes of each segment s_i cooperate to perform the following operation. Consider e_l^i, e_r^i, the edges whose deletion separated segment s_i from the rest of the path. Let also e_m^i be the minimum weight edge in this segment. If the weight of either e_l^i or e_r^i is larger than that of e_m^i then e_m^i is removed from the segment, and the largest of e_l^i or e_r^i is reinserted as a part of the segment. If two adjacent segments wish to swap the same separator (e_l^i equals e_r^{i+1}) for a lighter one, then the segment with the smallest minimum weight edge e_m^i wins and performs the swap. Clearly the above can be performed distributively in constant time (since k is a constant).

The above correction may create segments that contain less than k nodes, or segments that are at most three times as long as the original segments. (A segment that itself is not split may join both the segment at its left and its right, however then these two segments are cut in exchange for the separating edges.)

The rest of the algorithm needs no changes, except that we will have no segments that are longer than $15k+6$ for some constant k, so the special treatment of long segments is not necessary. The time complexity of the resulting procedure is $O(\log^* n)$ since in the current paper we assume that k is a constant.

Theorem 4.2. *For any tree T, and matching $M(T)$ computed as above, we have*

$$2w(M(T)) \geq \left(1 - \frac{1}{k+1}\right) M^*(T) .$$

Hence, the approximation ratio of the algorithm is $2/(1 - \frac{1}{k+1}) = 2\frac{k+1}{k}$, which equals $2 + \epsilon$ if we set $\epsilon = \frac{2}{k}$.

Proof. First we compute $\sum_{S \in S(P)} w(S)$ for an arbitrary path P. As Algorithm DOM-Partition(k) returns segments of length at least $k+1$ (if the path has length at least $k+1$, otherwise no edges are cut), the number of cut edges from P is at most $|P|/(k+1)$. Swapping the originally cut edges by lower weight ones does not change the number of cut edges. Since in every original segment we

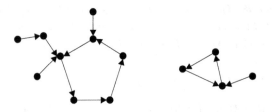

Fig. 2. Examples of simple graphs

swapped the minimum weight edge for a removed edge, the expected weight of the cut edges (after the swap) is at most $w(P)/(k+1)$. Hence

$$\sum_{S \in S(P)} w(S) = (1 - \frac{1}{k+1})w(P) \ .$$

Next we consider the weight of the matching returned for P. Similar to the proof of Th. 3.10, and using only Lemma 3.3 to bound $w(M(S))$ by $\frac{1}{2}w(S)$ (the proof of that lemma not only works in the expected case but also in the deterministic case, because all segments have length less than $15k+6$) we have

$$w(M(P)) \geq \sum_{S \in S(P)} \frac{1}{2}w(S) \ .$$

This equals

$$\frac{1}{2} \sum_{S \in S(P)} w(S) \ ,$$

and using the above result we see that

$$w(M(P)) \geq \frac{1}{2}(1 - \frac{1}{k+1})w(P) \ .$$

Combining the matching for all paths constructed for the tree, and by Lemma 3.1, the theorem follows.   □

5   Regular and Almost Regular Graphs

We now show how the algorithms from the previous sections can be used to compute a constant approximation for the maximum matching of unweighted, arbitrary regular graph and almost regular graphs.

In order to describe the algorithm we need the following notations. We denote by "directed simple" a directed graph were each node has only one outgoing edge, in other words the out degree of the node is 1. Note that the number of directed edges in a directed-simple graph is n and therefore there is at least one directed cycle. See Fig. 2.

The algorithm works in three phases. In the first phase we generate a directed-simple spanning subgraph, and in the second phase we transform the directed-simple spanning subgraph into a collection of disjoint paths. Finally, in the third phase, we can use one of the algorithms of the previous sections to compute the maximum matching for the graph (by assigning weight 1 to all edges). We make sure that the number of remaining edges after the second phase is a constant fraction of the number of nodes n. Because the maximum matching of an unweighted graph is never larger than $\frac{n}{2}$, the resulting maximum matching is a constant approximation.

A natural way to construct a directed-simple spanning subgraph is for each node to select a random neighbour from the d neighbours uniformly. After this step, we have n directed edges. Denote the union of all these directed edges by $G'(V, E')$. Let V_i' be the set of all the nodes in G' that have a degree i (in-degree plus out-degree) and $\overline{V'}_i$ be the set of all the nodes in G' of degree bigger than or equal to i. The next lemma estimates the size of $V_1', V_2', \overline{V'}_3$.

Lemma 5.1. *Let G' be the directed graph constructed as above. Then*

$$E[|V_1'|] = (1 - \frac{1}{d})^d n \leq \frac{1}{e} n$$

$$E[|V_2'|] = ((1 - \frac{1}{d})^{d-1})n \geq \frac{1}{e} n$$

$$E[|\overline{V'}_3|] = n - E[|V_2'|] - E[|V_1'|]$$

Proof. First we compute the probability for a node to have a degree 2. $P[d'(v) = 2] = \frac{d}{d}(1 - \frac{1}{d})^{d-1} = (1 - \frac{1}{d})^{d-1} \geq \frac{1}{e}$. The probability of a node to be of degree 1 is $P[d'(v) = 1] = (1 - \frac{1}{d})^d \leq \frac{1}{e}$. Now we use the linearity of the expectation and the lemma follows.

The next corollary shows that for a regular graph the expected size of V_2' is linear.

Corollary 5.2. *For all $d \geq 2$*

$$n/e \geq E[|V_1'|] \geq n/4$$

$$n/2 \geq E[|V_2'|] \geq n/e$$

$$0.416n \geq E[|\overline{V'}_3|] \geq (1 - 1/e - 1/2)n = 0.132n$$

When we remove the direction from the edges we get that each connected component in G' is a union of paths, and it ends in a cycle.

Let v be a node in G' s.t $d(v') > 2$. From the definition of G' it follows that there is only one edge that is outgoing from the node (this is the edge that was chosen by the node) and this node has more than one ingoing edge. In the next time step we transform the graph G' into the graph G'' by randomly removing all the extra incoming edges except one. After this step all the nodes in $\overline{V'}_3$ have out-degree at most 1 and in-degree equal to 1. Note that the out-degree can actually become 0 if the outgoing edge happens to be incoming to another node v'' with degree $d(v'') > 2$ who removes this edge to reduce its in-degree to 1. Clearly after this step $E[|V_1''|] + E[|V_2''|] \geq E[|V_3'|] > 0.132n$.

Because all these nodes have degree 1 or 2, we have at least as many edges in the graph as well. Now, we assign weights 1 to all remaining edges, and run one of the algorithms from the previous section to compute a maximum weighted matching for the remaining graph. This yields a matching of size at least $(E[|V_1''|] + E[|V_2''|])/3 \geq \frac{0.132}{3}n$. The reason we divide by 3 is that we may have a cycle of length 3.

Since a maximum matching of the original unweighted regular graph is never larger than $\frac{n}{2}$, it follows that we have a randomised algorithm that proximate the maximum matching with a in expected $\frac{1}{2}\frac{3}{0.132} = 11.36$ approximation ratio for a regular graph in a constant time.

Note that the same proof will work if the graph is not an r-regular graph but an almost r-regular graph. We say that a graph G is an α-d-regular graph if $\frac{\Delta}{\delta} < \alpha$, where Δ is the maximal degree and δ is the minimal degree. The next lemma replaces lemma 5.1 for α-d regular graphs.

Lemma 5.3. *Let G be an α-d-regular graph then*

$$E[|V_1'|] \leq ne^{-1/\alpha}$$

$$E[|V_2'|] \leq ne^{-1/\alpha}\frac{\Delta}{\Delta - 1}$$

$$E[|\overline{V}'_3|] = n - E[|V_2'|] - E[|V_1'|]$$

If $\Delta = 2$ then our graph is a forest with some cycles. Using the results from section 3.5 it follows that for a graph without cycles we have a $2 + \epsilon$ approximation. Sine a cycle is very close to a path we can use the same idea for cycles, and get a $2 + \epsilon$ approximation for graphs with $\Delta \leq 2$. So we may assume that $\Delta \geq 3$. The next corollary shows that for $1/\log(5/2) = 1.09136 > \alpha$-$d$-regular graph the expected size of V_3' is linear.

Corollary 5.4. *for all $\delta \geq 2$ and $1.09136 > \alpha$,*

$$E[|V_3'|] \geq n\left(1 - e^{-1/\alpha}\frac{2\Delta - 1}{\Delta - 1}\right) = \Omega(n) .$$

Since the size of V_3' is linear in n we can apply the algorithm from the previous section and get a constant approximation which depends on α.

Corollary 5.5. *Let $G(n, p)$ when $p > \frac{\log(n)}{n}$ be a random graph. Then our approximation algorithm for matching is a constant approximation.*

From the previous corollary it may seem our algorithm always computes a matching with a constant approximation to the maximum matching. The next example shows that this is not the case. Let K_n be a clique of size n. Let C_{n^2} be a cycle that contains n^2 nodes. We connect all the node in C_{n^2} to all the nodes in K_n.

Note that the number of node in this graph is $|V| = n^2 + n$, and the number of edges is $n^2 + n^3$. The order of the maximum matching is $O(n^2)$. However, the expected size of the matching that our algorithm computes is $2n$. So in this case the approximation ratio of the algorithm is $O(\sqrt{|V|})$.

6 Conclusions

We have presented efficient distributed algorithms that compute good approximations for the maximum weighted matchings for arbitrary weighted trees. Equally good algorithms for general graphs that compute constant approximations in sub-logarithmic time are not known. We have shown why our approach of constructing paths fails in the general case. Different techniques therefore seem required to handle arbitrary graphs efficiently.

References

[Avi83]   AVIS, D. A survey of heuristics for the weighted matching problem. *Networks* **13** (1983), 475–493.

[CHS02]   CHATTOPADHYAY, S., HIGHAM, L., AND SEYFFARTH, K. Dynamic and self-stabilizing distributed matching. In *21st PODC* (Monterey, CA, USA, 2002), ACM Press, pp. 290–297.

[Gab90]   GABOW, H. Data structures for weighted matching and nearest common ancestors with linking. In *1th SODA* (San Fransisco, Ca., USA, 1990), ACM Press, pp. 434–443.

[GPS87]   GOLDBERG, A. V., PLOTKIN, S., AND SHANNON, G. Parallel symmetry breaking in sparse graphs. In *19th STOC* (New York City, NY, USA, 1987), ACM Press.

[Hoe04]   HOEPMAN, J.-H. Simple distributed weighted matchings, 2004. eprint cs.DC/0410047.

[II86]   ISRAELI, A., AND ITAI, A. A fast and simple randomized parallel algorithm for maximal matching. *Inf. Proc. Letters* **22** (1986), 77–80.

[KS00]   KARAATA, M., AND SALEH, K. A distributed self-stabilizing algorithm for finding maximal matching. *Computer Systems Science and Engineering* **3** (2000), 175–180.

[KP98]   KUTTEN, S., AND PELEG, D. Fast distributed construction of k-dominating sets and applications. *Journal of Algorithms* **28**, 1 (1998), 40–66.

[MV80]   MICALI, S., AND VAZIRANI, V. An $O(\sqrt{V}E)$ algorithm for finding maximum matching in general graphs. In *21st FOCS* (Syracuse, NY, USA, 1980), IEEE Comp. Soc. Press, pp. 17–27.

[Pre99]   PREIS, R. Linear time 1/2-approximation algorithm for maximum weighted matching in general graphs. In *16th STACS* (Trier, Germany, 1999), C. Meinel and S. Tison (Eds.), LNCS 1563, Springer, pp. 259–269.

[UC00]   UEHARA, R., AND CHEN, Z. Parallel approximation algorithms for maximum weighted matching in general graphs. *Inf. Proc. Letters* **76** (2000), 13–17.

[WW04]   WATTENHOFER, M., AND WATTENHOFER, R. Distributed weighted matching. In *18th DISC* (Amsterdam, the Netherlands, 2004), R. Guerraoui (Ed.), LNCS 3274, Springer, pp. 335–348.

Approximation Strategies for Routing Edge Disjoint Paths in Complete Graphs

Adrian Kosowski[*]

Gdańsk University of Technology
Department of Algorithms and System Modeling
kosowski@sphere.eti.pg.gda.pl

Abstract. The paper deals with the well known Maximum Edge Disjoint Paths Problem (MaxEDP), restricted to complete graphs. We propose an off-line 3.75-approximation algorithm and an on-line 6.47-approximation algorithm, improving earlier 9-approximation algorithms due to Carmi, Erlebach and Okamoto (Proceedings WG'03, 143–155). Next, it is shown that no on-line algorithm for the considered problem is ever better than a 1.50-approximation. Finally, the proposed approximation techniques are adapted for other routing problems in complete graphs, leading to an off-line 3-approximation (on-line 4-approximation) for routing with minimum edge load, and an off-line 4.5-approximation (on-line 6-approximation) for routing with a minimum number of WDM wavelengths.

1 Introduction

The fundamental networking problem of establishing point-to-point connections between pairs of nodes in order to handle communication requests has given rise to numerous path routing problems in graph theory. The topology of the network is modeled in the form of a graph whose vertices correspond to nodes, while edges represent direct physical connections between nodes. This paper deals with the well established problem of handling the maximum possible number of communication requests without using a single physical link more than once, known as the *Maximum Edge Disjoint Paths Problem* (MaxEDP). We focus on the construction of approximation algorithms for the *NP*-hard MaxEDP problem in complete graphs, which are used to model networks with direct connections between all pairs of nodes. Two basic algorithmic approaches are considered — *off-line algorithms*, which compute a routing for a known set of requests provided at input, and *on-line algorithms*, which have to handle requests individually, in the order in which they appear.

Problem definition. The physical architecture of the network is given in the form of an undirected graph $G = (V, E)$, where V denotes the set of nodes, while

---

[*] Research supported by the State Committee for Scientific Research (Poland) Grant No. 4 T11C 047 25.

P. Flocchini and L. Gąsieniec (Eds.): SIROCCO 2006, LNCS 4056, pp. 130–142, 2006.

E represents the set of connections between them. A sequence of edges $P = (e_1, e_2, \ldots, e_l) \in E^l$, such that $e_i = \{v_i, v_{i+1}\}$ for some two vertices $v_i, v_{i+1} \in V$, is called a *path* of length $l = |P|$ in G, with endpoints v_1 and v_{l+1}. The symbol $P_{\{u,v\}}$ is used to denote any path in G with endpoints $u, v \in V$. A pair of paths P_1 and P_2 is called *conflicting* if there exists an edge $e \in E$ such that $e \in P_1$ and $e \in P_2$. For a given set of paths R in graph G, the *conflict graph* $Q(R)$ is a simple graph with vertex set R and edges connecting all pairs of vertices corresponding to paths from set R which conflict in G.

An *instance* I in network G is defined as any multiset of pairs $\{u, v\}$, $u, v \in V$, $u \neq v$, such that each element of I represents a single *communication request* between a pair of nodes. An equivalent representation of instance I may be given in the form of the *instance multigraph* $H(I) = (V, I)$, where communication requests are treated as edges of $H(I)$. A *routing* R of instance I in network G is a multiset of paths in G, such that there is a one-to-one correspondence between paths $P_{\{u,v\}} \in R$ and elements $\{u, v\} \in I$. The set of all routings of instance I is denoted as $\mathcal{R}(I)$. For use in further considerations, we define the following parameters for any routing R:

- *dilation* $\mathbf{d}(R)$, defined as the length of the longest path in routing R: $\mathbf{d}(R) = \max_{P \in R} |P|$,
- *edge load* $\pi(R)$, given by the formula: $\pi(R) = \max_{e \in E} |\{P \in R : e \in P\}|$.

A routing R is said to consist of *edge disjoint paths* if $\pi(R) = 1$, or equivalently, if conflict graph $Q(R)$ has no edges. A formal definition of the MAXEDP problem, expressed in these terms, is given below.

Maximum Edge Disjoint Paths Problem [MaxEDP]

Input: Instance I in graph G.

Solution: A set of pairwise edge-disjoint paths R_{OPT}, such that $R_{\text{OPT}} \in \mathcal{R}(I_{\text{OPT}})$ for some instance $I_{\text{OPT}} \subseteq I$.

Goal: Maximise the cardinality of R_{OPT}.

Notation. Throughout the paper, the complete graph with vertex set V is denoted K_V. Unless otherwise stated, we will assume that the MAXEDP problem is considered for the instance I in complete graph $G = K_V = (V, E)$. The optimal solution to the MAXEDP problem is some routing $R_{\text{OPT}} \in \mathcal{R}(I_{\text{OPT}})$ $(I_{\text{OPT}} \subseteq I)$, while approximation algorithms yield a solution denoted as $R_S \in \mathcal{R}(I_S)$ $(I_S \subseteq I)$, of not greater cardinality than R_{OPT}. Approximation ratios are understood in terms of upper bounds on the ratio $\frac{|I_{\text{OPT}}|}{|I_S|}$. The number of elements of a set or multiset, and also the length of a path, is written as $|P|$. The symbols Δ_H and χ'_H are used to denote the maximum vertex degree and the chromatic index of multigraph H, respectively.

State-of-the-art results. In the case of general networks G, the MAXEDP problem is closely related to a family of unsplittable flow problems. In consequence MAXEDP is *NP*-hard, difficult to approximate in polynomial time within

Table 1. New complexity results for the MAXEDP problem in complete graphs

Instance restriction	Off-line complexity		On-line complexity	
$\Delta_{H(I)} \leq \frac{|V|}{12}$	$O(|V|^3)$	Prop. 3	$O(|V|)$ per request	Cor. 3
$|I| < |V|$	$O(|V|)$	Prop. 2	$O(|V|)$ per request	Cor. 3
$|I| < k|V|$, const $k > 0$	$O(|V|^3)$	Thm. 2	not approx. within	Thm. 8
$|I| < |V|^s$, const $s \in (1, 2)$	$NPH, PTAS$	Thm. 3, 4	1.50 for $|I| \geq 3|V|$	Thm. 8
general case	3.75-approximation	Thm. 1	6.47-approximation	Thm. 7

a constant factor, and difficult to approximate within a factor of $O(\log^{\frac{1}{3}-\varepsilon} |E|)$, for any $\varepsilon > 0$ (unless $NP \subseteq ZPTIME(n^{\text{poly} \log n})$, [1]). The variant of MAXEDP defined for directed graphs is even difficult to approximate within $O(|E|^{\frac{1}{2}-\varepsilon})$, for any $\varepsilon > 0$ [12]. Both the directed and undirected version are approximable within a factor of $O(|E|^{\frac{1}{2}})$ [15].

When graph G is the complete graph K_V, the MAXEDP problem, though remaining NP-hard, becomes approximable within a constant factor. The best known approximation ratio was equal to 9 both in the off-line and on-line model of computation, due to Carmi, Erlebach and Okamoto [4]. A comparison of known approximation algorithms is provided in Table 2 at the end of the paper.

Our contribution and outline of the paper. In Section 2 we deal with the off-line MAXEDP problem in complete graphs, providing a 3.75-approximation algorithm based on the simple combinatorial concept of edge-coloring. Moreover, we show that for instances with significantly fewer than $|V|^2$ requests, the problem is either polynomially solvable, or admits a polynomial time approximation scheme. For the on-line version of the problem, in Section 3 we provide a 6.47-approximation algorithm, and show that no algorithm is better than 1.50-approximate, even for restricted instances. A summary of the most important new results concerning the MAXEDP problem is given in Table 1. Finally, in Section 4 we discuss the application of similar approximation techniques to other routing problems in complete graphs, and remark on their implementation in a distributed setting.

2   The Off-Line MaxEDP Problem in Complete Graphs

In the off-line routing model, it is assumed that all pairs of vertices forming the routed instance are initially known and all paths are determined by the routing algorithm at the same time.

2.1   Preliminaries: Bounds on Solution Cardinality

Factors in a multigraph. Let F_v be a set of nonnegative integers defined for each vertex $v \in V$. An F-*factor* in multigraph $H = (V, I)$ is a set of edges of H such that the number of edges from this set which are incident to vertex v

belongs to F_v. An $[a,b]$-*factor* is defined as an F-factor such that each set F_v consists of all integers from the range $[a,b]$. An $[a,b]$-factor with the maximum number of edges may be found efficiently by reduction to a minimum weighted perfect matching problem.

Proposition 1 ([16],[11]). *The problem of finding an $[a,b]$-factor with the maximum possible number of edges in multigraph $H = (V,I)$ can be solved in $O(|I|^3)$ time.*

Let I be an instance in graph K_V. Consider an instance I_{OPT} yielding an optimal solution to the MAXEDP problem for instance I. It is immediately evident that any vertex $v \in V$ can belong to at most $\deg_{K_V} v = |V| - 1$ requests of I_{OPT}, hence I_{OPT} is a $[0, |V| - 1]$-factor in $H(I)$ and we have the following bound.

Corollary 1. *The cardinality of the optimal solution to the MAXEDP problem for I is bounded from above by the size of the maximum $[0, |V|-1]$-factor in $H(I)$.*

Instances admitting an edge-disjoint routing. It is interesting to note that relatively wide classes of instances can be entirely routed using edge disjoint paths and in polynomial time. A short characterisation of two classes useful in further considerations is given below.

Proposition 2. *If $|I| < |V|$, then the entire instance I can be routed in K_V by edge disjoint paths, and a solution $R_{\mathrm{OPT}} \in \mathcal{R}(I)$ to the MAXEDP problem, such that $\mathrm{d}(R_{\mathrm{OPT}}) \leq 2$, can be determined in $O(|V|)$ time.*

Proof. The proof is constructive and proceeds by induction with respect to $|V|$. For $|V| = 2$, we have $|I| \leq 1$ and the proposition is obviously true. Next, let $|V| > 2$ be fixed and let $u \in V$ be a vertex belonging to the smallest number of requests in I, i.e. such that u is of minimal degree in $H(I)$. Since $|I| < |V|$, it is evident that $\deg_{H(I)} u = 0$ or $\deg_{H(I)} u = 1$. In the former case, we select an arbitrary request $\{v_1, v_2\} \in I$, and return the solution to the MAXEDP problem for I in K_V in the form of path $(\{v_1, u\}, \{u, v_2\})$ added to the solution to MAXEDP for instance $I \setminus \{\{v_1, v_2\}\}$ in complete graph $K_{V \setminus \{u\}}$. Thus $|R_{\mathrm{OPT}}| = 1 + (|I| - 1) = |I|$ by the inductive assumption. In the latter case, let $\{u, v\} \in I$ be the only request involving vertex u. The sought routing then consists of the single-edge path $(\{u, v\})$ added to the solution to MAXEDP for instance $I \setminus \{\{u, v\}\}$ in $K_{V \setminus \{u\}}$. The described approach may easily be implemented in the form of an algorithm with $O(|V|)$ time complexity. $\qquad\square$

Observe that the thesis of Proposition 2 does not hold if $|I| = |V|$ (it suffices to consider an instance composed of $|V|$ requests between a fixed pair of vertices). Nevertheless, if $|I| \in O(|V|)$ the problem can be solved in polynomial time (see Theorem 2).

Proposition 3. *If $\Delta_{H(I)} \leq \frac{|V|}{12}$, then the entire instance I can be routed in K_V by edge disjoint paths, and a solution $R_{\mathrm{OPT}} \in \mathcal{R}(I)$ to the MAXEDP problem, such that $\mathrm{d}(R_{\mathrm{OPT}}) \leq 2$, can be determined in $O(|V||I|)$ time.*

Proof. First, let us observe that the size of any instance I fulfilling the assumptions of the theorem is bounded by $|I| \leq \frac{|V|}{2} \cdot \frac{|V|}{12}$. The sought routing $R_{\text{OPT}} \in \mathcal{R}(I)$ consisting of edge disjoint paths can be formed by sequentially assigning paths to requests from I (in arbitrary order), in such a way as to preserve the following conditions:

1. The length of any path added to R_{OPT} is at most 2.
2. Each vertex of graph K_V is the center of at most $\frac{|V|}{12}$ paths.

It suffices to show that the described construction of routing R_{OPT} is always possible. Suppose that at some stage of the algorithm R_{OPT} fulfills conditions 1 and 2, and the next considered request is $\{v_1, v_2\}$. Vertex v_1 is the endpoint of at most $\frac{|V|}{12} - 1$ paths and the center of at most $\frac{|V|}{12}$ paths already belonging to R_{OPT}, thus at least $\frac{3|V|}{4}$ edges of K_V incident to v_1 do not belong to any path of R_{OPT}. The same is true for vertex v_2. Thus we immediately have that the set U of vertices connected to both v_1 and v_2 by edges unused in R_{OPT} is of cardinality $|U| \geq \frac{3|V|}{4} + \frac{3|V|}{4} - |V| = \frac{|V|}{2}$. Since routing R_{OPT} currently consists of fewer than $|I| \leq \frac{|V|}{2} \cdot \frac{|V|}{12}$ paths, by the pigeonhole principle there must exist a vertex $u \in U$ such that u is the center of fewer than $\frac{|V|}{12}$ paths from R_{OPT}. Therefore the request $\{v_1, v_2\}$ may be fulfilled by adding path $(\{v_1, u\}, \{u, v_2\})$ [1] to routing R_{OPT}, thus preserving the assumptions of the construction, which completes the proof. $\square$

2.2 An Off-Line 3.75-Approximation Algorithm

Theorem 1. *There exists a 3.75-approximation algorithm for the* MaxEDP *problem in complete graphs with* $O(|I|^3)$ *runtime. The dilation of the returned solution is not greater than 2.*

Proof. Let I be an arbitrary instance in complete graph K_V, and let $I_{\text{OPT}} \subseteq I$ be a subset of the considered instance whose routing is an optimal solution to the MaxEDP problem. We denote by $H^* = (V, I^*)$ a multigraph $H^* \subseteq H(I)$ with the maximum possible number of edges, such that $\Delta_{H^*} < |V|$. Since the edge set of multigraph H^* is in fact a maximum $[0, |V| - 1]$-factor in $H(I)$, by Proposition 1 multigraph H^* can be determined in $O(|I|^3)$ time. Moreover, by Corollary 1 we have $|I_{\text{OPT}}| \leq |I^*|$.

We will now show that there exists an algorithm with $O(|I|^3)$ runtime which finds a routing $R_S \in \mathcal{R}(I_S)$ composed of edge disjoint paths, such that $I_S \subseteq I^* \subseteq I$ and the obtained solution is a 3.75-approximation of the optimal MaxEDP solution, $|I_S| \geq \frac{|I^*|}{3.75} \geq \frac{|I_{\text{OPT}}|}{3.75}$. Instance I_S is constructed as a subset of the edge set of multigraph H^*. Since $\Delta_{H^*} < |V|$, by a well known result due to Shannon [10], the chromatic index χ'_{H^*} is bounded by $\chi'_{H^*} \leq \frac{3\Delta_{H^*}}{2} < \frac{3|V|}{2}$, and an edge coloring of multigraph H^* using not more than $\frac{3|V|}{2}$ colors can be obtained in $O(|I|^3)$ time. Without loss of generality we may assume that colors are labelled

---

[1] Throughout the paper, we assume that edges of the form $\{v, v\}$ which appear in notation when enumerating edges of paths should be treated as nonexistent.

with integers from the range $\{1,\ldots,\frac{3|V|}{2}\}$, in such a way that a color with a smaller label is never assigned to fewer edges than a color with a larger label. Let I_C denote the subset of edges from I^* colored with colors from the range $\{1,\ldots,|V|\}$. Due to the adopted ordering of the color labels, we immediately have $|I_C| \geq \frac{2}{3}|I^*|$. For each edge $\{v_1, v_2\} \in I_C$, let $c_{\{v_1,v_2\}}$ denote the color assigned to this edge, which is an integer from the range $\{1,\ldots,|V|\}$, and as such may be treated as an identifier of some vertex in graph K_V (see Fig. 1 for an exemplary illustration).

Let us now consider routing R_C of instance I_C in graph K_V, defined as follows: $R_C = \{(\{v_1, c_{\{v_1,v_2\}}\}, \{c_{\{v_1,v_2\}}, v_2\}) : \{v_1, v_2\} \in I_C\}$. No vertex of H^* may ever be incident to two edges from I_C of the same color, therefore each edge $\{v_1, v_2\}$ of graph K_V belongs to at most two paths of routing R_C — one path, in which v_1 is an end vertex and v_2 is a central vertex (an edge color in I_C), and another path in which the functions of vertices v_1 and v_2 are reversed. Routing R_C thus fulfills the following conditions: $d(R_C) \leq 2$ and $\pi(R_C) \leq 2$. Consequently, each path of R_C may only conflict with at most two other paths, and the conflict graph $Q(R_C)$ is of degree bounded by $\Delta_{Q(R_C)} \leq 2$. Graph $Q(R_C)$ is thus a set of isolated vertices, paths and cycles. Notice that the three vertex cycle C_3 is a connected component of $Q(R_C)$ only if some three paths form a triangle, i.e. $P_1, P_2, P_3 \in R_C$ and $P_1 = (\{v_1, v_3\}, \{v_3, v_2\})$, $P_2 = (\{v_2, v_1\}, \{v_1, v_3\})$, $P_3 = (\{v_3, v_2\}, \{v_2, v_1\})$, for some three vertices $v_1, v_2, v_3 \in V$. Such a structure may however be easily eliminated by removing paths P_1, P_2, P_3 from R_C and replacing them by the following three paths: $P_1' = (\{v_1, v_2\})$, $P_2' = (\{v_2, v_3\})$, $P_3' = (\{v_3, v_1\})$, which satisfy the same set of requests and whose conflict graph consists of three isolated vertices.

The sought suboptimal solution R_S to the MaxEDP problem is now obtained by indicating a maximum independent set R_S in conflict graph $Q(R_C)$. Graph $Q(R_C)$ has $|R_C|$ vertices, and once all cycles C_3 have been eliminated the independent set R_S consists of at least $\frac{2}{5}|R_C|$ vertices (or equivalently, $|I_S| \geq \frac{2}{5}|I_C|$). Therefore, we finally obtain the following bound:

$$\frac{|I_{\text{OPT}}|}{|I_S|} \leq \frac{|I^*|}{|I_S|} = \frac{|I^*|}{|I_C|}\frac{|I_C|}{|I_S|} \leq \frac{3}{2}\cdot\frac{5}{2} = 3.75$$

which completes the proof of the approximation ratio of the designed algorithm.

□

It is interesting to note that although the off-line MaxEDP problem in complete graphs is NP-hard even for relatively small instances (Theorem 3), the conjecture that it is APX-hard still remains open [4], and the only inapproximability result concerns the on-line problem (Theorem 8). In fact, in the following subsection we show that for all instances of sufficiently bounded size, the off-line MaxEDP problem is not APX-hard.

2.3   Problem Complexity for Bounded Instances

We now deal with the MaxEDP problem restricted to instances I such that $|I| < |V|^s$ for some $s < 2$, and study the increasing difficulty of the problem with the increase of the bound on $|I|$.

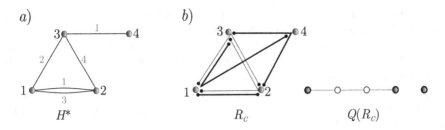

Fig. 1. Construction of an approximate solution to the MAXEDP problem for instance $I = \{\{1, 2\}, \{1, 2\}, \{1, 3\}, \{2, 3\}, \{3, 4\}\}$ in complete graph K_4: a) an edge coloring of multigraph H^* (in the considered case $I_C = I^* = I$), b) a routing R_C of instance I_C in graph K_4 and its conflict graph $Q(R_C)$ (the independent set of paths forming the sought routing R_S is marked in bold)

Theorem 2. *An optimal solution to the* MAXEDP *problem in complete graphs can be determined in* $O(|V|^3)$ *time if the size of the input instance is bounded by* $|I| \le k|V|$, *for any constant value of parameter* $k > 0$.

Proof. Let $T \subseteq V$ be defined as the set of all vertices belonging to more than $\frac{|V|}{24}$ requests, $T = \{v \in V : \deg_{H(I)} v > \frac{|V|}{24}\}$. Suppose that $|V| \ge 1248k$ (the problem for all smaller graphs may be solved by exhaustive search).

Property. The size of the solution $R_{\text{OPT}} \in \mathcal{R}(I_{\text{OPT}})$ *to the* MAXEDP *problem for instance* I *remains unchanged even if paths need not be disjoint with respect to edges from the edge set* E^* *of subgraph* $K_{V \setminus T} \subseteq K_V$. Indeed, let R^* be a routing of a maximal possible instance $I^* \subseteq I$ such that no edge from $E \setminus E^*$ belongs to more than one path of R^*. We create an instance I^{**} in graph $K_{V \setminus T}$ by successively considering all paths $P \in R^*$, and adding to I^{**} a request consisting of the first and the last vertex from $V \setminus T$ which appears in P. We now proceed to establish that it is possible to reroute instance I^{**} in $K_{V \setminus T}$ using edge disjoint paths, leading to the conclusion that $I_{\text{OPT}} = I^*$ is a valid solution to MAXEDP in K_V. Let $v \in V \setminus T$ be arbitrarily chosen. By definition of set T, we have $\deg_{H(I^*)} v \le \deg_{H(I)} v \le \frac{|V|}{24}$. Since each vertex $v \in V \setminus T$ is obviously connected to at most $|T|$ vertices from T, we immediately have $\deg_{H(I^{**} \setminus I^*)} v \le |T|$. Combining the last two inequalities we obtain $\deg_{H(I^{**})} v \le \frac{|V|}{24} + |T|$. Since $2k|V| \ge 2|I| = \sum_{v \in V} \deg_{H(I)} v \ge \sum_{v \in T} \deg_{H(I)} v \ge \frac{|V|}{24}|T|$, we have $|T| \le 48k$. Taking into account the assumption $|V| \ge 1248k$, we finally obtain $\deg_{H(I^{**})} v \le \frac{|V|}{24} + |T| \le \frac{|V|}{24} + 48k = \frac{1}{12}(\frac{|V|+1248k}{2} - 48k) \le \frac{1}{12}|V \setminus T|$, which means that by Proposition 3 instance I^{**} can be routed in $K_{V \setminus T}$ by means of edge disjoint paths, closing the proof of the property.

Let $I' \subseteq I$ denote the set of requests from I with at least one vertex in T, and let $I'_{\text{OPT}} \subseteq I'$ be a maximal subset of I' which can be routed by edge disjoint paths in K_V. The instance $I'_{\text{OPT}} \cup (I \setminus I') \subseteq I$ is therefore a maximal subset of I which can be routed by paths conflicting only within the edge set of graph $K_{V \setminus T}$, and by the proven Property such a routing can be converted to a correct solution to the MAXEDP problem for I. The problem of finding $I_{\text{OPT}} \subseteq I$ is thus reduced to finding $I'_{\text{OPT}} \subseteq I'$. Furthermore, when considering instance I' all vertices from set $V \setminus T$ may be regarded as indistinguishable (after once more relaxing the edge disjointness condition within $K_{V \setminus T}$). Thus graph K_V may be reduced to the multigraph G' formed by connecting each of the vertices of K_T to one additional vertex u using exactly $|V \setminus T|$ edges. In order to solve the MAXEDP problem for instance I' in G', we consider all possible arrangements of paths in the edge set of K_T, taken over all routings of all subsets of instance I'. Note that the number of such arrangements is bounded, since $|T| \in O(1)$. For a fixed arrangement of paths in the edge set of K_T, the MAXEDP problem for instance I' in G' can be easily reduced to the MAXEDP problem for a related instance in the multistar $G' \setminus K_T$. The latter problem can in turn be solved in $O(|I|^3) = O(|V|^3)$ time, using a generalisation of a technique from [7] (the solution proceeds by reduction to the problem of finding a maximal $[0, |V \setminus T|]$-factor in a multigraph). This procedure determines the complexity of the entire algorithm; the final rerouting step within $K_{V \setminus T}$ only requires $O(|V||I|) = O(|V|^2)$ time by Proposition 3.   □

Theorem 3. *The* MAXEDP *problem in complete graphs is NP-hard even for instances of size bounded by $|I| \leq |V|^s$, for any value of parameter $s > 1$.*

Proof (sketch). The proof proceeds by reduction from the MAXEDP problem in complete graphs with cardinality restriction $|I| \leq |V|^2$, which was shown to be *NP*-hard in [8]. Let $s = 1 + \varepsilon$, $\varepsilon > 0$. Consider an arbitrary subset of vertices $V' \subseteq V$ of cardinality equal to at most $|V|^\varepsilon$. Let I' be any instance in $K_{V'}$. We define instance I in K_V as follows: $I = I' \cup \{\{u, v\} : u \in V', v \in V \setminus V'\}$; for sufficiently large $|V|$ we have $|I| \leq |V|^s$. The proof is complete when we observe that an optimal solution R_{OPT} to the MAXEDP problem for instance I in graph K_V is always equal to the union of two sets of paths: the set of all one-edge paths connecting vertices from $K_{V'}$ with vertices from $K_{V \setminus V'}$, and some optimal solution R'_{OPT} to the MAXEDP problem for instance I' in graph $K_{V'}$. In particular, we have: $|R_{\text{OPT}}| = |R'_{\text{OPT}}| + |V'|(|V| - |V'|)$.   □

Theorem 4. *The* MAXEDP *problem in complete graphs admits a polynomial time approximation scheme for instances of size bounded by $|I| \leq |V|^s$, for any value of parameter $s < 2$.*

Proof (sketch). Let $|I| = |V|^s$, where $s = 2 - \varepsilon$, $\varepsilon > 0$. The proof is in essence similar to that of Theorem 2. We adopt the same definition of set T, obtaining $|T| \leq 48|V|^{1-\varepsilon}$. In all considerations we assume $|V| \geq 1248^{\frac{1}{\varepsilon}}$, so that the Property stated in the proof of Theorem 2 also holds in this case. By this property, any subset of instance I such that each vertex from T is the endpoint of at most $|V \setminus T|$ paths can be routed in K_V using edge disjoint paths. This implies that any maximal $[0, |V \setminus T|]$-factor in $H(I)$ is a suboptimal solution I_S to the considered MAXEDP problem. On the other hand, the cardinality of the optimal solution $|I_{\text{OPT}}|$ is bounded from above by the size of the maximal $[0, |V| - 1]$-factor in $H(I)$ by Corollary 1. The sizes of the considered factors in $H(I)$ are closely related, which leads to the following bound: $\frac{|I_{\text{OPT}}|}{|I_S|} \leq \frac{|V|}{|V| - 2|T|} \leq \frac{1}{1 - 96|V|^{-\varepsilon}}$. Thus, for any

$\delta > 0$ the considered approach achieves an approximation ratio of $1 + \delta$ provided $|V| > (96(1 + \max\{12, \delta^{-1}\}))^{\frac{1}{\epsilon}}$, whereas the problem may be optimally solved by exhaustive search for all smaller values of $|V|$.   □

A summary of the main results of the section is given in Table 1.

3   The On-Line MaxEDP Problem in Complete Graphs

On-line algorithms for the MAXEDP problem, which are considered in this paper, are treated as a special case of greedy algorithms. We assume that successive requests from instance I appear sequentially at input, becoming known to the algorithm only once the previous request has been processed. The decision taken at every step as to whether some path fulfilling the current request should be added to the constructed edge disjoint routing R_S is inadvertent and impossible to change at a later stage of the algorithm. Approximation ratios are calculated with respect to the best possible solution R_{OPT} in the off-line model.

3.1   An On-Line 6.47-Approximation Algorithm

A slight modification of the approximation algorithm provided for the off-line case (Theorem 1) allows for its on-line operation. In the considered approach, the algorithm sequentially processes requests from instance I, treating them as edges of multigraph $H(I)$, and at every step attempts to color the edge using a color from the range $\{1, \ldots, |V|\}$. A generalization of this problem was recently considered by Favrholdt and Nielsen [9], under the name of the *maximum k-edge-colorable subgraph problem* for a multigraph. They stated that any fair on-line algorithm (i.e. an algorithm which always colors an edge, if only a color from the range $\{1, \ldots, k\}$ is available) leads to a $\frac{1}{2\sqrt{3}-3}$-approximation of the solution. In fact, the obtained result was significantly stronger; we shall reformulate it here for easier use in further considerations.

Theorem 5 ([9]). *For any multigraph $H = (V, I)$, any fair on-line algorithm for the k-edge-colorable subgraph problem labels a subset of edges $I_C \subseteq I$ with colors $\{1, \ldots, k\}$, such that $|I_C| \geq (2\sqrt{3} - 3)|I^{**}|$, where I^{**} denotes a maximal $[0, k]$-factor in H.*

In particular, the above theorem holds for $k = |V|$, thus using the notation from Theorem 1 we may write $|I_C| \geq (2\sqrt{3} - 3)|I^*|$. As the coloring proceeds, the sought routing R_S may be incrementally constructed using an on-line independent set algorithm applied to graph $Q(R_C)$. Since graph $Q(R_C)$ only consists of cycles, paths and isolated vertices, we obtain $|I_S| \geq \frac{1}{3}|I_C|$. Combining the obtained relations leads to the bound:

$$\frac{|I_{\text{OPT}}|}{|I_S|} \leq \frac{|I^*|}{|I_S|} = \frac{|I^*|}{|I_C|} \frac{|I_C|}{|I_S|} \leq \frac{1}{2\sqrt{3}-3} \cdot 3 < 6.47$$

which may be expressed by means of the following statement.

Corollary 2. *There exists a 6.47-approximation algorithm for the on-line* MAX-EDP *problem in complete graphs, requiring $O(|V|)$ time to process a single request. The dilation of the returned solution is not greater than 2.*

In fact, the algorithm resulting from the above considerations can be written in much simpler form, as described in the next subsection.

3.2 Performance Analysis of the BGA Algorithm

The *bounded length greedy algorithm* (BGA) is an on-line strategy for the MAX-EDP problem, introduced in [13]. The basic principle of its operation is that at every step an attempt is made to route the current request by the shortest possible path P which does not contain any of the edges already belonging to R_S, and to add P to the solution R_S provided $|P| \leq L$, where L is a fixed parameter of the algorithm. The computed routing R_S therefore fulfills the bound $\mathsf{d}(R_S) \leq L$. The BGA strategy was last studied by Carmi, Erlebach and Okamoto [4], who bounded its approximation ratio for $L = 4$ using an unsplittable flow technique.

Theorem 6 ([4]). *The* BGA *strategy with $L = 4$ is a 9-approximation on-line algorithm for the* MAXEDP *problem in complete graphs.*

However, it is interesting to note that further bounding of the parameter L may lead to algorithms for which a better approximation ratio can be proven.

Theorem 7. *The* BGA *strategy with $L = 2$ is a 6.47-approximation on-line algorithm for the* MAXEDP *problem in complete graphs.*

Proof (sketch). The proof is based on the observation that each step of BGA with $L = 2$ combines the properties of an on-line algorithm for the edge-colorable subgraph problem with those of an on-line independent set algorithm, thus implementing an approach very similar to that described in Subsection 3.1. A request $\{u, v\}$ can only be routed using BGA by a path $P = (\{u, w\}, \{w, v\})$ of length at most 2 via some vertex $w \in V$ if edge $\{u, v\}$ of multigraph $H(I)$ can be labeled with color $w \in \{1, \ldots, |V|\}$, and if path P does not conflict with any paths previously added to R_S. The only difference is that the $|V|$-edge-colorable subgraph of $H(I)$ implicitly found by the BGA algorithm need not correspond to that obtained by means of any fair algorithm, since in a step of BGA an edge of $H(I)$ is not colored whenever any color assignment is possible, but only when assigning a color contributes to the size of the resultant solution R_S. Careful analysis shows that this does not affect the overall approximation ratio which remains equal to 6.47 (Corollary 2). □

A further interesting property of the BGA strategy with parameter $L = 2$ is that it finds an edge disjoint routing of the whole instance I in the cases considered in Propositions 2 and 3.

Corollary 3. *If $\Delta_{H(I)} \leq \frac{|V|}{12}$, or $|I| \leq |V| - 1$, then the entire instance I can be routed in G_V by edge disjoint paths, and an optimal solution such that $\mathsf{d}(R_{\mathrm{OPT}}) \leq 2$ is always determined by the* BGA *strategy with $L = 2$.*

3.3 Inapproximability Results

Whereas the complexity of finding a solution to the off-line MAXEDP problem in complete graphs still remains open, we now show that the on-line version is not approximable within a constant factor for sufficiently large instances.

Theorem 8. *There does not exist any on-line approximation algorithm for the* MAXEDP *problem in complete graphs with an approximation ratio smaller than* 1.50, *even when considering instances of size* $|I| < k|V|$, *for any* $k \geq 3$.

Proof. By contradiction, suppose that some on-line MAXEDP algorithm A has an approximation ratio not worse than 1.50. Given any graph K_V, let instance I begin with $|V| - 1$ requests of the form $\{u, v\}$, for some two distinguished vertices $u, v \in V$. At this point the routing R_S obtained by algorithm A consists of p paths, where $p \geq \frac{2}{3}(|V| - 1)$ (otherwise the instance is ended, and we have $|R_{\text{OPT}}| = |V| - 1 > 1.50|R_S|)$. Instance I is now completed by presenting a further $2(|V| - 2)$ requests of the form $\{u, w\}$ and $\{v, w\}$, taken over all vertices $w \in V \setminus \{u, v\}$. Since the number of paths which end in any vertex (in particular, u or v) cannot exceed $|V| - 1$, the total number of paths eventually belonging to R_S is bounded by $|R_S| \leq p + 2((|V| - 1) - p) \leq \frac{4}{3}(|V| - 1)$, whereas $|R_{\text{OPT}}| = 2(|V| - 2) + 1 = 2(|V| - 1) - 1$, hence the ratio $\frac{|R_{\text{OPT}}|}{|R_S|}$ cannot be smaller than 1.50 for arbitrarily large values of $|V|$.   □

Even in the on-line model, the gap remaining between the 1.50 inapproximability result of Theorem 8 and the 6.47-approximation algorithm from Theorem 7 is quite substantial. A partial attempt to bridge it may be performed by considering the inapproximability of specific classes of on-line algorithms. For example, the BGA algorithm and similar strategies are never better than 3-approximate for certain classes of instances [4].

4 Final Conclusions

The technique adopted in the proof of Theorem 1 — which may basically be thought of as *routing by edge coloring* — provides efficient approximation

Table 2. A comparison of presented approximation algorithms for the MAXEDP problem in complete graphs with previous results (updated from [4])

Principle of operation	Model	Approximation ratio	Dilation	Reference
Shortest-path-first variant of BGA	off-line	54		[8], 2001
Set tripartition	off-line	27		[8], 2001
BGA with $L = 4$	on-line	17	≤ 4	[13], 2002
BGA with $L = 4$	on-line	9	≤ 4	[4], 2003
BGA with $L = 2$	on-line	6.47	≤ 2	Thm. 7
Routing by edge coloring	off-line	3.75	≤ 2	Thm. 1

algorithms for a number of routing problems in complete graphs and similar extremely dense topologies. When applying this approach, the approximation ratio may vary depending on the considered problem, and is usually given in the form of the product of two parameters $M_1 \cdot M_2$, where M_1 denotes the relative loss in the first phase of the algorithm (determining an edge coloring), and M_2 is the relative loss in the second phase (post-processing the edge coloring).

For the MaxEDP problem, the applied techniques constitute a substantial improvement on earlier results (Table 2). We now give two more examples of routing problems for which fixed-ratio approximation algorithms can be similarly obtained.

The edge load routing problem. For a given instance I in graph K_V, we consider the problem of finding a routing $R_{\mathrm{OPT}} \in \mathcal{R}(I)$, such that edge load $\pi(R_{\mathrm{OPT}})$ is the minimum possible [2, 3]. In order to construct an approximation approach with respect to $\pi(R_S)$ within K_V, observe that multigraph $H(I)$ can always be efficiently edge-colored with at most $1.5(|V| - 1)\,\pi(R_{\mathrm{OPT}})$ colors in the off-line model, or $2(|V| - 1)\,\pi(R_{\mathrm{OPT}})$ colors in the on-line model. By applying a similar approach as that in the proof of Theorem 2, it is easy to see that the instance corresponding to any $(|V| - 1)$-edge-colorable subgraph of $H(I)$ can always be routed with load at most 2, both in the off-line and the on-line model. Thus we have $M_2 = 2$ and $M_1 = 1.5$ (off-line) or $M_1 = 2$ (on-line), finally obtaining an off-line 3-approximation algorithm and an on-line 4-approximation algorithm for edge load routing in complete graphs.

The WDM wavelength count routing problem. This modification of the edge load routing problem is of special importance from the point of view of application in so called *all-optical wavelength division multiplexing* (WDM) networks [2, 5, 6]. For a given instance I in graph K_V, the sought routing $R_{\mathrm{OPT}} \in \mathcal{R}(I)$ should minimize the value of a parameter called *WDM wavelength count* $\mathtt{w}(R_{\mathrm{OPT}})$, defined as the chromatic number of conflict graph $Q(R_{\mathrm{OPT}})$. The proposed construction of an approximation algorithm with respect to $\mathtt{w}(R_S)$ is nearly the same as for bounded edge load, the only difference being that in the second stage of the algorithm $(|V| - 1)$-edge-colorable subgraphs of $H(I)$ can always be routed using 3 wavelengths. Therefore in this case we have $M_2 = 3$ and $M_1 = 1.5$ (off-line) or $M_1 = 2$ (on-line), yielding an off-line 4.5-approximation algorithm and an on-line 6-approximation algorithm for the considered problem.

Finally, let us remark on a general property of the approximate solutions obtained using the proposed approach: in all cases the dilation is bounded by a value of 2. Using paths with at most 1 intermediary node between the communicating pair of endpoints is advantageous from the point of view of resource usage, and additionally simplifies the routing process. Indeed, if the on-line version of the routing algorithm is considered in a distributed setting, each node can independently decide whether it may participate in the routing of a given communication request. Thus each request can be processed in $O(1)$ synchronous rounds, achieving a time-optimal routing process.

Acknowledgement. The author would like to express his gratitude to the anonymous referees for their numerous helpful comments and suggestions for the improvement of this paper.

References

1. M. Andrews, L. Zhang, Hardness of the undirected edge-disjoint paths problem. *Proc. STOC'05* (2005), 276–283.
2. B. Beauquier, J.C. Bermond, L. Gargano, P. Hell, S. Pèrennes and U. Vaccaro, Graph problems arising from wavelength routing in all-optical networks. *Proc. WOCS'97* (1997), Geneve, Switzerland.
3. J. Białogrodzki, Path Coloring and Routing in Graphs. In: *Graph Colorings*, M. Kubale ed., Contemporary Math. 352, AMS (2004), USA, 139–152.
4. P. Carmi, T. Erlebach, Y. Okamoto, Greedy edge-disjoint paths in complete graphs. *Proc. WG'03*, LNCS 2880 (2003), 143–155.
5. S. Choplin, L. Narayanan, J. Opatrny, Two-Hop Virtual Path Layout in Tori. *Proc. SIROCCO'04*, LNCS 3104 (2004), 69–78.
6. T. Erlebach, K. Jansen, The complexity of path coloring and call scheduling. *Theoret. Comp. Sci.* 255 (2001), 33–50.
7. T. Erlebach, K. Jansen, The Maximum Edge-Disjoint Paths Problem in Bidirected Trees. *SIAM J. Discret. Math.* 14 (2001), 326–355.
8. T. Erlebach, D. Vukadinović, New results for path problems in generalized stars, complete graphs, and brick wall graphs. *Proc. FCT'01*, LNCS 2138 (2001), 483–494.
9. L.M. Favrholdt, M.N. Nielsen, On-line edge-coloring with a fixed number of colors. *Algorithmica* 35 (2003), 176–191.
10. S. Fiorini, R.J. Wilson: *Edge-Colourings of Graphs*, Pittman (1977), USA.
11. H.N. Gabow, Data structures for weighted matching and nearest common ancestors with linking. *Proc. SODA'90* (1990), 434–443.
12. V. Guruswami *et al*, Near-optimal hardness results and approximation algorithms for edge-disjoint paths and related problems. *J. Comput. Syst. Sci.* 67 (2003), 473–496.
13. P. Kolman, C. Scheideler, Improved bounds for the unsplittable flow problem. *Proc. SODA'02* (2002), 184–193.
14. B. Ma, L. Wang, On the inapproximability of disjoint paths and minimum Steiner forest with bandwidth constraints, *J. Comput. Syst. Sci.* 60 (2000), 1–12.
15. A. Srinivasan, Improved approximations for edge-disjoint paths, unsplittable flow, and related routing problems, *Proc. FOCS'97* (1997), 416–425.
16. W.T. Tutte, A short proof of the factor theorem for finite graphs. *Canad. J. Math.* 6 (1954), 347–352.

Short Labels by Traversal and Jumping[*]

Nicolas Bonichon, Cyril Gavoille, and Arnaud Labourel

Laboratoire Bordelais de Recherche en Informatique, Université Bordeaux 1
{bonichon, gavoille, labourel}@labri.fr

Abstract. In this paper, we propose an efficient implicit representation of caterpillars and binary trees with n vertices. Our schemes, called *Traversal & Jumping*, assign to vertices of the tree distinct labels of $\log_2 n + O(1)$ bits, and support constant time adjacency queries between any two vertices by using only their labels. Moreover, all the labels can be constructed in $O(n)$ time.

1 Introduction

The two basic ways of representing a graph are adjacency matrices and adjacency lists. The latter representation is space efficient for sparse graphs, but adjacency queries require searching in the list, whereas matrices allow fast queries to the price of a super-linear space. Another technique, called *implicit representation* or *adjacency labeling scheme*, consists in assigning labels to each vertex such that adjacency queries can be computed alone from the labels of the two involved vertices without any extra information source. The goal is to minimize the maximum length of a label associated with a vertex while keeping fast adjacency queries.

Adjacency labeling schemes, introduced by [Bre66, BF67], have been investigated by [KNR88, KNR92]. They construct for several families of graphs adjacency labeling schemes with $O(\log n)$-bit labels. In particular, for trees the scheme consists in: 1) choosing an arbitrary prelabeling of the n vertices, a permutation of $\{1, \ldots, n\}$; 2) choosing a root; and 3) setting the label of a vertex to be the pair formed by its prelabel and the prelabel of its parent. The adjacency test checks whether the prelabel for one vertex equals the parent prelabel of the other vertex. Such labels are of $2 \lceil \log n \rceil$ bits[1], whereas $\lceil \log n \rceil$ bits are clearly necessary since labels must be different.

Improving the label length of this straightforward scheme is not an easy task. It has been however improved in a non trivial way by [AKM01] to $1.5 \log n + O(\log \log n)$ bits, and more recently to $\log n + O(\log^* n)$ bits[2] [AR02], leaving open the question of whether trees enjoy a labeling scheme with $\log n + O(1)$ bit labels.

---

[*] The three authors are supported by the project "GeoComp" of the ACI Masses de Données.
[1] All the logarithms are in base two.
[2] Log$^*n$ denotes the number of times log should be iterated to get a constant.

P. Flocchini and L. Gąsieniec (Eds.): SIROCCO 2006, LNCS 4056, pp. 143–156, 2006.

1.1   Related Work

Motivated by applications in XML search engines, and distributed applications as peer-to-peer networks or network routing, several other distributed data-structures with optimal $O(\log n)$-bit labels, have been developed.

For instance, routing in trees [FG01, TZ01], near-shortest path routing in specific networks [BG05, DL02, DL04], distance queries for interval, circular-arc, and permutation graphs [BG05, GP03a], etc. have $O(\log n)$-bit distributed data-structures. And, specifically for several queries on trees, we have: nearest common ancestor [AGKR04] with $O(\log n)$-bit labels, ancestry [AAK+05] with $\log n + O(\sqrt{\log n})$ bit labels, and small distance queries and other related functions with $\log n + \Theta(\log \log n)$ bit labels [KM01, ABR05]. Interestingly, it is shown in [ABR05] that for sibling queries in trees of maximum degree Δ, $\log n + \Theta(\log \log \Delta)$ bit labels are necessary and sufficient. A survey on labeling schemes can be founded in [GP03b]. All these schemes achieve labeling of length[3] $\log n + \omega(1)$.

To our best knowledge, for reasonably large families of graphs, no distributed data-structure is known to have an optimal label size up to an additive constant. In particular, for adjacency queries in trees, the current lower bound is $\log n$ and the upper bound is $\log n + O(\log^* n)$ [AR02]. This latter scheme, based on a recursive decomposition of the tree in $\Theta(\log^* n)$ levels, has adjacency query time of $\Omega(\log^* n)$.

1.2   Our Contributions

In this paper we present adjacency labeling schemes for caterpillars (i.e., a tree whose nonleaf vertices induce a path), and binary trees with n vertices. Both schemes assign distinct labels of $\log n + O(1)$ bits, and support constant time adjacency queries. Moreover, all the labels can be constructed in $O(n)$ time. We observe that the recursive scheme of [AR02] for general trees does not simplify for caterpillars or binary trees. The worst-case label length remains $\log n + O(\log^* n)$ and the adjacency query time $\Omega(\log^* n)$.

As far as we know, this is the first $\log n + O(1)$ bit adjacency labeling supporting constant query time for a family of trees including trees with an arbitrary numbers of arbitrary degree vertices (caterpillars). The technique, called *Traversal & Jumping*, is interesting on its own, and we believe that it might be extended to larger families of graphs, and to other queries.

1.3   Outline of Techniques

To introduce our labeling technique, let us consider an n-vertex caterpillar whose path is $x_1, \ldots, x_k$ and where the j-th leaf of x_i is denoted by $y_{i,j}$.

The first naive approach consists in labeling each vertex x_i with the pair $(i, 0)$ and $y_{i,j}$ with $(i, 1)$ but this scheme does not respect the uniqueness condition for the labels. A correct scheme can be obtained by using labels (i, j) for $y_{i,j}$. The

---

[3] $f(n) = \omega(g(n))$ if and only if $g(n) = o(f(n))$.

adjacency test is then trivial. This labeling, which is a variant of the tree labeling scheme presented above, is not efficient since every pair of nonnegative integers (i, j) with $i + j \leqslant n$ will be assigned by the scheme to some caterpillars. There are at least $(n/2)^2$ such pairs, yielding some labels of at least $2 \log n - O(1)$ bits.

A second less trivial labeling (with distinct labels) assigns to each vertex $y_{i,j}$ the pair (r_i, j) where r_i is the rank of the number of leaves of x_i, so that less bits are used for r_i if x_i has many leaves, leaving room for the index j. Because $j \leqslant n/r_i$, the label length reduced now to $\lceil \log r_i \rceil + \lceil \log(n/r_i) \rceil \leqslant \log n + O(1)$. However, the fields of the pair (r_i, j) have variable length, so $\log \log \min\{r_i, j\} + O(1)$ bits are required to code the position of the two values of the pair. Moreover, this scheme does not give adjacency for two nodes in the path. Anyway, as all possible pairs (r_i, j) can occur, this required an extra information of $\log \log \sqrt{n}$ bits (in the worst-case $\min\{r_i, j\} \geqslant \sqrt{n}$), yielding labels of length at least $\log n + \log \log n - O(1)$.

A third solution is to apply some recursive decomposition, as in [AR02]. However, any decomposition in a non-constant levels produces labels with a non-constant number of fields, yielding a label length of $\log n + \omega(1)$ bits, furthermore with $\omega(1)$ adjacency query time. Labelings with $\log n + O(1)$ bits require new ideas.

Roughly speaking, the Traversal & Jumping technique consists in:

1. Selecting a suitable traversal of the tree (or of the graph);
2. Associating with each vertex x some information $C(x)$;
3. Performing the traversal and assign the labels with increasing but non necessarily consecutive numbers to the vertices.

Intuitively, the adjacency test between x and y is done on the basis of $C(x)$ and $C(y)$. Actually, the jumps achieved in Step 3 are done by selecting an interval associated with each vertex in which its label must be. It is important to note that the intervals are ordered in the same way as the corresponding vertices in the traversal. Moreover, all vertex intervals must be disjoint. The position of the label of x in its interval is tuned in order to encode $C(x)$ in the label in a self-extracting way. In general, the information $C(x)$ determines the intervals length of all the neighbours of x which are after in the traversal.

The main difficulty is to design the minimal information $C(x)$ and to tune the jumps, i.e., the interval length. The maximum label length is simply determined by the value of the last label assigned during the traversal.

This technique fundamentally differs from previous schemes, in which a label is essentially viewed as a unique prelabel of $\lceil \log n \rceil$ bits plus some small extra fields, inevitably leading to labels of $\log n + \omega(1)$ bits. On the contrary, Traversal & Jumping abandons this representation, and uses the full range of values $[0, O(n)]$ to get labels of length $\log n + O(1)$.

Section 2 presents the scheme for caterpillars, and Section 3 for binary trees. We propose further works in Section 4.

1.4   Preliminaries

We assume a RAM model of computation with $\Omega(\log n)$-bit words. In this model, standard arithmetic operations on words of $O(\log n)$ bits can be done in constant time. These include additions, comparisons, binary masks, shifting, MSB and LSB (returning respectively the position of the most and least significant bit of a word).

Given a binary string A, we denote by $|A|$ its length, and for a binary string B, $A \circ B$ denotes the concatenation of A followed by B.

Given an $x \in \mathbb{N}$, we denote by $\lg x = \log \max \{x, 1\}$, and by $\mathrm{bin}(x)$ its standard binary representation. We have $|\mathrm{bin}(x)| = \lfloor \lg x \rfloor + 1$. We denote by $\mathrm{val}(w)$ the integer x such that $w = \mathrm{bin}(x)$. When it is clear from the context, we confuse w and $\mathrm{val}(w)$. We also denote by $\lceil x \rceil_2 = 2^{\lceil \lg x \rceil}$.

A *code* is a set of words, and a code is *suffix-free* if no words of the code is the ending of another one. A basic property of suffix-free codes it that they can be composed, by the concatenation of a fixed number of fields, to form new suffix-free codes. A simple suffix-free code is defined by $\mathrm{code}_0(x) = 1 \circ 0^x$, where 0^x is the binary string composed of x zeros. This code extends to more succinct codes defined recursively by $\mathrm{code}_{i+1}(x) = \mathrm{bin}(x) \circ \mathrm{code}_i(|\mathrm{bin}(x)| - 1)$ for every $i \geqslant 0$. It is easy to check that, for every $i \geqslant 0$, code_i is suffix-free. E.g., $\mathrm{code}_0(5) = 100000$, $\mathrm{code}_1(5) = 101\,100$, and $\mathrm{code}_2(5) = 101\,10\,10$.

If a word w has $\mathrm{code}_i(x)$ as suffix, then x can be extracted from w in $O(i)$ time (in particular with the use of LSB to extract the length of code_0). In the sequel, any integer sequence $x_1, \ldots, x_k$ can be stored as a suffix $\mathrm{code}_i(x_1) \circ \cdots \circ \mathrm{code}_i(x_k)$, and can be extracted in $O(ik)$ time.

In this paper, we will essentially use code_i for $i \in \{0, 1, 2\}$. We check that for every $x \in \mathbb{N}$, $|\mathrm{code}_0(x)| = x + 1$, $|\mathrm{code}_1(x)| = 2 \lfloor \lg x \rfloor + 2$, and $|\mathrm{code}_2(x)| = \lfloor \lg x \rfloor + 2 \lfloor \lg \lfloor \lg x \rfloor \rfloor + 3$.

Claim. Let w be a word, and z an integer. One can compute in constant time an integer $x \in [z, z + 2^{|w|})$ such that w is a suffix of $\mathrm{bin}(x)$.

Proof. Observe that, for all strings A and B, $\mathrm{val}(A \circ B) = \mathrm{val}(A) \cdot 2^{|B|} + \mathrm{val}(B)$, and that $\mathrm{val}(A) < 2^{|A|}$.

Let $u = \lfloor z/2^{|w|} \rfloor$ and $v = z \bmod 2^{|w|}$, so that $z = u \cdot 2^{|w|} + v$. Set $b = 0$ if $\mathrm{val}(w) \geqslant v$, and set $b = 1$ otherwise. The integer x is defined by $\mathrm{bin}(x) = \mathrm{bin}(u + b) \circ w$, that clearly contains w as suffix. Note that x can be computed in constant time using shifts, masks and MSB (in particular MSB is used to compute $|w|$ from w).

It remains to check that $x \in [z, z + 2^{|w|})$. We have $x = (u + b) \cdot 2^{|w|} + \mathrm{val}(w) = z - v + b \cdot 2^{|w|} + \mathrm{val}(w)$.

If $b = 0$, then $x = z - v + \mathrm{val}(w) \leqslant z + \mathrm{val}(w) < z + 2^{|w|}$. For $b = 0$, $\mathrm{val}(w) \geqslant v$, thus $x \geqslant z$.

If $b = 1$, then $x = z - v + 2^{|w|} + \mathrm{val}(w) > z + \mathrm{val}(w) \geqslant z$ since $v < 2^{|w|}$. For $b = 1$, $\mathrm{val}(w) < v$, thus $z - v + 2^{|w|} + \mathrm{val}(w) < z + 2^{|w|}$.

2   Caterpillars

A leaf is a vertex of degree one, and an inner vertex is a nonleaf vertex. A tree is a *caterpillar* if the subgraph induced by its inner vertices is a path.

Theorem 1. *The family of caterpillars with n vertices enjoys an adjacency labeling scheme with labels of length at most $\lceil \log n \rceil + 6$ bits, supporting constant time adjacency query. Moreover, all the labels can be constructed in $O(n)$ time.*

2.1   Description of the Labeling Scheme

Consider a caterpillar G of n vertices. We denote by $X = \{x_1, \ldots, x_k\}$ the inner vertices of G (ordered along the path). For every i, let $Y_i = \{y_{i,1}, \ldots, y_{i,d_i}\}$ be the set of leaves attached to x_i, with $d_i = 0$ if $Y_i = \varnothing$.

The traversal used in our scheme is a prefix traversal of the caterpillar rooted at x_1 where the vertices of Y_i are traversed before the vertex x_{i+1}. According to this traversal, the inner vertex x_i stores necessary information to determine the adjacency with the vertices of $Y_i \cup \{x_{i+1}\}$. The leaves do not store any specific information in their label.

With each inner vertex x_i, we associate an interval of length p_i, for some suitable integer p_i, in which its label $\ell(x_i)$ must be. For some technical reasons, impose that $p_i = 2^{t_i+3}$ with t_i is an integer $\geqslant 0$. With the set of the labels of Y_i we associate an interval of same length: $(\ell(x_i), \ell(x_i) + p_i]$. In this interval $\ell(y_{i,j}) = \ell(x_i) + j$. Finally, the interval associated with vertex x_{i+1} is $(\ell(x_i) + p_i, \ell(x_i) + p_i + p_{i+1}]$.

The information encoded by x_i is the ordered pair (t_i, t_{i+1}). To encode this information, we propose the following suffix-free code:

$$C(x_i) = \mathrm{code}_0(t_i + 3 - |\mathrm{code}_1(t_{i+1})|) \circ \mathrm{code}_1(t_{i+1}) .$$

Three conditions on p_i (and so on t_i) have to be satisfied to ensure that the code is valid. The value p_i must be large enough to encode the information, large enough so that all the labels of the vertices of Y_i can be placed in the interval $(\ell(x_i), \ell(x_i) + p_i]$, and $p_i \geqslant 8$. The following relation ensures such conditions:

$$t_i = \max\{|\mathrm{code}_1(t_{i+1})| - 3, \lceil \lg d_i \rceil - 3, 0\}, \text{ with } t_{k+1} = 0 .$$

So, given $\ell(x_i)$, $\ell(x_{i+1})$ is computed applying Claim 1.4 with $w = C(x_{i+1})$ and $z = \ell(x_i) + p_i$.

One can remark that the value of t_i depends on the value of t_{i+1}. The computation of the labels can be done with two traversals of the caterpillar. The value of the t_i is computed from a traversal of the path from x_k to x_1. A second traversal (a prefix one starting from x_1) computes the labels of the vertices. Each traversal takes $O(n)$ time. Finally, an additional bit is added to the labels to determine if the vertex is an inner vertex: $\ell'(x_i) = 1 \circ \ell(x_i)$ and $\ell'(y_{i,j}) = 0 \circ \ell(y_{i,j})$.

2.2   Adjacency Test

Lemma 1. *For every pair of vertices u and v, the adjacency between u and v can be computed in constant time from $\ell'(u)$ and $\ell'(v)$.*

Proof. Looking at the first bit of $\ell'(u)$ and $\ell'(v)$ one can check whether u and v belongs to X or to $Y = \bigcup_{i=1}^{i=k} Y_i$. Because two leaves cannot be adjacent, let us assume that $u \in X$ with $u = x_i$.

In constant time, we can compute $\ell(x_i)$, t_i and t_{i+1} from $\ell'(x_i)$, decoding $C(x_i)$. Recall that p_i and p_{i+1} can be directly deduced from t_i and t_{i+1}. There are two cases to consider:

- Case 1: $v \in Y$ (the first bit of the label is 0). By construction, the labels of vertices of Y_i and only these belong to the interval $(\ell(x_i), \ell(x_i) + p_i]$. Since the length of the labels is $O(\log n)$ (see Lemma 2), this test can be performed in constant time.
- Case 2: $v \in X$ (the first bit of the label is 1). Let $v = x_j$, and w.l.o.g. assume that $j > i$ (we simply check whether $\ell'(v) > \ell'(u)$).
  By construction if $j = i+1$, then $\ell(x_i) + p_i < \ell(x_j) \leqslant \ell(x_i) + p_i + p_{i+1}$. This interval may contain other labels (labels of vertices of Y_{i+1}), but the only label of inner vertex if $\ell(x_{i+1})$. This test can also be performed in constant time.

2.3 Label Length

Lemma 2. *The length of the labels is at most $\lceil \log n \rceil + 6$.*

Proof. First, let us show by induction the following property (P_m):

$$(P_m): \quad \sum_{i=m}^{k} p_i \leqslant 8 \left(\sum_{i=m}^{k} \lceil d_i + 1 \rceil_2 \right) - p_m .$$

(P_k) is true since $d_k > 0$ and $\sum_{i=k}^{i=k} p_i = p_k = \max\{8, \lceil d_k + 1 \rceil_2\} \leqslant 8 \sum_{i=k}^{k} \lceil d_i + 1 \rceil_2 - p_k$. Assume that (P_m) is true for some $m \in [2, k]$, and let us show (P_{m-1}):

Applying the induction hypothesis:

$$\sum_{i=m-1}^{k} p_i \leqslant 8 \left(\sum_{i=m}^{k} \lceil d_i + 1 \rceil_2 \right) - p_m + p_{m-1} .$$

There are three cases to consider:

- Case 1: $\lceil d_{m-1} \rceil_2 \leqslant 8$ and $2^{|\mathrm{code}_1(t_m)|} \leqslant 8 \Rightarrow p_{m-1} = 8$.

$$\sum_{i=m-1}^{k} p_i \leqslant 8 \left(\sum_{i=m}^{k} \lceil d_i + 1 \rceil_2 \right) - p_m + 8$$

Since $\lceil d_{m-1} + 1 \rceil_2 \geqslant 1$:

$$\sum_{i=m-1}^{k} p_i \leqslant 8 \left(\sum_{i=m-1}^{k} \lceil d_i + 1 \rceil_2 \right) - p_m$$

Since $p_m \geqslant p_{m-1}$:

$$\sum_{i=m-1}^{k} p_i \leqslant 8 \left(\sum_{i=m-1}^{k} \lceil d_i + 1 \rceil_2 \right) - p_{m-1}$$

- Case 2: $\lceil d_{m-1} \rceil_2 \geqslant 2^{|code_1(t_m)|} \Rightarrow p_{m-1} = \lceil d_{m-1} \rceil_2$.

$$\sum_{i=m-1}^{k} p_i \leqslant 8 \left(\sum_{i=m}^{k} \lceil d_i + 1 \rceil_2 \right) - p_k + \lceil d_{m-1} \rceil_2$$

$$\leqslant 8 \left(\sum_{i=m-1}^{k} \lceil d_i + 1 \rceil_2 \right) - p_k - 7 \lceil d_{m-1} + 1 \rceil_2$$

$$\leqslant 8 \left(\sum_{i=m-1}^{k} \lceil d_i + 1 \rceil_2 \right) - p_{m-1}$$

- Case 3: $\lceil d_{m-1} \rceil_2 < 2^{|code_1(t_m)|}$ and $8 < 2^{|code_1(t_m)|} \Rightarrow p_{m-1} = 2^{|code_1(t_m)|}$.

$$\sum_{i=m-1}^{k} p_i \leqslant 8 \left(\sum_{i=m}^{k} \lceil d_i + 1 \rceil_2 \right) - p_m + 2^{|code_1(t_m)|} \ .$$

Since $p_{m-1} = 2^{|code_1(t_m)|} = 2^{2 \lceil \log((\log(p_m)-3)+1) \rceil} \leqslant \frac{1}{2} p_m$:

$$\sum_{i=m-1}^{k} p_i \leqslant 8 \left(\sum_{i=m-1}^{k} \lceil d_i + 1 \rceil_2 \right) - \frac{1}{2} p_m$$

$$\leqslant 8 \left(\sum_{i=m-1}^{k} \lceil d_i + 1 \rceil_2 \right) - p_{m-1} \ .$$

So (P_m) is true for any positive $m \leqslant k$. Hence:

$$\sum_{i=1}^{k} p_i \leqslant 8 \left(\sum_{i=1}^{k} \lceil d_i + 1 \rceil_2 \right) \ .$$

The maximum label length is determined by the label of the last leaf of x_k, say $y_{k,j}$. We can bound $\ell(y_{k,j})$ by:

$$\ell(y_{k,j}) \leqslant 2 \sum_{i=1}^{k} p_i \leqslant 2^4 \sum_{i=1}^{k} \lceil d_i + 1 \rceil_2 \leqslant 2^5 \sum_{i=1}^{k} (d_i + 1) \leqslant 2^5 n \ .$$

The length labels $\ell(y_{k,j})$ is at most $\lceil \log n \rceil + 5$. The effective labels, $\ell'(v)$, use one more bit. So the label length is at most $\lceil \log n \rceil + 6$.

3   Binary Trees

Theorem 2. *The family of binary trees with n vertices enjoys an adjacency labeling scheme with labels of length at most $\log n + O(1)$ bits, supporting constant time adjacency query. Moreover, all the labels can be constructed in $O(n)$ time.*

3.1   Description of the Labeling Scheme

Let T be a rooted binary tree with n vertices. For any vertex v, let T_v denote the subtree of T rooted at v. Let r_T be the root of T. We denote by v^- and v^+ respectively the left and the right child of v (if exist). We assume that the children of every inner vertex v are ordered such that the weight of v^- is at most the weight of v^+, i.e., $|V(T_{v^-})| \leqslant |V(T_{v^+})|$. For the shake of the proof, we assume that v^- exists for every inner vertex v, possibly by completing the tree with some extra vertices. Note that this at most double the size of the tree.

The traversal considered in our scheme is a prefix traversal of T in which v^- is visited before v^+, for every inner vertex v. Let $s(v)$ be the length of the interval assigned to v, and let $p(v)$ be the length of the interval of values assigned to the labels of vertices of T_v (see Fig. 1). In the scheme, the interval associate with v is at the beginning of the interval of T_v (on Fig. 2 arrows $s(v)$ and $p(v)$ are aligned on the left). The interval of T_{v^-} is at distance $s(v)$ from $\ell(v)$ (i.e., the difference of the left boundaries of the intervals is $s(v)$). The interval of T_{v^+} begins at distance $s(v) + q(v^-)$ from $\ell(v)$ (cf. Fig. 2), for a suitable length $q(v^-)$.

In addition, our scheme imposes that $s(v)$ is a power of 2, and that $q(v^-)$ is a square. More precisely, $s(v) = 2^{m(v)}$ for some integer $m(v) \geqslant 0$, and $q(v^-) = \lceil \sqrt{p(v^-)} \rceil^2$. Observe that for every v, $q(v) = p(v) + O(\sqrt{p(v)})$, and that $q(v)$ can be encoded with half many bits than for $p(v)$.

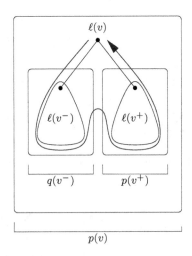

Fig. 1. Traversal of the tree

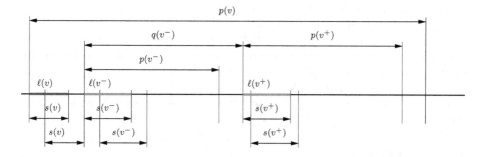

Fig. 2. Description of the labeling

To compute the adjacency with its children, vertex v stores a single bit 0 if it is a leaf, and the quadruple $(s(v^-), s(v^+), q(v^-), 1)$ if it is inner. To encode this information, we propose the following suffix-free code:

$$C(v) = \begin{cases} 0 & \text{if } v \text{ is a leaf} \\ \text{code}_1(m(v^-)) \circ \text{code}_1(m(v^+)) \circ \text{code}_2(\sqrt{q(v^-)}) \circ 1 & \text{otherwise} \end{cases}$$

We set $s(v) = 2^{|C(v)|-1}$, i.e., $m(v) = |C(v)| - 1$. To compute the labels we need first to compute $s(v)$ and $p(v)$ for each vertex v of T. This is done in linear time with a postfix traversal considering the recursive relation:

$$p(v) = \begin{cases} 2 & \text{if } v \text{ is a leaf} \\ q(v^-) + p(v^+) + 2s(v) & \text{otherwise} \end{cases}$$

Then, the labels can be computed in linear time with a traversal of T, and applying Claim 1.4 with $w = C(v)$.

3.2   Adjacency Test

Lemma 3. *Let any pair of vertices v and u, the adjacency of v and u can be computed in constant time from $\ell(v)$ and $\ell(u)$.*

Proof. W.l.o.g., we can consider that $\ell(v) < \ell(u)$. To test adjacency between v and u, we use the following conditions:
  v and u are adjacent if and only if y is inner and:

  – either $\ell(u) \in [\ell(v) + s(v), \ell(v) + s(v) + s(v^-))$ (in this case $u = v^-$);
  – or $\ell(u) \in [\ell(v) + s(v) + q(v^-), \ell(v) + s(v) + q(v^-) + s(v^+))$ (in this case $u = v^+$)

We can remark that this test can be computed in constant time from the labels of the vertices. Indeed, $s(v)$, $s(v^-)$, $s(v^+)$, and $q(v^-)$ can be extracted in constant time from $\ell(u)$.

It remains to prove the validity of this test. If v is a leaf (and $\ell(v) < \ell(u)$), v and u cannot be adjacent. To check it, it suffices to extract the last bit of

$\ell(v)$. Now assume v is inner. By construction, if $u = v^-$ then $\ell(u) \in [\ell(v) + s(v), \ell(v) + s(v) + s(v^-))$. In the same way, if $u = v^+$ then $\ell(u) \in [\ell(v) + s(v) + q(v^-), \ell(v) + s(v) + q(v^-) + s(v^+))$. Moreover, $\ell(v^-)$ is the only label in $[\ell(v) + s(v), \ell(v) + s(v) + s(v^-))$ because, in the construction we make a jump from $\ell(v)$ to $\ell(v^-)$ of length $s(v)$ and we make another jump from $\ell(v^-)$ to $\ell(v^{--})$ (if exists) of length at least $s(v^-)$.

With the same argument, we prove that $\ell(v^+)$ is the only label in $[\ell(v) + s(v) + q(v^-), \ell(v) + s(v) + q(v^-) + s(v^+))$.

3.3 Label Length

Lemma 4. *The length of the labels is at most $\log n + O(1)$.*

Proof. Let B_n be the family of all binary trees of at most n vertices such that every leaf has a sibling (i.e., v^- and v^+ exist for every inner vertex v).

For every tree $T \in B_n$ with $n \geqslant 3$:

$$p(r_T) = q(r_T{}^-) + p(r_T{}^+) + 2s(r_T) \quad \text{where}$$
$$\log s(r_T) = |C(r_T)| - 1 = |\text{code}_1(m(r_T{}^-))| + |\text{code}_1(m(r_T{}^+))|$$
$$+ |\text{code}_2(\lceil \sqrt{p(r_T{}^-)} \rceil)|$$
$$= |\text{code}_1(\log s(r_T{}^-))| + |\text{code}_1(\log s(r_T{}^+))| + |\text{code}_2(\lceil \sqrt{p(r_T{}^-)} \rceil)|$$

Let $P(n) = \max_{T \in B_n} p(r_T)$, and $S(n) = \max_{T \in B_n} s(r_T)$. Because every label assigned to T ranges in $[0, P(n))$, our goal is to upper bound $P(n)$ by $O(n)$.

Let $i = |V(T_{r_T^-})|$ be the weight of the left subtree. We have $p(r_T{}^-) \leqslant P(i)$ and $p(r_T{}^+) \leqslant P(n-1-i)$. Similarly, $s(r_T{}^-) \leqslant S(i)$ and $s(r_T{}^+) \leqslant S(n-1-i)$. By the ordering of the children, observe that $i \leqslant n-1-i$, i.e., i can only range in $I = \{1, \ldots, \lfloor (n-1)/2 \rfloor\}$.

From previous equations we derive (recall that $q(v^-) = \lceil \sqrt{p(v^-)} \rceil^2$):

$$P(n) = \max_{i \in I} \left\{ \lceil \sqrt{P(i)} \rceil^2 + P(n-i-1) + 2S \right\} \quad \text{where}$$

$$\log S = |\text{code}_1(\log S(i))| + |\text{code}_1(\log S(n-i-1))| + |\text{code}_2(\lceil \sqrt{P(i)} \rceil)|$$

with $P(1) = 2$ and $S(1) = 1$. Note that for $x \in \mathbb{N}$, $2^{\lg x} \leqslant x + 1$. The two first terms of $\log S$ can be bounded by:

$$|\text{code}_1(\log S(i))| = 2 \lfloor \lg \log S(i) \rfloor + 2, \text{ and thus}$$
$$2^{|\text{code}_1(\log S(i))|} \leqslant 4 \log^2 S(i) + 4, \text{ and similarly}$$
$$2^{|\text{code}_1(\log S(n-i-1))|} \leqslant 4 \log^2 S(n-i-1) + 4.$$

By construction, $p(v) \geqslant 2s(v)$ for every v, thus $S(n) \leqslant P(n)/2$ for every n. So, bounding $(x-1)^2 \leqslant x^2 - 1$, we obtain:

$$2^{|\text{code}_1(\log S(i))|} \leqslant 4 \log^2 P(i) \quad \text{and} \quad 2^{|\text{code}_1(\log S(n-i-1))|} \leqslant 4 \log^2 P(n-i-1).$$

Let $u = \sqrt{P(i)}$. We have $u \geqslant 1$. The third term of $\log S$ can be bounded by:

$$|\mathrm{code}_2(\lceil u \rceil)| = \lfloor \lg \lceil u \rceil \rfloor + 2 \lfloor \lg \lfloor \lg \lceil u \rceil \rfloor \rfloor + 3, \text{ and thus}$$
$$2^{|\mathrm{code}_2(\lceil u \rceil)|} \leqslant 8 \cdot (\lceil u \rceil + 1) \cdot (\lfloor \lg \lceil u \rceil \rfloor^2 + 1) \leqslant 16u(\log^2 u + 2).$$

Therefore,

$$S \leqslant 4 \cdot \log^2 P(i) \cdot \log^2 P(n - i - 1)) \cdot 16\sqrt{P(i)} \cdot (\log^2 \sqrt{P(i)} + 2)$$
$$\leqslant 16\sqrt{P(i)} \cdot (\log^2 P(i) + 8) \cdot \log^2 P(i) \cdot \log^2 P(n - i - 1)$$

One can check that for $x \geqslant 1$, $\lceil \sqrt{x} \rceil^2 \leqslant x + 2\sqrt{x}$. Hence:

$$P(n) \leqslant \max_{i \in I} \left\{ P(i) + 2\sqrt{P(i)} + P(n - i - 1) + 2S \right\}$$
$$\leqslant \max_{i \in I} \left\{ P(i) + P(n - i - 1) + 34\sqrt{P(i)} \log^4 P(i) \log^2 P(n - i - 1) \right\}$$

In particular, we deduce that $\exists \alpha, \beta, \gamma, \delta \in \mathbb{R}^+, \delta < 1/2 < \gamma < 1$ and $\delta + \gamma < 1$ such that:

$$P(n) \leqslant \max_{i \in I} \left\{ P(i) + P(n - i - 1) + \alpha P(i)^\gamma P(n - i - 1)^\delta + \beta \right\}.$$

The following claim shows that $P(n) = O(n)$, and so the label length is $\log n + O(1)$.

Claim. Let $P(n)$ be a sequence. If there are $\alpha, \beta, \gamma, \delta \in \mathbb{R}^+, \delta < \gamma < 1, \delta + \gamma < 1$ such that $P(n) \leqslant \max_{i \in I} \left\{ P(i) + P(n - i - 1) + \alpha P(i)^\gamma P(n - i - 1)^\delta + \beta \right\}$ and $P(1) > 0$, then $P(n) = O(n)$.

Proof. Let a and b be two positive constants we will determine later. Let us prove by induction the property (Q_n):

$$(Q_n): \quad P(n) \leqslant an - bn^{\gamma + \delta}.$$

Q_1 is true if and only if a and b satisfy $P(1) \leqslant a - b$. Assume that Q_i is true for $i < n$.

$$P(n) \leqslant \max_{i \in I} \left\{ an - bi^{\gamma + \delta} - b(n - i - 1)^{\gamma + \delta} + \alpha(an)^\gamma (a(n - i - 1))^\delta - (a - \beta) \right\}$$
$$\leqslant an - (a - \beta) + \max_{i \in I} \left\{ h(n, i) + f(n, i) \right\}$$

with $h(n, i) = -bi^{\gamma + \delta} - b(n - i - 1)^{\gamma + \delta}$ and $f(n, i) = a^{\gamma + \delta} n^\gamma (n - i - 1)^\delta$.

In order to bound $\max_{i \in I} h(n, i)$, we compute:

$$\frac{\partial}{\partial i} h(n, i) = bi^{\gamma + \delta}(n - i - 1)^{\gamma + \delta} \left(\frac{i^{1 - \gamma - \delta} - (n - i - 1)^{1 - \gamma - \delta}}{i(n - i - 1)} \right)$$

For $i \in I$, $\frac{\partial}{\partial i} h(n, i) \leqslant 0$ because $i \leqslant n - i - 1$ and $\gamma > \delta$. So, we obtain:

$$\max_{i \in I} h(n, i) \leqslant -bn^{\gamma + \delta} \leqslant -2^{\gamma + \delta} b \left(\frac{n}{2} \right)^{\gamma + \delta}$$

In order to bound $\max_{i \in I} f(n, i)$, we compute:

$$\frac{\partial}{\partial i} f(n, i) \;=\; a^{\gamma+\delta} i^{\gamma} (n - i - 1)^{\delta} \left(\frac{\gamma(n - i - 1) - \delta i}{i(n - i - 1)} \right).$$

For $i \in I$, $\frac{\partial}{\partial i} f(n, i) \geqslant 0$ because $i \leqslant n - i - 1$ and $\gamma > \delta$. So:

$$\max_{i \in I} f(n, i) \;\leqslant\; a^{\gamma+\delta} \left(\frac{n}{2} \right)^{\gamma+\delta}$$

and thus,

$$P(n) \;\leqslant\; an - \left(2^{\gamma+\delta} b - a^{\gamma+\delta} \right) n^{\gamma+\delta} - (a - \beta).$$

The two constants must fulfill the following equalities:

$$\begin{cases} P(1) \leqslant a - b \\ 2^{\gamma+\delta} b - a^{\gamma+\delta} \alpha \geqslant b \\ a - \beta \geqslant 0 \end{cases}$$

For instance, it suffices to choose b such that:

$$a - P(1) \;\geqslant\; b \;\geqslant\; a^{\gamma+\delta} \alpha \quad \text{with } a \geqslant \beta$$

which is possible for a large enough since $\gamma + \delta < 1$.

This completes the proof of Lemma 4.

4   Conclusion

The unsolved *implicit graph representation conjecture* of [KNR88, KNR92] asks whether every hereditary[4] family of graphs with $2^{O(n \log n)}$ labeled graphs of n vertices enjoys a $O(\log n)$-bit adjacency labeling scheme. This is motivated by the fact that every family with at least $2^{cn \log n}$ labeled graphs of n vertices requires adjacency labels of at least $c \log n$ bits.

Our schemes suggest that, at least for trees, labels of $\log n + O(1)$ bits may be possible. Therefore, we propose to prove or to disprove the following:

Every hereditary family of graphs with at most $n! 2^{O(n)} = 2^{n \log n + O(n)}$ labeled graphs of n vertices enjoys an adjacency labeling scheme with labels of $\log n + O(1)$ bits.

We observe that several well-known families of graphs are concerned by this proposition: trees, planar graphs, bounded treewidth graphs, graphs of bounded genus, graphs excluding a fixed minor (cf. [NRTW05] for counting such graphs). Proving the latter conjecture appears to be hard, e.g., the best upper bound for planar graphs is only $3 \log n + O(\log^* n)$.

---

[4] That is a family of graphs closed under induced subgraph taking.

References

[AAK⁺05]   Serge Abiteboul, Stephen Alstrup, Haim Kaplan, Tova Milo, and Theis Rauhe. Compact labeling schemes for ancestor queries. *SIAM Journal on Computing*, 2005.

[ABR05]   Stephen Alstrup, Philip Bille, and Theis Rauhe. Labeling schemes for small distances in trees. *SIAM Journal on Discrete Mathematics*, 19(2):448–462, 2005.

[AGKR04]   Stephen Alstrup, Cyril Gavoille, Haim Kaplan, and Theis Rauhe. Nearest common ancestors: A survey and a new algorithm for a distributed environment. *Theory of Computing Systems*, 37:441–456, 2004.

[AKM01]   Serge Abiteboul, Haim Kaplan, and Tova Milo. Compact labeling schemes for ancestor queries. In 12th *Symposium on Discrete Algorithms (SODA)*, pages 547–556. ACM-SIAM, January 2001.

[AR02]   Stephen Alstrup and Theis Rauhe. Small induced-universal graphs and compact implicit graph representations. In 43rd *Annual IEEE Symposium on Foundations of Computer Science (FOCS)*, pages 53–62. IEEE Computer Society Press, November 2002.

[BF67]   Melvin A. Breuer and Jon Folkman. An unexpected result on coding the vertices of a graph. *Journal of Mathematical Analysis and Applications*, 20:583–600, 1967.

[BG05]   Fabrice Bazzaro and Cyril Gavoille. Localized and compact data-structure for comparability graphs. In 16th *Annual International Symposium on Algorithms and Computation (ISAAC)*, volume 3827 of Lecture Notes in Computer Science, pages 1122–1131. Springer, December 2005.

[Bre66]   Melvin A. Breuer. Coding the vertexes of a graph. *IEEE Transactions on Information Theory*, IT-12:148–153, 1966.

[DL02]   Feodor F. Dragan and Irina Lomonosov. New routing schemes for interval graphs, circular-arc graphs, and permutation graphs. In 14th *IASTED International Conference on Parallel and Distributed Computing and Systems (PDCS)*, pages 78–83, November 2002.

[DL04]   Feodor F. Dragan and Irina Lomonosov. On compact and efficient routing in certain graph classes. In 15th *Annual International Symposium on Algorithms and Computation (ISAAC)*, volume 3341 of Lecture Notes in Computer Science, pages 402–414. Springer, December 2004.

[FG01]   Pierre Fraigniaud and Cyril Gavoille. Routing in trees. In Fernando Orejas, Paul G. Spirakis, and Jan van Leeuwen, editors, 28th *International Colloquium on Automata, Languages and Programming (ICALP)*, volume 2076 of Lecture Notes in Computer Science, pages 757–772. Springer, July 2001.

[GP03a]   Cyril Gavoille and Christophe Paul. Optimal distance labeling schemes for interval and circular-arc graphs. In G. Di Battista and U. Zwick, editors, 11th *Annual European Symposium on Algorithms (ESA)*, volume 2832 of Lecture Notes in Computer Science, pages 254–265. Springer, September 2003.

[GP03b]   Cyril Gavoille and David Peleg. Compact and localized distributed data structures. *Journal of Distributed Computing*, 16:111–120, May 2003. PODC 20-Year Special Issue.

[KM01]   Haim Kaplan and Tova Milo. Short and simple labels for small distances and other functions. In 7th *International Workshop on Algorithms and Data Structures (WADS)*, volume 2125 of Lecture Notes in Computer Science, pages 32–40. Springer, August 2001.

[KNR88]   Sampath Kannan, Moni Naor, and Steven Rudich. Implicit representation of graphs. In 20th *Annual ACM Symposium on Theory of Computing (STOC)*, pages 334–343. ACM Press, May 1988.

[KNR92]   Sampath Kannan, Moni Naor, and Steven Rudich. Implicit representation of graphs. *SIAM Journal on Discrete Mathematics*, 5:596–603, 1992.

[NRTW05] Serguei Norine, Neil Robertson, Robin Thomas, and Paul Wollan. Proper minor-closed families are small. *Journal of Combinatorial Theory, Series B*, 2005. To appear.

[TZ01]   Mikkel Thorup and Uri Zwick. Compact routing schemes. In 13th *Annual ACM Symposium on Parallel Algorithms and Architectures (SPAA)*, pages 1–10. ACM Press, July 2001.

An Optimal Rebuilding Strategy for a Decremental Tree Problem

Nicolas Thibault and Christian Laforest

Tour Evry 2, LaMI/IBISC, Université d'Evry, 523 place des terrasses,
91000 EVRY France
{nthibaul, laforest}@lami.univ-evry.fr

Abstract. This paper is devoted to the following *decremental* problem. Initially, a graph and a distinguished subset of vertices, called *initial group*, are given. This group is connected by an initial tree. The decremental part of the input is given by an *on-line* sequence of withdrawals of vertices of the initial group, removed on-line one after one. The goal is to keep connected each successive group by a tree, satisfying a *quality* constraint: The maximum distance (called *diameter*) in each constructed tree must be kept in a given range compared to the best possible one. Under this quality constraint, our objective is to minimize the *number of critical stages* of the sequence of constructed trees. We call "critical" a stage where the current tree is rebuilt. We propose a strategy leading to at most $O(\log i)$ critical stages (i is the number of removed members). We also prove that there exist situations where $\Omega(\log i)$ critical stages are necessary to *any algorithm* to maintain the quality constraint. Our strategy is then worst case optimal in order of magnitude.

A lot of works have been devoted to the construction of trees spanning a given set of vertices in a graph. For example the *Steiner tree* problem, where the goal is to span a set (called *group*) of distinguished vertices (called *members*) with a minimum weight tree, has been extensively studied. As the problem is NP-complete, numerous *approximation algorithms* have been designed (see [1, 3] for example). In [8], Waxman was the first author to present the on-line version of this problem in which vertices to add in, or to remove from, the current group *revealed* one by one (see [2] references on on-line problems). In this first paper, he divides the problem into two categories: A model in which "heavy" changes of the current tree are not allowed and a model in which changes are allowed. Then, Imase and Waxman proposed in [4] two different strategies corresponding to the two models above. In the first one the tree is just incremented or decremented and the degradation of weight is evaluated, whereas in the second one they allow changes in the current tree to maintain a guaranty on the weight. At each stage, they prove that they construct with the first strategy a tree whose weight is at a logarithmic ratio compared to the optimal one (i.e. the weight of a Steiner tree of the current group), and that they construct with the second strategy a tree whose weight is at a constant ratio compared to the optimal one. They give for the second strategy an upper bound of $O(\sqrt{i})$ on the number of elementary

P. Flocchini and L. Gąsieniec (Eds.): SIROCCO 2006, LNCS 4056, pp. 157–170, 2006.
© Springer-Verlag Berlin Heidelberg 2006

changes per stage (where i is the number of new members). However, the tree can potentially be changed at each stage; this means that each stage is potentially what we call later a *critical stage*.

In [6], a very similar on-line Steiner tree problem with a delay constraint from one node to the others is studied. But the authors only evaluate their method with simulations, and they give no upper bound for the different competitive ratios. Note that in [4, 6], only the number of elementary changes is taken into account to measure the level of damage due to the allowed changes in the current tree.

In this paper we are concerned with a decremental group problem where the members to remove are revealed on-line one by one. However, we do not focus on the same objective function (the weight of the tree) but on a different measure: The *diameter* of the current group induced by the current tree. Note that we consider here a model in which changes are allowed because it can easily be shown that *any* on-line algorithm without critical stage *cannot* guarantee a constant competitive ratio (for the diameter objective function we consider in this paper). That is why we fix here a "relative budget", called *quality constraint*, on the diameter and we propose an algorithm minimizing the number of critical stages necessary to guarantee this budget constraint at each stage. Note also that we have proved that our algorithm leads to a constant number of elementary changes per stage in average (but we do not give in this paper the definitions and the proof associated to this problem because of the limitation on the number of pages).

A motivation for such model and objective function is the construction of connection structures for groups of members in networks. An important QoS parameter is *latency* that is expressed here in terms of maximum distance between users. This maximum distance must be guaranteed (our *quality constraint*). However, this must be done by minimizing the number of critical stages since they induce perturbations in communications in the current group (implying many re-routing operations between members in the current tree).

In Section 1 we describe more formally our problem. More precisely, in Subsection 1.1 we describe and motivate the constraints (namely the *tree* and *quality* constraints) that must be satisfied at each stage. In Subsection 1.2 we give the definitions of a *critical stage*. In Subsection 2.1 we propose an algorithm satisfying the construction constraints (in Section 2.2). In Subsection 2.3 we prove that our algorithm leads to at most $O(\log i)$ critical stages (where i is the number of removed members). We prove in Section 3 that our strategy is worst case optimal in order of magnitude for the number of critical stages criterion by constructing a scenario in which at least $\Omega(\log i)$ critical stages are necessary for *any* algorithm to satisfy the *quality* constraint. These results show that our algorithm is worst case optimal for the number of critical stages.

1 Definitions and Notations

Let $G = (V, E, w)$ be any connected weighted graph representing a network. V is the set of vertices (modeling the nodes of the network), E the set of edges (modeling the set of physical links) and w a positive weight function of the edges (modeling the length of the edges). We denote by $d_G(u, v)$ the *distance between u and v in G*, i.e. the sum of the weights of the edges of a minimum weight path between u and v in G.

Definition 1 (Diameter of a group M). *Let $G = (V, E, w)$ be a graph and let $M \subseteq V$ be a group. We denote the* diameter of M in G *by*

$$D_G(M) = \max\{d_G(u, v) \; : \; u, v \in M\}.$$

1.1 Construction Constraints

In our problem, the graph $G = (V, E, w)$ and an *initial group* $M_0 \subseteq V$ are given (with $M_0 \neq \varnothing$). For example, in a meeting on network (called net-meeting) this initial group M_0 represents the set of members present at the beginning of the meeting. A structure, noted $T_0 = (V_0, E_0)$, must be created to connect the *members* of M_0 (T_0 spans M_0 in $G : M_0 \subseteq V_0 \subseteq V$, $E_0 \subseteq E$).

However, members may leave the meeting. These members must be removed from the current group (we underline that they are **not** removed from the underlying graph G). Let $m_0 = |M_0|$ be the size of the initial group. Let $u_1, u_2, \ldots, u_i, \ldots$ ($i \leq m_0 - 1$) be the sequence of members to remove. For every i, $1 \leq i \leq m_0 - 1$, we denote by $M_i = M_{i-1} \backslash \{u_i\}$ the ith group, and by $m_i = |M_i|$ its size. Thus, starting from the initial connection structure T_0 for M_0, at each *stage* of withdrawal i, the member u_i is removed by updating the current structure T_{i-1} (spanning M_{i-1}) to obtain T_i spanning M_i.

Note that as the members to remove are revealed one by one, we are in an *on-line* model. It means that we do not know the future: Neither in which order the members are removed, nor what is the set of members to remove. Hence, each stage can potentially be the last one; this explains why we are interested by giving guarantees *at each stage*.

We need the following definition that presents the best possible connection tree for the group M_k, minimizing the diameter parameter.

Definition 2 (Optimal tree). *Let $G = (V, E, w)$ be a graph. For every i, $0 \leq i \leq m_0 - 1$, we denote by T_i^* a tree satisfying*

$$D_{T_i^*}(M_i) = min\{D_T(M_i) : T \text{ tree spanning } M_i\}.$$

We are now ready to give the two constraints that each current structure T_i must satisfy.

- **The *tree* constraint:** For every i, $0 \leq i \leq m_0 - 1$, T_i must be a *tree*, spanning M_i, in which all leaves are in M_i (we call that a *pruned* tree).
- **The *quality* constraint:** Let $c \geq 1$ be any fixed constant representing the *required level of quality*. For every i, $0 \leq i \leq m_0 - 1$, we must have $D_{T_i}(M_i) \leq c \cdot D_{T_i^*}(M_i)$.

As in a net-meeting the current structure T_i is used to support the communications between members of M_i, the *tree constraint* is set in order to simplify the mechanisms of routing and duplication of information in T_i. Indeed, there is only one route between any pair of members in a tree; moreover as there is no cycle, a simple flooding process can be used to broadcast information from any member. This flooding naturally ends at the leaves that are members (because trees are pruned); there is no need of costly process to control it.

The *quality constraint* of level c is set to guarantee that the induced diameter of the current group in T_i is not too large compared to the best possible diameter in T_i^* (at most c times the best possible diameter).

In the rest of the paper we say that an algorithm solves our problem if, for any on-line sequence of successive groups $M_0 \supset \cdots \supset M_i$, it returns a sequence of trees $T_0, \ldots, T_i$ (T_i spanning M_i) satisfying the *tree* and *quality* constraints.

1.2   The Criterion to Minimize

In this subsection we present the cost associated with any algorithm satisfying the tree and quality constraints. We first need the following definitions.

Definition 3 (Critical stage). *Let $\mathscr{A}$ be an algorithm solving our problem. At stage i, $1 \leq i \leq m_0 - 1$, Algorithm $\mathscr{A}$ builds $T_i = (V_i, E_i)$ from $T_{i-1} = (V_{i-1}, E_{i-1})$. Stage i is a* critical stage *if $E_i \nsubseteq E_{i-1}$.*

We distinguish critical stages from other stages since they generate a lot of perturbations. Indeed, if i is a critical stage, the communication routes in T_{i-1} between members already in the current group M_{i-1} have to be changed in T_i. Potentially all the routing tables of the connecting nodes must be modified. This generates a heavy traffic to update them. Moreover the current communications between members of M_{i-1} initiated before the changes may be interrupted. That is why the number of critical stages must be minimized.

On the other hand, the withdrawal of a member by just removing useless branches in the tree generates only local changes and is not considered as a critical stage (since in this case $E_i \subseteq E_{i-1}$). The update of the routing can just be done by broadcasting the information of the departure of the leaving member in the new tree T_i. This does not create any re-routing between the other members. The aim of this paper is to minimize the total number $\sharp CS(T_0, \ldots, T_i)$ of critical stages while respecting the *tree* and the *quality* constraints.

2   Our Algorithm CS

2.1   Definition of Algorithm CS (Critical Stages)

To define Algorithm CS, we need the following algorithm, called MD for Minimum Distance. We denote by MD(M) Algorithm MD applied to group M of size m to find a particular group $M(r^*)$ of size $\lfloor \frac{m}{2} \rfloor + 1$ and what we call its associated *root*.

Algorithm MD(M)

1. For each $r \in M$, sort the m vertices of M by non decreasing value of their distance to r:
$r, u_1^r, \ldots, u_{m-1}^r$ $(d_G(r, u_1^r) \leq d_G(r, u_2^r) \leq \cdots \leq d_G(r, u_{m-1}^r))$.
Let $M(r) = \left\{ r, u_1^r, \ldots, u_{\lfloor \frac{m}{2} \rfloor}^r \right\}$.

2. Return r^* and its associated group $M(r^*)$ such that

$$d_G\left(r^*, u_{\lfloor \frac{m}{2} \rfloor}^{r^*}\right) = \min\left\{ d_G\left(r, u_{\lfloor \frac{m}{2} \rfloor}^r\right) \ : \ r \in M \right\}$$

Note that for all $r \in M$, the vertices $u_1^r, \ldots, u_{m-1}^r$ can be sorted by non decreasing value of $d_G(r, u_k^r)$ and the associated group $M(r)$ can be constructed in polynomial time by using Dijkstra's algorithm. Thus, Algorithm MD(M) finds $M(r^*)$ and its associated root r^* in polynomial time.

The main idea of Algorithm CS is to define particular stages numbers, called *rebuilding stages* during which we (totally) reconstruct the current tree (to match the quality constraint). Between two successive *rebuilding stages*, a member is leaving by just removing the dead branches of the current tree (in order to maintain at each stage a pruned tree to satisfy the *tree* constraint).

The following sequence (a_k) defines the rebuilding stages of our algorithm: $m_{a_0} = m_0$ is the size of the initial group M_0 and for every a_k $(k \geq 1)$, $m_{a_k} = \left\lfloor \frac{m_{a_{k-1}}}{2} \right\rfloor$ is the size of the group M_{a_k}.

Algorithm CS

 - Initially, at stage $a_0 = 0$:
CS builds a shortest path tree spanning the first group M_0, rooted in $r_0 \in M_0$, where $r_0 \in M_0$ is the root found by MD(M_0).
 - After the last rebuilding stage a_k:
Let M_{a_k+j} be the current group and let u_{a_k+j} be the jth member revealed to be removed since the last rebuilding stage a_k.
   • If $m_{a_k+j} > \left\lfloor \frac{m_{a_k}}{2} \right\rfloor$ (corresponding to $j < m_{a_k} - \left\lfloor \frac{m_{a_k}}{2} \right\rfloor$):
Update the tree $T_{a_k+j-1} = (V_{a_k+j-1}, E_{a_k+j-1})$ by pruning potential useless branches.
We obtain the pruned tree $T_{a_k+j} = (V_{a_k+j}, E_{a_k+j})$ spanning M_{a_k+j} satisfying $E_{a_k+j} \subseteq E_{a_k+j-1}$.
   • Otherwise, we have $m_{a_k+j} = \left\lfloor \frac{m_{a_k}}{2} \right\rfloor$ (corresponding to $j = m_{a_k} - \left\lfloor \frac{m_{a_k}}{2} \right\rfloor$):
This is a rebuilding stage and we have $m_{a_k+j} = \left\lfloor \frac{m_{a_k}}{2} \right\rfloor = m_{a_{k+1}}$.
Break the current tree and construct $T_{a_{k+1}}$, a shortest path tree spanning $M_{a_{k+1}}$, rooted in $r_{a_{k+1}}$ (where $r_{a_{k+1}} \in M_{a_{k+1}}$ is the root found by MD($M_{a_{k+1}}$)). Thus, a_{k+1} is the new last rebuilding stage.

The rebuilding stages of CS can be critical stages (because the current tree is broken and rebuilt). The other stages are non critical because the algorithm only removes from the current tree useless branches to obtain the new tree.

Note that this algorithm is polynomial because it uses Algorithm MD (MD is polynomial) and because updating a tree by removing useless branches can be done in polynomial time.

Note also that by construction, at each stage, the *tree* constraint is satisfied. Section 2.2 shows that it also respects the *quality* constraint for a level of quality $c = 4$.

2.2 CS Respects the *Quality* Constraint

Theorem 1 shows that CS respects the *quality* constraint with a level of quality $c = 4$.

Theorem 1. *Let $G = (V, E, w)$ be a graph. For any sequence of withdrawals, at every stage i, $0 \leq i \leq m_0 - 1$ (i is the number of removed members), let T_i^* be an optimal (off-line) tree spanning M_i for the diameter. CS respects the* quality *constraint with a level of quality $c = 4$, i.e. for every i, $0 \leq i \leq m_0 - 1$, we have*

$$D_{T_i}(M_i) \leq 4 D_{T_i^*}(M_i).$$

Proof.

- If i is a stage of rebuilding. In this case, $i = a_k$. Let $u_0, v_0 \in M_{a_k}$ be such that $d_{T_{a_k}}(u_0, v_0) = D_{T_{a_k}}(M_{a_k})$ (where T_{a_k} is the tree spanning M_{a_k} rooted in r^* built by CS at stage a_k). We have

$$\begin{aligned}
D_{T_{a_k}}(M_{a_k}) = d_{T_{a_k}}(u_0, v_0) &\leq d_{T_{a_k}}(u_0, r^*) + d_{T_{a_k}}(r^*, v_0) \\
&\text{(by triangular inequality)} \\
= d_G(u_0, r^*) + d_G(r^*, v_0) &\leq 2 D_G(M_{a_k}) \\
&\text{(because } T_{a_k} \text{ is a shortest path tree rooted} \\
&\text{in } r^* \text{ and because } u_0, v_0, r^* \in M_{a_k}) \\
\leq 2 D_{T_{a_k}^*}(M_{a_k}) &\leq 4 D_{T_{a_k}^*}(M_{a_k}) \\
&\text{(because for every tree } T \text{ spanning} \\
&\text{a group } M, D_G(M) \leq D_T(M))
\end{aligned}$$

- Otherwise a_k is not a stage of rebuilding. Let j, $1 \leq j < m_{a_k} - \lfloor \frac{m_{a_k}}{2} \rfloor$ be the number of removed vertices after the last rebuilding, happening at stage a_k (i.e. j is such that $m_{a_k+j} \geq \lfloor \frac{m_{a_k}}{2} \rfloor + 1$). Let $M(r^*) = \{r^*, u_1^{r^*}, \ldots, u_{\lfloor \frac{m_{a_k}}{2} \rfloor}^{r^*}\}$ be the set returned by MD(M_{a_k}). As $M_{a_k+j} \subset M_{a_k}$ (by definition of the sequence of withdrawals) and $M(r^*) \subseteq M_{a_k}$ with $m_{a_k+j} \geq \lfloor \frac{m_{a_k}}{2} \rfloor + 1$ and $|M(r^*)| = \lfloor \frac{m_{a_k}}{2} \rfloor + 1$, we have $M_{a_k+j} \cap M(r^*) \neq \emptyset$. Thus, there exists $v \in M_{a_k+j} \cap M(r^*)$. As $v \in M(r^*)$, $v = r^*$ or $v = u_l^{r^*}$, with $l \leq \lfloor \frac{m_{a_k}}{2} \rfloor$. As $r^*, u_1^{r^*}, \ldots, u_{\lfloor \frac{m_{a_k}}{2} \rfloor}^{r^*}$ are sorted by non decreasing value of their distance to r^* (see definition of Algorithm MD), we have

$$d_G(r^*, v) \leq d_G\left(r^*, u_{\lfloor \frac{m_{a_k}}{2} \rfloor}^{r^*}\right) \tag{1}$$

Moreover, as Algorithm $\mathsf{MD}(M_{a_k})$ finds r^* and $M(r^*)$ such that $d_G\left(r^*, u^{r^*}_{\lfloor\frac{m_{a_k}}{2}\rfloor}\right) = \min\left\{d_G\left(r, u^r_{\lfloor\frac{m_{a_k}}{2}\rfloor}\right) : r \in M_{a_k}\right\}$, for every $r^0 \in M_{a_k+j} \subset M_{a_k}$ we have

$$d_G\left(r^*, u^{r^*}_{\lfloor\frac{m_{a_k}}{2}\rfloor}\right) \leq d_G\left(r^0, u^{r^0}_{\lfloor\frac{m_{a_k}}{2}\rfloor}\right) \tag{2}$$

As $m_{a_k+j} \geq \lfloor\frac{m_{a_k}}{2}\rfloor + 1$ and as $r^0, u^{r^0}_1, \ldots, u^{r^0}_{m_{a_k}-1}$ are sorted by non decreasing value of their distance to r^0, there exists $u^{r^0}_l \in M_{a_k+j}$ with $\lfloor\frac{m_{a_k}}{2}\rfloor \leq l \leq m_{a_k+j} - 1$ such that

$$d_G\left(r^0, u^{r^0}_{\lfloor\frac{m_{a_k}}{2}\rfloor}\right) \leq d_G\left(r^0, u^{r^0}_l\right) \tag{3}$$

By (1), (2), (3) and as r^0 and $u^{r^0}_l$ are in M_{a_k+j}, by definition of the diameter, we obtain

$$\exists v \in M_{a_k+j} \cap M(r^*) : d_G(r^*, v) \leq D_G(M_{a_k+j}) \tag{4}$$

Let $u^0 \in M_{a_k+j}$ and $v^0 \in M_{a_k+j}$ be such that $d_{T_{a_k+j}}(u^0, v^0) = D_{T_{a_k+j}}(M_{a_k+j})$ (where T_{a_k+j} is the tree spanning M_{a_k+j} built by CS at stage $a_k + j$). We have

$$
\begin{aligned}
D_{T_{a_k+j}}(M_{a_k+j}) = d_{T_{a_k+j}}(u^0, v^0) &= d_{T_{a_k}}(u^0, v^0) \\
&\qquad \text{(since, by definition of Algorithm CS,} \\
&\qquad \text{we have } T_{a_k+j} \subseteq T_{a_k}) \\
&\leq d_{T_{a_k}}(u^0, r^*) + d_{T_{a_k}}(r^*, v^0) \\
&\qquad \text{(by triangular inequality)} \\
&= d_G(u^0, r^*) + d_G(r^*, v^0) \\
&\qquad \text{(because } T_{a_k} \text{ is a shortest path tree rooted in } r^*) \\
&\leq d_G(u^0, v) + d_G(v, r^*) + d_G(r^*, v) + d_G(v, v^0) \\
&\qquad \text{(by triangular inequality, using vertex } v \text{ of (4))} \\
&\leq 4D_G(M_{a_k+j}) \\
&\qquad \text{(because } v \in M_{a_k+j}, u^0 \in M_{a_k+j}, \\
&\qquad v^0 \in M_{a_k+j} \text{ and by (4))} \\
&\leq 4D_{T^*_{a_k+j}}(M_{a_k+j}) \\
&\qquad \text{(because for every tree } T \text{ spanning} \\
&\qquad \text{a group } M, D_G(M) \leq D_T(M))
\end{aligned}
$$

In conclusion, for every i, $0 \leq i \leq m_0 - 1$, we obtain $D_{T_i}(M_i) \leq 4D_{T^*_i}(M_i)$. $\quad\square$

2.3 CS Leads to $O(\log i)$ Critical Stages

Theorem 2. *Let $G = (V, E, w)$ be a graph. For any sequence of withdrawals, let $T_0, \ldots, T_i$ $(0 \le i \le m_0 - 1)$ be the sequence of trees constructed by CS. We have*

$$\sharp CS(T_0, \ldots, T_i) \le \lfloor \log_2(2i) \rfloor = O(\log i)$$

Proof. Two cases may occur:

- If $i < m_0 - \lfloor \frac{m_0}{2} \rfloor$, by definition of CS, there is no rebuilding stage. Thus, $CS(T_0, \ldots, T_i) = 0$.
- Otherwise, $i \ge m_0 - \lfloor \frac{m_0}{2} \rfloor \ge \frac{m_0}{2}$ and we obtain

$$m_0 \le 2i \tag{5}$$

Moreover, by definition of the sequence (a_k) and CS, if there are p rebuildings (that are critical stages), then p is such that

$$m_{a_{p+1}} < m_0 - i \le m_{a_p} \Rightarrow m_0 - i \le \frac{m_0}{2^p} \quad \text{(by definiton of sequence } (a_k)\text{,}$$
$$\forall k, \, m_{a_k} \le \frac{m_0}{2^k})$$

$$\Rightarrow m_0 - i \le \frac{2i}{2^p} \quad \text{(by (5))}$$

$$\Rightarrow 1 \le \frac{2i}{2^p} \quad \text{(by definition, } i \le m_0 - 1)$$

$$\Rightarrow p \le \lfloor \log_2(2i) \rfloor \quad \text{(because } p \text{ is an integer)}$$

$$\Rightarrow p \le O(\log i) \qquad \qquad \square$$

3 Lower Bound for the Number of Critical Stages of Any Algorithm

In this section, we prove that for *any* algorithm respecting the *tree* constraint and the *quality* constraint, for any sufficiently large i, there exists a particular sequence of withdrawals leading to at least $\Omega(\log i)$ critical stages. To prove that, we describe the graph G in Section 3.1. Then, we define the particular on-line sequence of withdrawals in Section 3.2 and prove the main result in Section 3.3.

3.1 Description of the Graph G

Let $k, d, 0 \le k \le d$ and $3 \le p$ be any integer. We define graphs $G_k^p = (V_k^p, E_k^p, w_k^p)$ recursively on k as follows:

- $G_0^p = (V_0^p, E_0^p, w_0^p)$ is the cycle of length p such that $\forall e \in E_0^p, \, w_0^p(e) = 2^d$.
- $\forall k, 1 \le k \le d$, we define $G_k^p = (V_k^p, E_k^p, w_k^p)$ as follows. $\forall v \in V_{k-1}^p$, let $C_v = (V_v^C, E_v^C, w_v^C)$ be a cycle of length p such that $v \in V_v^C$, $(V_v^C \backslash \{v\}) \cap V_{k-1}^p = \emptyset$ and $w_v^C(e) = \frac{2^{d-k}}{p^k}$.

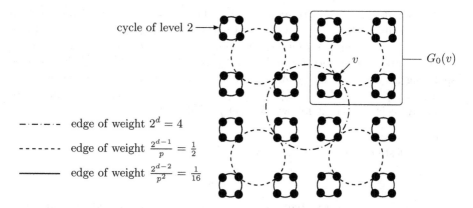

cycle of level 2

$G_0(v)$

v

- — - — edge of weight $2^d = 4$
- — — — edge of weight $\frac{2^{d-1}}{p} = \frac{1}{2}$
- ——— edge of weight $\frac{2^{d-2}}{p^2} = \frac{1}{16}$

Fig. 1. The graph G_2^4

$G_k^p = (V_k^p, E_k^p, w_k^p)$ is the graph such that:

- $V_k^p = V_{k-1}^p \cup \bigcup_{v \in V_{k-1}^p} V_v^C$
- $E_k^p = E_{k-1}^p \cup \bigcup_{v \in V_{k-1}^p} E_v^C$
- $\forall e \in E_{k-1}^p, w_k^p(e) = w_{k-1}^p(e)$ and $\forall e \in \bigcup_{v \in V_{k-1}^p} E_v^C, w_k^p(e) = w_v^C(e)$.

See Figure 1 for an illustration of G_k^p (with $k = 2$ and $p = 4$). We can now define the graph $G = (V, E, w)$. Let $c \geq 1$ be the constant corresponding to the required level of quality and let d be a positive integer sufficiently large such that $i \leq \left| V_d^{\lceil 6c+2 \rceil} \right| - 1$ (where i is the number of removed vertices and $V_d^{\lceil 6c+2 \rceil} = M_0$ is the initial group). We set $G = G_d^{\lceil 6c+2 \rceil}$.

Definition of a Cycle of Level k

We say that a cycle $C = (V^C, E^C, w)$ (subgraph of G) is of level k ($0 \leq k \leq d$) if each edge $e \in E^C$ has weight $w(e) = \frac{2^{d-k}}{p^k}$. See Figure 1 for an illustration of such cycle (Note that G_2^4 is too small to be a possible graph of the form $G_d^{\lceil 6c+2 \rceil}$, but this is just an illustration).

Definition of the Subgraphs $G_k(v)$

Let v be any vertex of the graph G ($v \in V$). Let k be the smallest index such that $C_k = (V_k^C, E_k^C, w)$ is the cycle of level k containing v ($v \in V_k^C$). We define $G_k(v) = (V_k(v), E_k(v), w)$ the subgraph induced by every vertices and edges which can be reached from vertex v by going through edges of weight strictly less than $\frac{2^{d-k}}{p^k}$ (i.e. by going through edges of cycles of level strictly more than k). See Figure 1 for an illustration of such subgraph.

3.2   Definition of the Sequence of Withdrawals $M_0 \supset \cdots \supset M_i$

Let A be *any* online algorithm respecting the tree and quality of level c constraints. We use an *adaptive adversary* to define the sequence of withdrawals in the graph $G = (V, E, w)$ defined above.

We first define a generic sequence of withdrawals of vertices. Note that we do not specify each elementary stage of withdrawal, but only the "main" stages interesting for our analysis (stages of the form $i = \alpha(k, b)$). For every $k \geq 0$, for every $b \in \{0, 1\}$, for every $i = \alpha(k, b)$ $(0 \leq \alpha(k, b) \leq m_0 - 1)$, let T_i be the tree spanning M_i constructed by Algorithm A at stage i. Note that at each stage, we have $\alpha(k, b) = |M_0| - |M_{\alpha(k,b)}|$ $(0 \leq \alpha(k, b) \leq m_0 - 1)$. The sequence of withdrawals is defined as follows. We set $p = \lceil 6c + 2 \rceil$.

Basic Cases
- At stage $\alpha(0, 0) = 0$, we have

$$M_{\alpha(0,0)} = V$$

As $T_{\alpha(0,0)}$ is a tree spanning $M_{\alpha(0,0)}$, it is necessarily made up of, amongst other things, all the edges of the cycle $C_0 = (V_0^C, E_0^C, w)$, except one edge e_0. Let v_0^1 and v_0^2 be the two vertices connected by e_0. The adaptive adversary now removes (one by one) from $M_{\alpha(0,0)}$ all the vertices in $\bigcup_{v \in V_0^C \setminus \{v_0^1, v_0^2\}} V_1(v)$ in order to obtain $M_{\alpha(0,1)}$.

- At stage $\alpha(0, 1)$, we have

$$M_{\alpha(0,1)} = V_1(v_0^1) \cup V_1(v_0^2)$$

The adaptive adversary now removes (one by one) from $M_{\alpha(0,1)}$ all the vertices in $V_1(v_0^1)$ in order to obtain $M_{\alpha(1,0)}$ (note that the adversary chooses arbitrarily to remove all the vertices in $V_1(v_0^1)$ rather than in $V_1(v_0^2)$).

Main Cases
- At stage $\alpha(k, 0)$. Let $C_k = (V_k^C, E_k^C, w)$ be the cycle of level k such that $V_k^C \subset M_{\alpha(k-1,1)}$. We have

$$M_{\alpha(k,0)} = \bigcup_{v \in V_k^C} V_{k+1}(v)$$

As $T_{\alpha(k,0)}$ is a tree spanning $M_{\alpha(k,0)}$, it is necessarily made up of, amongst other things, all the edges of the cycle C_k, except one edge e_k. Let v_k^1 and v_k^2 be the two vertices connected by e_k. The adaptive adversary now removes (one by one) from $M_{\alpha(k,0)}$ all the vertices in $\bigcup_{v \in V_k^C \setminus \{v_k^1, v_k^2\}} V_{k+1}(v)$ in order to obtain $M_{\alpha(k,1)}$.
- At stage $\alpha(k, 1)$, we have

$$M_{\alpha(k,1)} = V_{k+1}(v_k^1) \cup V_{k+1}(v_k^2)$$

The adaptive adversary now removes (one by one) from $M_{\alpha(k,1)}$ all the vertices in $V_{k+1}(v_k^1)$ in order to obtain $M_{\alpha(k+1,0)}$ (note that the adversary chooses arbitrarily to remove all the vertices in $V_{k+1}(v_k^1)$ rather than in $V_{k+1}(v_k^2)$).

We specify with $\alpha(k,b)$ only the "main" stages of the sequence of withdrawals, corresponding to the stages where the adaptive adversary has to make a choice. Indeed, between two successive "main" stages $\alpha(k,0)$ and $\alpha(k,1)$ (resp. $\alpha(k,1)$ and $\alpha(k+1,0)$), the vertices are removed one by one in any order. Note that we stop removing vertices after the last "main" stage, when exactly i vertices have been removed. See Figure 2 for an illustration of the six first "main" stages $\alpha(0,0)$, $\alpha(0,1)$, $\alpha(1,0)$, $\alpha(1,1)$, $\alpha(2,0)$ and $\alpha(2,1)$, where the successive trees are built by an arbitrary algorithm (Note that G_2^4 is too small to be a possible graph of the form $G_d^{\lceil 6c+2 \rceil}$, but this figure is just an illustration of a sequence of withdrawals).

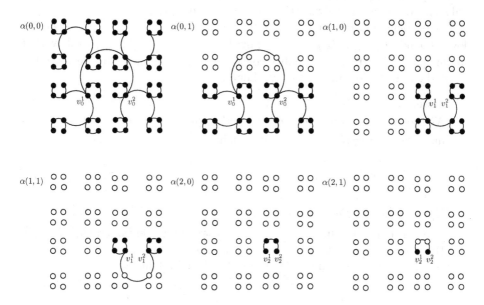

Fig. 2. Illustration of the sequence of withdrawals on graph G_2^4

3.3   Any Algorithm Leads to $\Omega(\log i)$ Critical Stages

Lemmas 1 and 2 are preliminary technical results (Lemma 1 is trivial. A proof can be found in [5]).

Lemma 1. *Let $G = (V, E, w)$ be any graph. For every $M \subseteq V$, there exists a tree T^{off} spanning M such that*

$$D_{T^{\text{off}}}(M) \leq 2 D_G(M)$$

The following Lemma is central in our analysis. It describes sub-sequences of withdrawals where *at least* one rebuilding/critical stage occurs.

Lemma 2. *Let $c \geq 1$ be any constant (representing the required level of quality). For every $k \geq 0$, let $T^*_{\alpha(k,0)}, T^*_{\alpha(k,0)+1}, \ldots, T^*_{\alpha(k,1)}$ be the trees respectively spanning $M_{\alpha(k,0)}, M_{\alpha(k,0)+1}, \ldots, M_{\alpha(k,1)}$ optimal for the diameter and let $T_{\alpha(k,0)},$*

$T_{\alpha(k,0)+1}, \ldots, T_{\alpha(k,1)}$ *be any trees respectively spanning* $M_{\alpha(k,0)}, M_{\alpha(k,0)+1}, \ldots,$ $M_{\alpha(k,1)}$. *If for every* i, $\alpha(k,0) \leq i \leq \alpha(k,1)$, *we have* $D_{T_i}(M_i) \leq c \cdot D_{T_i^*}(M_i)$, *then*

$$\sharp CS(T_{\alpha(k,0)}, T_{\alpha(k,0)+1}, \ldots, T_{\alpha(k,1)}) \geq 1.$$

Proof. We prove Lemma 2 by contradiction. Suppose that there exists $k \geq 0$ such that for every i, $\alpha(k,0) \leq i \leq \alpha(k,1)$, the *quality* constraint is satisfied *and* there is no critical stage, i.e. there exists $k \geq 0$ such that for every i, $\alpha(k,0) \leq i \leq \alpha(k,1)$, we have $D_{T_i}(M_i) \leq c \cdot D_{T_i^*}(M_i)$ and $T_{\alpha(k,0)} \supseteq T_{\alpha(k,0)+1} \supseteq \cdots \supseteq T_{\alpha(k,1)}$.

These trees are made up of, amongst other things, all edges of the cycle $C_k \subset G$, except one edge, noted e_k. We insist on the fact that, because there is no critical stage, this edge e_k is always the same in all trees $T_{\alpha(k,0)}, T_{\alpha(k,0)+1}, \ldots, T_{\alpha(k,1)}$.

Let us focus now on stage $\alpha(k,1)$, where $M_{\alpha(k,1)} = V_{k+1}(v_k^1) \cup V_{k+1}(v_k^2)$. We lower bound $D_{T_{\alpha(k,1)}}(M_{\alpha(k,1)})$ and upper bound $D_{T_{\alpha(k,1)}^*}(M_{\alpha(k,1)})$ to show that at this particular stage, the quality constraint is not satisfied. This leads to the wanted contradiction and proves the Lemma.

- Lower bound of $D_{T_{\alpha(k,1)}}(M_{\alpha(k,1)})$

  As the two extremities v_k^1 and v_k^2 of the edge e_k are separated by a path made of $p - 1 = \lceil 6c + 1 \rceil$ edges of weight $\frac{2^{d-k}}{p^k}$ in $T_{\alpha(k,1)}$ we have

$$D_{T_{\alpha(k,1)}}(M_{\alpha(k,1)}) \geq (p-1)\frac{2^{d-k}}{p^k} \geq (6c+1)\frac{2^{d-k}}{p^k} \tag{6}$$

- Upper bound of $D_{T_{\alpha(k,1)}^*}(M_{\alpha(k,1)})$

  In order to upper bound $D_{T_{\alpha(k,1)}^*}(M_{\alpha(k,1)})$, we first upper bound $D_G(M_{\alpha(k,1)})$. By construction of the graph G, two cases may occur:
  1. If $k = d$, there is no cycle of level $k+1$ in G. Thus, we have

$$D_G(M_{\alpha(k,1)}) = w(e_k) = \frac{2^{d-k}}{p^k} \leq 3\frac{2^{d-k}}{p^k}$$

  2. If $k \leq d - 1$, we have

$$D_G(M_{\alpha(k,1)}) \leq D_G(V_{k+1}(v_k^1)) + d_G(v_k^1, v_k^2) + D_G(V_{k+1}(v_k^2))$$

$$\leq \sum_{e \in E_{k+1}(v_k^1)} w(e) + w(e_k) + \sum_{e \in E_{k+1}(v_k^2)} w(e)$$

(because for every graph or subgraph

$$G = (V, E, w), D_G(V) \leq \sum_{e \in E} w(e))$$

$$= \sum_{l=k+1}^{d} \frac{2^{d-l}}{p^l} p^{l-k} + \frac{2^{d-k}}{p^k} + \sum_{l=k+1}^{d} \frac{2^{d-l}}{p^l} p^{l-k}$$

$$\leq \frac{2}{p^k} \sum_{l=k+1}^{d} 2^{d-l} + \frac{2^{d-k}}{p^k} \leq 2\frac{2^{d-k}}{p^k} + \frac{2^{d-k}}{p^k} = 3\frac{2^{d-k}}{p^k}$$

Moreover, by Lemma 1, there exists a tree $T^{\text{off}}_{\alpha(k,1)}$ spanning $M_{\alpha(k,1)}$ such that $D_{T^{\text{off}}_{\alpha(k,1)}}(M_{\alpha(k,1)}) \leq 2D_G(M_{\alpha(k,1)})$. Thus, as $T^*_{\alpha(k,1)}$ is a tree spanning $M_{\alpha(k,1)}$ optimal for the diameter, we have

$$D_{T^*_{\alpha(k,1)}}(M_{\alpha(k,1)}) \leq D_{T^{\text{off}}_{\alpha(k,1)}}(M_{\alpha(k,1)}) \leq 2D_G(M_{\alpha(k,1)}) \leq 6\frac{2^{d-k}}{p^k} \quad (7)$$

By (6) and (7), we obtain

$$\frac{D_{T_{\alpha(k,1)}}(M_{\alpha(k,1)})}{D_{T^*_{\alpha(k,1)}}(M_{\alpha(k,1)})} \geq \frac{(6c+1)\frac{2^{d-k}}{p^k}}{6\frac{2^{d-k}}{p^k}} \geq c + \frac{1}{6} > c$$

This result contradicts the assumption that the *quality* constraint is satisfied. Thus, Lemma 2 is proved by contradiction.   □

The following Theorem shows that if the *tree* constraint and the *quality* constraint are satisfied, *any* algorithm leads to $\Omega(\log i)$ critical stages, where i is the number of removed vertices.

Theorem 3. *Let $c \geq 1$ be any constant. For any algorithm A, for every sufficiently large i, there exists a graph G_0, there exists $M_0 \supset \cdots \supset M_i$, such that if Algorithm A returns a sequence of trees $T_0, \ldots, T_i$ respectively spanning $M_0 \supset \cdots \supset M_i$ respecting the* quality constraint *of level c, then*

$$\sharp CS(T_0, \ldots, T_i) = \Omega(\log i)$$

Proof. Let $c \geq 1$ be any constant. We set $p = \lceil 6c + 2 \rceil$. Let i be the number of removed vertices. There exists d and G_0 (where G_0 is graph G, defined in Section 3.1), there exists $M_0 \supset \cdots \supset M_i$ (the sequence defined in Section 3.2) such that

$$\alpha(d-1,1) \leq i \leq \alpha(d,1) \leq |V| = p^d$$

Thus, we have $i \leq p^d \Rightarrow \log_p i \leq d \Rightarrow \log_p i \leq d$. And as $p = \lceil 6c + 2 \rceil$ is a constant, we have $d \geq \Omega(\log i)$. Moreover, by Lemma 2, we have

$$\begin{cases} \sharp CS(T_{\alpha(0,0)}, T_{\alpha(0,0)+1}, \ldots, T_{\alpha(0,1)}) \geq 1 \\ \sharp CS(T_{\alpha(1,0)}, T_{\alpha(1,0)+1}, \ldots, T_{\alpha(1,1)}) \geq 1 \\ \qquad\qquad\qquad \vdots \\ \sharp CS(T_{\alpha(d-1,0)}, T_{\alpha(d-1,0)+1}, \ldots, T_{\alpha(d-1,1)}) \geq 1 \end{cases}$$

$$\Rightarrow \sharp CS(T_{\alpha(0,0)}, \ldots, T_{\alpha(d-1,1)}) \geq d$$
$$\Rightarrow \sharp CS(T_0, \ldots, T_i) \geq d \qquad \text{(because } i \geq \alpha(d-1,1))$$
$$\Rightarrow \sharp CS(T_0, \ldots, T_i) \geq \Omega(\log i) \qquad \text{(because } d \geq \Omega(\log i)) \qquad \square$$

Theorem 2 and Theorem 3 show that Algorithm CS is worst case optimal in order of magnitude for the number of critical stages criterion.

4   Conclusion

We have proposed an algorithm, called CS, solving an on-line covering problem of members by respecting the following *quality* constraint: For each stage of withdrawal, the diameter between members induced by the built tree is at most a constant time the best possible value. Moreover, our algorithm is easy to use. Indeed, for a stage of withdrawal, either it breaks the tree and rebuilds a new one which is a tree of shortest paths (only $O(\log i)$ times, where i is the number of removed members), or it just updates the current tree by removing useless branches (in all the other cases).

Moreover, our algorithm is worst case optimal in order of magnitude for the number of critical stages: It leads to $O(\log i)$ critical stages and we showed that *any* algorithm leads to $\Omega(\log i)$ critical stages in the worst case. We also have proved that the number of elementary changes per stage (see equivalent definition in [4]) is constant in average. Due to space limitation, we do not include these results. Note that we only consider the decremental problem because the incremental problem (adding new members in the current tree) considering the diameter as quality constraint is trivial. Indeed, plugging each new member with a shortest path to the initial member leads to 0 critical stage with a level of quality $c = 2$. We also have results with another objective function than the diameter. Indeed, concerning the average distance between members of the groups, we proved similar results in [7] for the incremental version of the problem. We are now currently working on mixing additions and withdrawals.

Acknowledgments

The authors wish to thank the anonymous referees for their very useful comments.

References

1. G. Ausiello, P. Crescenzi, G. Gambosi, V. Kann, A. Marchetti Spaccamela, and M. Protasi, *Complexity and approximation*, Springer, 1999.
2. A. Borodin and R. El-Yaniv, *Online computation and competitive analysis*, Cambridge University press, 1998.
3. D. Hochbaum, *Approximation algorithms for NP-hard problems*, PWS publishing compagny, 1997.
4. M. Imase and B. Waxman, *Dynamic steiner tree problem*, SIAM J. Discr. Math., 4 (1991), pp. 369–384.
5. C. Laforest, *A good balance between weight and distances for multipoint trees*, in International Conference On Principles Of DIstributed Systems 2002, pp. 195–204.
6. S. Raghavan, G. Manimaran, and C. S. R. Murthy, *A rearrangeable algorithm for the construction of delay-constrained dynamic multicast trees*, IEEE/ACM (SIGCOMM), ACM Press, 7 (1999).
7. N. Thibault and C. Laforest, *An optimal rebuilding strategy for an incremental tree problem, submitted in 2004 to journal of interconnection networks*.
8. B. Waxman, *Routing of multipoint connections*, IEEE Journal on Selected Areas in Communications, 6 (1988), pp. 1617–1622.

Optimal Delay for Media-on-Demand with Pre-loading and Pre-buffering

Amotz Bar-Noy[1], Richard E. Ladner[2,*], and Tami Tamir[3]

[1] Computer and Information Science Department, Brooklyn College, 2900 Bedford
Avenue Brooklyn, NY 11210
amotz@sci.brooklyn.cuny.edu
[2] Department of Computer Science and Engineering, Box 352350, University of
Washington, Seattle, WA 98195
ladner@cs.washington.edu
[3] School of Computer Science, The Interdisciplinary Center, P.O.Box 167,
Herzliya 46150, Israel
tami@idc.ac.il

Abstract. Broadcasting popular media to clients is the ultimate scalable solution for media-on-demand. The simple solution of downloading and viewing the media from one channel cannot guarantee a reasonable startup delay for viewing with no interruptions. Two known techniques to reduce the delay are pre-loading and pre-buffering. In the former an initial segment of the media is already in the client buffer, and in the latter segments of the media are not transmitted in sequence and clients may pre-buffer later segments of the media before viewing them. In both techniques, the client should be capable to receive streams from channels at the same time of handling its own buffer and view the media from either one of the channels or the buffer.

In this paper we consider broadcasting schemes that combine pre-loading and pre-buffering. We present a complete tradeoff between (i) the size of the pre-loading; (ii) the maximal possible delay for an uninterrupted playback; (iii) the number of media; and (iv) the number of channels allocated per one media. For a given B the size of the pre-loading as a fraction of the media length, for m media, and for h channels per media, we first establish a lower bound for the minimal maximum delay, D, as a fraction of the movie length, for an uninterrupted playback of any media out of the m media. We then present an upper bound that approaches this lower bound when each media can be fragmented into many segments.

1 Introduction

Media-on-demand (MoD) is the demand by clients to read, listen, or view various types of media. In its simplest function, the clients would like to have an uninterrupted playback with as minimal as possible start-up delay. The subject of this paper is to reduce the maximal start-up delay for MoD systems that support

---

* The research was partially funded by NSF Grant No. CCF-0223578.

P. Flocchini and L. Gąsieniec (Eds.): SIROCCO 2006, LNCS 4056, pp. 171–181, 2006.

uninterrupted service. Our main objective is to achieve the smallest maximal start-up delay for given amount of two resources: the bandwidth of the system, and the client local memory.

There are two main types of systems that support MoD: unicast systems and broadcast systems. The former guarantees an immediate service as long as there are not too many clients. The latter can support many clients but cannot guarantee immediate service. For popular media, *broadcasting* is the ultimate scalable solution. In various broadcasting schemes, different parts of the media are transmitted on channels viewable to the clients. This paper considers the potential benefit of broadcasting schemes from using some of the client memory for storing in advance (pre-loading) parts of the media.

In the simplest implementation of MoD systems, clients who wish to view a movie[1], select the channel that would start broadcasting this movie the earliest after their request time. Movies are broadcast on one channel or several channels. Thus if h channels are allocated to a movie of length L time units, the maximal start-up delay is L/h units by starting a new transmission every L/h time units.

In recent years, more efficient broadcasting schemes that are based on pre-buffering were suggested. In these schemes, each movie is partitioned into segments, and the segments are transmitted on the channels in some order, not necessarily their order in the movie. The client is reading all the channels simultaneously, 'collecting' segments to its local memory, and is watching the segments of the movie in order - some directly from the channels and some from its own memory.

The above broadcasting schemes require customers to get the service through a set-top-box (STB) capable of storing locally the transmitted data. This requires that the STB is equipped with a local memory (disk). In fact, this technology is already available: digital VCRs offered by ReplayTV [19], TiVo [22], and UltimateTV [23], have capacities of at least 300 gigabytes, enabling the client to store hours of movies in perfect quality. The disk capacity can be used to store entire movies and also pre-buffered segments and pre-loaded segments of other movies. Usually the former type of movies will be non-popular movies where the latter type will be popular movies for which the broadcasting solution is more beneficial. In this paper, we consider broadcasting schemes that combine pre-loading and pre-buffering. That is, we assume that some prefix of the movie is stored at the client's machine, and therefore he or she should only receive the remainder of the movie. We present a complete tradeoff between (i) the size of the pre-loading; (ii) the maximal possible delay for an uninterrupted playback; (iii) the number of movies; and (iv) the number of channels allocated per movie.

For a given B the size of the pre-loading as a fraction of the movie length, for m movies, and for h channels per movie, we first establish a lower bound for the minimal maximum delay, D, as a fraction of the movie length, for an uninterrupted playback of any movie out of the m movies. We then present an upper bound that approaches this lower bound when each movie can be fragmented to many segments.

---

[1] For convenience, we use the terminology of movies in Video-on-Demand (VoD).

1.1 Model and Preliminaries

The system broadcasts m movies on h channels. Unless specified otherwise, assume that all m movies have the same length, L, normalized to be one time unit ($L = 1$). Each movie is partitioned into s segments of equal length. Segment size may range from a single bit (which is theoretically interesting) to the whole movie (in case of a single segment). The segments are indexed 1 to s in the order they should be viewed. The segments of the movies may be broadcast in any order on any channel. Assume that it takes one time slot to transmit or view a segment and thus, the length of the time slot is $1/s$. Assume further that all the channels are synchronized in the sense that the starting points for the time slots coincide in all of them. Clients may buffer or view segments from any channel since they may receive data from all of them. In other words, the receiving bandwidth of each client is h. This implies that a client buffers or views segment i the first time he or she can do so after the arrival time. Clients may buffer any number of segments before the viewing process begins.

The maximal possible delay of a client is denoted by D and is given as a fraction of the movie length. That is, if for example $D = 1/4$, no client will wait more than $1/4$ of a movie length till it can start an uninterrupted playback of the movie. Let $d = Ds$ denote the maximal delay measured as number of segments (time-slots). In the broadcasting schemes we present, the maximum delay is given in units of time-slots, thus we assume that D is a multiple of $1/s$.

The basic principle in all the schemes that use pre-buffering is that early segments should be broadcast more frequently than later segments. Intuitively, a client needs to watch the zth segment only $z - 1$ time-slots after it starts watching the movie, therefore, the zth segment, can be transmitted less often than earlier segments. Formally, in [2], optimal schemes that are based on pre-buffering are developed using the *windows scheduling* problem and the following is shown:

Theorem 1. *Let $\mathcal{S}$ be a schedule that broadcasts $s \geq 1$ segments for a movie on $h \geq 1$ channels. Then $\mathcal{S}$ guarantees a maximum start-up delay of $d > 0$ time-slots if and only if segment z is transmitted once in any window of $d + z - 1$ segments for each $1 \leq z \leq s$.*

Assume now that out of the s segments composing the movie, the first b are *preloaded* and are stored at the client's local machine (set top box), the other $s - b$ segments are transmitted on channels. Clearly, the client can always watch the first b segments with no delay. Consider the remainder of the movie as a complete (shorter) movie. Assume there exists a broadcasting scheme that enables any client to view this movie with delay at most d' (in number of segments units). The idea is to overlap the time the client watches the first b segments with the time it is waiting to the rest of the data. This would result in a delay $\max(0, d' - b)$. The challenge is to schedule the remaining $s - b$ segments on the broadcasting channels in a way that minimizes this term.

Example: Consider a single movie transmitted on a single channel. Assume that the client has at his local machine all but the last 5 segments, which are not pre-loaded, and are transmitted on the channel in the following (repeated) order:

$$[1, 3, 2, 4, 1, 5, 2, 3, 1, 4, 2, 5]$$

In this order, the segments $1, 2$ are transmitted every 4 slots and the segments $3, 4, 5$ are transmitted every 6 slots. Recall that the movie is partitioned into s segments, thus, these 5 segments are segments $s - 4, \ldots, s$ of the movie. The first $b = s - 5$ segments are pre-loaded and available to the client at any time (thus, $B = (s - 5)/s$). By Theorem 1 the above transmission of the last 5 segments guarantees a delay of at most $d' = 4$ slots for viewing with no interruptions the last 5 segments. Thus, if $s \geq 9$, or equivalently, $b \geq 4$ meaning that the client has at least the first 4 segments of the movie, then there is no delay at all. If $s < 9$, the delay with pre-loading is $4 - b$ slots which is $D = (4 - b)/(b + 5) = (9 - s)/s$ of the whole movie. We get the following tradeoff between B and D:

s	B	D
5	0	4/5
6	1/6	3/6
7	2/7	2/7
8	3/8	1/8
9	4/9	0
> 9	$(s-5)/s$	0

In particular, this means that in order to guarantee no delay the pre-loading size should be 4/9 of the movie length, and with no pre-loading the maximal delay is 4/5 of the movie length.

Table 1 provides a glossary of the notation used in the paper.

The lower bound and the matching broadcasting scheme we present assume that the client's memory stores only prefixes of movies. One might doubt that this is optimal, and suggest it might be better to store late parts of the movie

Table 1. Glossary of notations

notation	meaning
h	number of channels
m	number of different movies
$\rho = h/m$	the ratio between number of channels and number of movies.
s	number of segments per each movie.
B	the size of the pre-loading buffer as a fraction of the movie length.
$b = Bs$	the size of the pre-loading buffer as a number of segments
D	the maximal delay for an uninterrupted playback as a fraction of the movie length.
$d = Ds$	the maximal delay as a number of segments
d'	the maximal delay for the non pre-loaded part, as a number of segments

and broadcast earlier ones. The following Theorem should remove such doubts - it states that the best way to use an allocated amount B of memory to a movie is by storing (pre-loading) a prefix of size B of this movie.

Theorem 2. *For any broadcasting scheme that combines pre-loading and broadcasting, if memory of size B is allocated to a movie, then it is optimal to store from this movie a* prefix *of size B.*

Proof. The idea is similar to the optimality proof of the offline algorithm for caching - that evicts from the cache the page that will be requested last among the pages that are currently in the cache. Consider any broadcasting scheme S in which for some movie there exists a bit i that is not pre-loaded, while some bit $j > i$ is. Since j is pre-loaded, it is never transmitted by the scheme.

Consider the scheme S' in which bit i is pre-loaded and bit j is transmitted whenever bit i was transmitted in S. Clearly, the user will have bit i on time (from its memory) and bit j will be available at the time bit i was available in S. Since we assume that all clients read the movie in order, bit i is requested before bit j, therefore, by having bit j available in S' at the time i was available in S, the client's delay can only decrease.

1.2   Related Work

MoD systems, and in particular the solution of broadcasting, have been studied extensively in recent years. The paper [3] surveys broadcasting protocols and describes the development of these protocols, starting with Staggered broadcasting protocols, in which the movies are simply transmitted repeatedly on the channels (e.g., [4]), through Pyramid-based broadcasting protocols, in which movies are partitioned into segments and different segments are broadcast on different channels [24], and finally Harmonic broadcasting protocols in which segment i is allocated bandwidth proportional to $1/i$ (e.g., [8]).

The case when there is no pre-loading and pre-buffering may start only when clients start viewing the movie received much attention in the recent decade. The papers [7, 9] present a simple schedule of one movie on h channels by partitioning the movie into $2^h - 1$ segments. Their schedule implies a maximal start-up delay of $1/(2^h - 1)$ for a movie of length 1. This scheme is improved in the Pagoda scheme ([15]), the new Pagoda scheme ([11]), the Recursive Frequency-Splitting scheme ([21]), the Harmonic broadcasting scheme ([1]), and the Polyharmonic broadcasting scheme ([14]. In these schemes, the worst-case maximal delay asymptotically approaches $1/(e^b - 1)$ for total bandwidth b. Several papers, e.g., [6] have shown this bound on delay to be optimal.

Harmonic broadcasting is implemented in [1] by a reduction from the *window-scheduling* problem. Specifically, the movie is partitioned into s equal-sized segments that are scheduled on the channels such that the gap between any two consecutive appearances of segment i is at most i. For a given number of channels, the goal is to maximize s, and as a result, minimize the start-up delay (which is at most $1/s$). A schedule based on this principle is shown to approach the lower bound as $h \to \infty$. The papers [12, 13] also allow clients to start buffering segments before they start viewing the movie to achieve better results. However, they demonstrate the usefulness of this observation only for small examples. The paper [2] gives asymptotic matching upper and lower bounds on the maximal delay of a broadcasting scheme that uses pre-buffering.

The papers [18, 16] consider pre-loading, but only for the case of zero delay. The paper [18] does not allow pre-buffering before the clients start watching the movie whereas the paper [16] improves the results by allowing this feature. In another work on pre-loading [10], it is assumed that each client pre-loads segments of a different set of movies, according to the client's choice. Earlier works on pre-loading assume that the preloaded data is stored at a proxy server and not at the client's local machine [5, 20].

1.3   Contribution

We consider broadcasting schemes that combine pre-loading and pre-buffering. We present a complete tradeoff between (i) the size of the pre-loading; (ii) the maximal possible delay for an uninterrupted playback; (iii) the number of media; and (iv) the number of channels allocated per one media.

For a given B the size of the pre-loading as a fraction of the media length, for m media, and for h channels per media, we first establish a lower bound for the minimal maximum delay, D, as a fraction of the movie length, for an uninterrupted playback of any media out of the m media. We then present an upper bound that approaches this lower bound when each media can be fragmented to many segments.

2   A Lower Bound for the Maximal Delay

We first compute a lower bound for the maximal delay for a fixed $s \geq 1$ number of segments per movie. Then we calculate the general lower bound by letting s tend to infinity. For ease of presentation we assume that both $b = Bs$ and $d = Ds$ are integers.

Each client has the first $b = Bs$ segments of each movie in its buffer. Therefore, the channels need to broadcast only segments $b + 1, \ldots, s$. Since the maximal delay is d, segment i of each movie should be broadcast at least once in any window of size $d + i$ for $b + 1 \leq i \leq s$. That is, segment i consumes at least $1/(d + i)$ of a channel. Since the total number of channels is h and since there are m movies, it follows that

$$m \sum_{i=b+1}^{s} \frac{1}{i + d} \leq h .$$

This is equivalent to

$$\sum_{i=b+d+1}^{s+d} \frac{1}{i} \leq \rho .$$

Using the known bound on the harmonic number $H_n = \sum_{i=1}^{n}(1/i)$ implies

$$\ln\left(\frac{s+d}{b+d}\right) \leq \rho .$$

Since $b = Bs$ and $d = Ds$, this is equivalent to

$$\frac{1+D}{B+D} \leq e^\rho .$$

By manipulating the above inequality we get the lower bound for D given B

$$D \geq \frac{1 - Be^\rho}{e^\rho - 1} .$$

Equivalently, the lower bound for B given D is

$$B \geq \frac{1 - D(e^\rho - 1)}{e^\rho} .$$

In particular, when $B = 0$ the lower bound matches the known lower bound [6]

$$D \geq \frac{1}{e^\rho - 1} .$$

When $D = 0$ the lower bound for B is

$$B \geq \frac{1}{e^\rho} .$$

For example, in order to guarantee no delay for a single movie transmitted on a single channel the client must pre-load at least $1/e \approx 0.368$ of the movie.

3   Optimal Schedules

In this section we present an upper bound that approaches the lower bound from Section 2 when each movie can be fragmented into many segments. In the optimal schedule the last segments of each movie are transmitted in such a way that earlier segments are transmitted more often. Assume first a transition of a single movie, that is, $m = 1$. Consider a schedule of the numbers $[x..y]$ on h channels such that for any $x \leq i \leq y$, in each window of i consecutive slots, the number i appears at least once in one of the channels. For example

$$[4, 6, 5, 7, 4, 8, 5, 6, 4, 7, 5, 8]$$

is such a schedule for $h = 1$, $x = 4$, and $y = 8$.

Suppose we interpret the numbers $x, \ldots, y$ as segments $s - y + x, \ldots, s$ of the movie. This reflects a partition of the movie into s segments each of length $1/s$ of the movie length. The segments that are not transmitted should be stored at the client memory, thus, the pre-loading size is $b = s - y + x - 1$ which implies $B = (s - y + x - 1)/s$. Furthermore, the delay with pre-buffering is $D = (y + 1 - s)/s$. It follows that a viable range for s is from $y - x + 1$ to $y + 1$ (it might be that $s > y + 1$ but the delay never reduces below 0), and we get the following tradeoff between B and Dg:

s	B	D
$y - x + 1$	0	$x/(y - x + 1)$
$y - x + 2$	$1/(y - x + 2)$	$(x - 1)/(y - x + 2)$
$y - x + i$	$(i - 1)/(y - x + i)$	$(x - i + 1)/(y - x + i)$
y	$(x - 1)/y$	$1/y$
$y + 1$	$x/(y + 1)$	0
$> y + 1$	$(s + x - y - 1)/s$	0

In particular, this means that in order to guarantee no delay with this schedule the pre-loading size should be $x/(y + 1)$ of the movie length and with no pre-loading the maximal delay is $x/(y - x + 1)$ of the movie length.

The upper bound is achieved for such a schedule of the last segments, that is, for some range $[x..y]$, any $x \le i \le y$, is transmitted at least once in each window of i consecutive slots. This is a special instance of the *windows scheduling* problem studied in [2]. Consider the general case $s = y - x + i$ in which $B = (i-1)/(y-x+i)$ and $D = (x - i + 1)/(y - x + i)$. Assign $z = y + 1$ and $j = x - i + 1$. With these variables,

$$B = \frac{x - j}{z - j} \qquad D = \frac{j}{z - j}.$$

Further, assign $w = z/j$. It follows that

$$B = \frac{x/j - 1}{w - 1} \qquad D = \frac{1}{w - 1}.$$

As shown in [2], in the limit, for large values of s (and consequently large values of x, y), there exists a valid schedule of $[x..y]$ such that

$$\frac{x}{y - x + 1} \approx \frac{1}{e^h - 1}.$$

This implies that

$$x = \frac{y + 1}{e^h} \approx \frac{z}{e^h}.$$

Furthermore, the values of B and D as a function of w are

$$B = \frac{w/e^h - 1}{w - 1} \qquad D = \frac{1}{w - 1}.$$

Plugging $w = 1 + 1/D$ in the equality for B yields

$$B = D\left(\frac{1 + 1/D}{e^h} - 1\right) = \frac{D + 1}{e^h} - D = \frac{1}{e^h} - \left(1 - \frac{1}{e^h}\right)D.$$

Equivalently,

$$D = \frac{1/e^h - B}{1 - 1/e^h} = \frac{1 - e^h B}{e^h - 1}.$$

For the special case of $D = 0$ we have $(1 - e^h B) = 0$ or equivalently

$$B = \frac{1}{e^h} \ .$$

For the special case of $B = 0$ we have $1/e^h = (1 - 1/e^h)D$ or equivalently

$$D = \frac{1}{e^h - 1} \ .$$

The calculation for the general case of $m > 1$ is identical. For each of the m movies, segments $s - y + x, \ldots, s$ are transmitted with windows $x, \ldots, y$, respectively. Along the whole calculation it is possible to replace h by $\rho = h/m$. Note that all the above upper bounds match the lower bounds from Section 2.

4   Discussion

In this paper we showed a tradeoff between the size of the pre-loaded buffer and the guaranteed delay for an uninterrupted playback of movies. We first proved the optimal possible tradeoff and then demonstrated how to achieve it when a movie may be partitioned to many segments. In what follows we discuss several possible extensions.

Limiting the receiving bandwidth. In this paper we assumed that a client can buffer segments of the movie from all the channels. This means that the receiving bandwidth of a client is h times more than the playback bandwidth. Several papers explored the case where the receiving bandwidth is only r times the playback bandwidth for some $1 < r < h$ (e.g. [17]). However, no paper consider this case with the pre-loading capability.

Limited size buffers. Early works on this model assumed that the buffer size for the pre-buffered segments is bounded as a fraction of the movie length (see the survey [3]). Although it seems that the sky is the limit for cheap and large memory, this might not be the case for hand-held set top boxes. It is interesting therefore to investigate the tradeoff between the pre-loaded buffer size and the pre-buffered buffer size when their sum is bounded.

Movies with different popularity. The solution of broadcasting (in contrast to unicast) is suitable for popular media. However, even among popular media there are different levels of popularity. In particular, only a small number of movies is very popular at a specific time. It is very intriguing to see how the combination of pre-loading and pre-buffering can be used to provide smaller delay to the highly requested movies while increasing the maximal possible delay for less popular movies. The problem can be modelled as follows. Consider a system with m movies with different popularity. The popularity parameter of movie i is denoted by p_i such that $\sum_{i=1}^{m} \frac{1}{p_i} = 1$. The parameter p_i can be viewed as the probability that the next client's request is to watch movie i. Let D_i denote the maximal possible delay a broadcasting scheme guarantees for a movie i, then the goal is to minimize $\sum_{i=1}^{m} p_i D_i$. That is, the weighted maximal possible delay (also the expected maximal delay) of the whole system. In practice, especially since the popularity parameter varies drastically over time, it is not practical to assume that each movie has a specific popularity parameter and instead a simpler model may be addressed. The system distinguishes between the *hot* movies and the rest of the popular movies. There are various ways to ensure smaller delay to the hot movies, they can be transmitted more often, or a larger portion of these movies might be pre-loaded.

References

1. A. Bar-Noy and R. E. Ladner. Windows Scheduling Problems for Broadcast Systems. *SIAM Journal on Computing (SICOMP)*, 32(4):1091–1113, 2003.
2. A. Bar-Noy, R. E. Ladner, and T. Tamir. Scheduling techniques for media-on-demand. *Proc. of the 14-th Annual ACM-SIAM Symposium on Discrete Algorithms*, 791-800, 2003.
3. S. W. Carter, D. D. E. Long, and J. Pâris. Video-on-Demand Broadcasting Protocols. In *Multimedia Communications: Directions and Innovations (J. D. Gibson, Editors)*, Academic Press, San Diego, 179–189, 2000.
4. A. Dan, D. Sitaram, and P. Shahabuddin. Dynamic Batching Policies for an On-Demand Video Server. *ACM Multimedia Systems Journal*, 4(3):112–121, 1996.
5. D. Eager, M. Ferris and M. Vernon. Optimized regional caching for on-demand data delivery. In *Proc. 1999 Multimedia Computing and Networking Conference (MMCN'99)*, 1999.
6. L. Engebretsen and M. Sudan. Harmonic Broadcasting is Optimal. In *Proceedings of the 13th Annual ACM-SIAM Symposium on Discrete Algorithms (SODA)*, 431 − 432, 2002.
7. K. A. Hua, Y. Cai, and S. Sheu Exploiting Client Bandwidth for More Efficient Video Broadcast. In *Proceedings of the 7th International Conference on Computer Communication and Networks (ICCCN)*, 848–856, 1998.
8. L. Juhn and L. Tseng. Harmonic Broadcasting for Video-on-Demand Service. *IEEE Transactions on Broadcasting*, 43(3):268–271, 1997.
9. L. Juhn and L. Tseng. Fast Data Broadcasting and Receiving Scheme for Popular Video Service. *IEEE Transactions on Broadcasting*, 44(1):100–105, 1998.
10. J. F. Pâris. A Broadcasting Protocol for Video-on-Demand Using Optional Partial Preloading. In *Proceedings of the 11th International Conference on Computing*, vol.I, 319Ű-329, 2002.
11. J. Pâris. A Simple Low-Bandwidth Broadcasting Protocol for Video-on-Demand. In *Proceedings of the 8th International Conference on Computer Communications and Networks (IC3N)*, 118–123, 1999.
12. J. Pâris. A Fixed-Delay Broadcasting Protocol for Video-on-Demand. In *Proceedings of the 10th International Conference on Computer Communications and Networks (IC3N)*, 418–423, 2001.
13. J. Pâris. A Simple but Efficient Broadcasting Protocol for Video-on-Demand. In *Proceedings of the 24th International Performance of Computers and Communication Conference (IPCCC 2005)*, 167-Ű174, 2005.

14. J. Pâris, S. W. Carter, and D. D. E. Long. A Low Bandwidth Broadcasting Protocol for Video on Demand. In *Proceedings of the 7th International Conference on Computer Communications and Networks (IC3N)*, 690–697, 1998.

15. J. Pâris, S. W. Carter, and D. D. E. Long. A Hybrid Broadcasting Protocol for Video on Demand. In *Proceedings of the IS&T/SPIE Conference on Multimedia Computing and Networking (MMCN)*, 317–326, 1999.

16. J. Pâris and D. D. E. Long. The Case for Aggressive Partial Preloading in Broadcasting Protocols for Video-on-Demand. In *Proceedings of the IEEE International Conference on Multimedia and Expo (ICME)*, 113–116, 2001.

17. J. Pâris and D. D. E. Long. Limiting the Receiving Bandwidth of Broadcasting Protocols for Video-on-Demand. In *Proceedings of the Euromedia Conference*, 107-111, 2000.

18. J. Pâris, D. D. E. Long, and P. E. Mantey, Zero-Delay Broadcasting Protocols for Video-on-Demand. In *Proceedings of the 1999 ACM Multimedia Conference* pages 189-197,

19. ReplayTV. *http://www.replay.com*

20. S. Sen, J. Rexford, and D. Towsley. Proxy prefix caching for multimedia streams. In *Proceedings of the IEEE 18th Conference on Computer Communications (INFOCOM)*, 1310–1319, 1999.

21. Y. C. Tseng, M. H. Yang, and C. H. Chang. A Recursive Frequency-Splitting Scheme for Broadcasting Hot Video in VOD Service. *IEEE Transactions on Communications*, 50(8):1348–1355, 2002.

22. TiVo Technologies. *http://www.tivo.com*

23. UltimateTV. *http://www.ultimatetv.com*

24. S. Viswanathan and T. Imielinski. Metropolitan Area Video-on-Demand Service Using Pyramid Broadcasting. *ACM Multimedia Systems*, 4(3):197–208, 1996.

Strongly Terminating Early-Stopping k-Set Agreement in Synchronous Systems with General Omission Failures

Philippe Raïpin Parvédy, Michel Raynal, and Corentin Travers

IRISA, Université de Rennes 1, Campus de Beaulieu, 35042 Rennes, France
{praipinp, raynal, ctravers}@irisa.fr

Abstract. The k-set agreement problem is a generalization of the consensus problem: considering a system made up of n processes where each process proposes a value, each non-faulty process has to decide a value such that a decided value is a proposed value, and no more than k different values are decided. It has recently be shown that, in the crash failure model, $\min(\lfloor \frac{f}{k} \rfloor + 2, \lfloor \frac{t}{k} \rfloor + 1)$ is a lower bound on the number of rounds for the non-faulty processes to decide (where t is an upper bound on the number of process crashes, and f, $0 \leq f \leq t$, the actual number of crashes).

This paper considers the k-set agreement problem in synchronous systems where up to $t < n/2$ processes can experience general omission failures (i.e., a process can crash or omit sending or receiving messages). It first introduces a new property, called *strong termination*. This property is on the processes that decide. It is satisfied if, not only every non-faulty process, but any process that neither crashes nor commits receive omission failures decides. The paper then presents a k-set agreement protocol that enjoys the following features. First, it is strongly terminating (to our knowledge, it is the first agreement protocol to satisfy this property, whatever the failure model considered). Then, it is *early deciding and stopping* in the sense that a process that either is non-faulty or commits only send omission failures decides and halts by round $\min(\lfloor \frac{f}{k} \rfloor + 2, \lfloor \frac{t}{k} \rfloor + 1)$. To our knowledge, this is the first early deciding k-set agreement protocol for the general omission failure model. Moreover, the protocol provides also the following additional *early stopping* property: a process that commits receive omission failures (and does not crash) executes at most $\min(\lceil \frac{f}{k} \rceil + 2, \lfloor \frac{t}{k} \rfloor + 1)$ rounds. It is worth noticing that the protocol allows each property (strong termination vs early deciding/stopping vs early stopping) not to be obtained at the detriment of the two others.

The combination of the fact that $\min(\lfloor \frac{f}{k} \rfloor + 2, \lfloor \frac{t}{k} \rfloor + 1)$ is lower bound on the number of rounds in the crash failure model, and the very existence of the proposed protocol has two very interesting consequences. First, it shows that, although general omission failure model is more severe than the crash failure model, both models have the same lower bound for the non-faulty processes to decide. Second, it shows that, in the general omission failure model, that bound applies also the processes that commit only send omission failures.

Keywords: Agreement problem, Crash failure, Strong Termination, Early decision, Early stopping, Efficiency, k-set agreement, Message-passing system, Receive omission failure, Round-based computation, Send omission failure, Synchronous system.

P. Flocchini and L. Gąsieniec (Eds.): SIROCCO 2006, LNCS 4056, pp. 182–196, 2006.
© Springer-Verlag Berlin Heidelberg 2006

1   Introduction

Context of the paper k-set and consensus problems. The k-set agreement problem generalizes the uniform consensus problem (that corresponds to the case $k = 1$). It has been introduced by S. Chaudhuri who, considering the crash failure model, investigated how the number of choices (k) allowed to the processes is related to the maximum number (t) of processes that can be faulty (i.e., that can crash) [7]. The problem can be defined as follows. Each of the n processes (processors) defining the system starts with its own value (called "proposed value"). Each process that does not crash has to decide a value (termination), in such a way that a decided value is a proposed value (validity) and no more than k different values are decided (agreement)[1].

k-set agreement can trivially be solved in crash-prone asynchronous systems when $k > t$ [7]. A one communication step protocol is as follows: (1) $t + 1$ processes are arbitrarily selected prior to the execution; (2) each of these processes sends its value to all processes; (3) a process decides the first value it receives. Differently, it has been shown that there is no solution in these systems as soon as $k \le t$ [5, 17, 31]. (The asynchronous consensus impossibility, case $k = 1$, was demonstrated before, using a different technique [11]. A combinatorial characterization of the tasks which are solvable in presence of one process crash is presented in [3]). Several approaches have been proposed to circumvent the impossibility to solve the k-set agreement problem in process crash prone asynchronous systems (e.g., probabilistic protocols [22], or unreliable failure detectors with limited scope accuracy [16, 21, 32]).

The situation is different in process crash prone synchronous systems where the k-set agreement problem can always be solved, whatever the value of t with respect to k. It has also been shown that, in the worst case, the lower bound on the number of rounds (time complexity measured in communication steps) is $\lfloor t/k \rfloor + 1$ [8]. (This bound generalizes the $t + 1$ lower bound associated with the consensus problem [1, 2, 10, 20]. See also [4] for the case $t = 1$.)

Early decision. Although failures do occur, they are rare in practice. For the uniform consensus problem ($k = 1$), this observation has motivated the design of early deciding synchronous protocols [6, 9, 19, 30], i.e., protocols that can cope with up to t process crashes, but decide in less than $t + 1$ rounds in favorable circumstances (i.e., when there are few failures). More precisely, these protocols allow the processes to decide in $\min(f + 2, t + 1)$ rounds, where f is the number of processes that crash during a run, $0 \le f \le t$, which has been shown to be optimal (the worst scenario being when there is exactly one crash per round) [6, 18][2].

In a very interesting way, it has very recently been shown that the early deciding lower bound for the k-set agreement problem in the synchronous crash failure model is $\lfloor f/k \rfloor + 2$ for $0 \le \lfloor f/k \rfloor \le \lfloor t/k \rfloor - 2$, and $\lfloor f/k \rfloor + 1$ otherwise [12]. This lower bound,

---

[1] A process that decides and thereafter crashes is not allowed to decide one more value, in addition to the k allowed values. This is why k-set agreement generalizes *uniform consensus* where no two processes (be they faulty or not) can decide different values. Non-uniform consensus allows a faulty process to decide a value different from the value decided by the correct processes. The non-uniform version of the k-set agreement problem has not been investigated in the literature.

[2] More precisely, the lower bound is $f + 2$ when $f \le t - 2$, and $f + 1$ when $f = t - 1$ or $f = t$.

not only generalizes the corresponding uniform consensus lower bound, but also shows an "inescapable tradeoff" among the number t of crashes tolerated, the number f of actual crashes, the degree k of coordination we want to achieve, and the best running time achievable [8]. As far as the time/coordination degree tradeoff is concerned, it is important to notice that, when compared to consensus, k-set agreement divides the running time by k (e.g., allowing two values to be decided halves the running time).

Related work. While not-early deciding k-set agreement protocols for the synchronous crash failure model (i.e., protocols that always terminate in $\lfloor t/k \rfloor + 1$ rounds) are now well understood [2, 8, 20], to our knowledge, so far only two early deciding k-set agreement protocols have been proposed [13, 27] for that model. The protocol described in [13] assumes $t < n - k$, which means that (contrarily to what we could "normally" hope) the number of crashes t that can be tolerated decreases as the coordination degree k increases. The protocol described in [27], which imposes no constraint on t (i.e., $t < n$), is based on a mechanism that allows the processes to take into account the actual pattern of crash failures and not only their number, thereby allowing the processes to decide in much less than $\lfloor f/k \rfloor + 2$ rounds in a lot of cases (the worst case being only when the crashes are evenly distributed in the rounds with k crashes per round). We have recently designed an early deciding k-set agreement protocol for the synchronous send (only) omission failure model [28].

Content of the paper. This paper investigates the k-set agreement problem in synchronous systems prone to general omission failures and presents a corresponding protocol. This failure model lies between the crash failure model and the Byzantine failure model [24]: a faulty process is a process that crashes, or omits sending or receiving messages [14, 25]. This failure model is particularly interesting as it provides the system designers with a realistic way to represent input or output buffer overflow failures of at most t processes [14, 25]. The proposed protocol enjoys several noteworthy properties.

– The usual termination property used to define an agreement problem concerns only the correct processes: they all have to decide. This requirement is tied to the problem, independently of a particular model. Due to the very nature of the corresponding faults, there is no way to force a faulty process to decide in the crash failure model. It is the same in the Byzantine failure model where a faulty process that does not crash can decide an arbitrary value.

  The situation is different in the general omission failure model where a faulty process that does not crash cannot have an arbitrary behavior. On one side, due to the nature of the receive omission failures committed by a process, there are runs where that process can forever be prevented from learning that it can decide a value without violating the agreement property (at most k different values are decided). So, for such a process, the best that can be done in the general case is either to decide a (correct) value, or halt without deciding because it does not know whether it has a value that can be decided. On the other side, a process that commits only send omission failures receives all the messages sent to it, and should consequently be able to always decide a correct value.

  We say that a protocol is *strongly terminating* if it forces to decide all the processes that neither crash nor commit receive omission failures (we call them the *good* processes; the other processes are called *bad* processes). This new termination criterion is both theoretically and practically relevant: it extends the termination property to all the processes that are committing only "benign" faults. The proposed protocol is strongly terminating[3].

- Although, as discussed before, early decision be an interesting property, some early-deciding (consensus) protocols make a difference between early decision and early stopping: they allow a correct process to decide in $\min(f + 2, t + 1)$ but stop only at a later round (e.g., [9]). Here we are interested in early-deciding protocols in which a process decides and stops during the very same round. More precisely, the protocol has the following property:

    - A good process decides and halts by round $\min(\lfloor \frac{f}{k} \rfloor + 2, \lfloor \frac{t}{k} \rfloor + 1)$.

    So, when $\lfloor \frac{f}{k} \rfloor \leq \lfloor \frac{t}{k} \rfloor - 2$, the protocol has the noteworthy property to extend the $\lfloor \frac{f}{k} \rfloor + 2$ lower bound for a correct process to decide (1) from the crash failure model to the general omission failure model, and (2) from the correct processes to all the good processes.

    As noticed before, it is not possible to force a bad process to decide. So, for these processes the protocol "does its best", namely it ensures the following early stopping property:

    - No process executes more than $\min(\lceil \frac{f}{k} \rceil + 2, \lfloor \frac{t}{k} \rfloor + 1)$ rounds.

    Let us notice that it is possible that a bad process decides just before halting. Moreover, when $f = x\, k$ where x is an integer (which is always the case for consensus), or when there is no fault ($f = 0$), a bad process executes no more rounds than a good process. In the other cases, it executes at most one additional round.

- Each message carries a proposed value and two boolean arrays of size n (sets of process identities). This means that, if we do not consider the size of the proposed values (that does not depends on the protocol), the bit complexity is upper bounded by $O(n^2 f/k)$ per process.

The design of a protocol that satisfies, simultaneously and despite process crashes and general omission faults, the agreement property of the k-set problem, strong termination, early decision and stopping for the good processes and early stopping for the bad processes is not entirely obvious, as these properties are partly antagonistic. This is due to the fact that agreement requires that no more than k distinct values be decided (be the deciding processes correct or not), strong termination requires that, in addition to the correct processes, a well defined class of faulty processes decide, and early stopping requires the processes to halt as soon as possible. Consequently the protocol should not prevent processes from deciding at different rounds, and so, after it has decided, a process can appear to the other processes as committing omission failures, while it is actually correct. Finally, the strong termination property prevents the elimination from the protocol of a faulty process that commits only send omission failures as soon as it has been discovered faulty, as that process has to decide a value if it does not crash later. A major difficulty in the design of the protocol consists in obtaining simultaneously all these properties and not each one at the price of not satisfying one of the others.

---

[3] None of the uniform consensus protocols for the synchronous general omission failure model that we are aware of (e.g., [25, 26]) is strongly terminating.

General transformations from a synchronous failure model to another synchronous failure model (e.g., from omission to crash) are presented in [23]. These transformations are general (they are not associated with particular problems) and have a cost (simulating a round in the crash failure model requires two rounds in the more severe omission failure model). So, they are not relevant for our purpose.

When instantiated with $k = 1$, the protocol provides a new uniform consensus protocol for the synchronous general omission failure model. To our knowledge, this is the first uniform consensus protocol that enjoys strong termination and directs all the processes to terminate by round $\min(f + 2, t + 1)$. Let us finally observe that the paper leaves open two problems for future research. The first consists in proving or disproving that $\lceil \frac{f}{k} \rceil + 2$ is a tight lower bound for a bad process to stop when $f = k\,x + y$ with x and y being integers and $0 < y < k$ (we think it is). The second problem concerns t: is $t < n/2$ a lower bound to solve the strongly terminating early stopping k-set problem? (Let us remark that the answer is "yes" for $k = 1$ [23, 30].)

k-set protocol can be useful to allocate shareable resources. As an example, let us consider the allocation of broadcast frequencies in communication networks (this example is taken from [20]). Such a protocol allows processes to agree on a small number of frequencies for broadcasting large data (e.g., a movie). As the communication is broadcast based, the processes can receive the data using the same frequency.

Roadmap. The paper consists of 6 sections. Section 2 presents the computation model and gives a definition of the k-set agreement problem. To underline its basic design principles and make its understanding easier, the protocol is presented incrementally. Section 3 presents first a strongly terminating k-set agreement protocol. Then, Section 5 enriches this basic protocol to obtain a strongly terminating, early stopping k-set agreement protocol. Formal statements of the properties (lemmas and theorems) of both protocols are provided in Section 4 and Section 6, respectively. Due to the page limitation, the full proofs of these properties do not appear in this paper. The interested reader can find them in a companion technical report [29] available on-line.

2   Computation Model and Strongly Terminating k-Set Agreement

2.1   Round-Based Synchronous System

The system model consists of a finite set of processes, namely, $\Pi = \{p_1, \ldots, p_n\}$, that communicate and synchronize by sending and receiving messages through channels. Every pair of processes p_i and p_j is connected by a channel denoted (p_i, p_j). The underlying communication system is assumed to be failure-free: there is no creation, alteration, loss or duplication of message.

The system is *synchronous*. This means that each of its executions consists of a sequence of *rounds*. Those are identified by the successive integers $1, 2$, etc. For the processes, the current round number appears as a global variable r that they can read, and whose progress is managed by the underlying system. A round is made up of three consecutive phases:

- A send phase in which each process sends messages.
- A receive phase in which each process receives messages. The fundamental property of the synchronous model lies in the fact that a message sent by a process p_i to a process p_j at round r, is received by p_j at the same round r.
- A computation phase during which each process processes the messages it received during that round and executes local computation.

2.2   Process Failure Model

A process is *faulty* during an execution if its behavior deviates from that prescribed by its algorithm, otherwise it is *correct*. A *failure model* defines how a faulty process can deviate from its algorithm [15]. We consider here the following failure models:

- Crash failure. A faulty process stops its execution prematurely. After it has crashed, a process does nothing. Let us observe that if a process crashes in the middle of a sending phase, only a subset of the messages it was supposed to send might actually be sent.
- Send Omission failure. A faulty process crashes or omits sending messages it was supposed to send [14].
- General Omission failure. A faulty process crashes, omits sending messages it was supposed to send or omits receiving messages it was supposed to receive (receive omission) [25].

It is easy to see that these failure models are of increasing "severity" in the sense that any protocol that solves a problem in the General Omission (resp., Send Omission) failure model, also solves it in the (less severe) Send Omission (resp., Crash) failure model [15]. This paper considers the General Omission failure model. As already indicated, n, t and f denote the total number of processes, the maximum number of processes that can be faulty, and the actual number of processes that are faulty in a given run, respectively ($0 \leq f \leq t < n/2$).

As defined in the introduction, *good* processes are the processes that neither crash nor commit receive omission failures. A *bad* process is a process that commits receive omission failures or crashes. So, given a run, each process is either good or bad. A good process commits only "benign" failures, while a bad process commits "severe" failures.

2.3   Strongly Terminating k-Set Agreement

The problem has been informally stated in the Introduction: every process p_i *proposes* a value v_i and each correct process has to *decide* on a value in relation to the set of proposed values. More precisely, the k-set agreement problem is defined by the following three properties:

- Termination: Every correct process decides.
- Validity: If a process decides v, then v was proposed by some process.
- Agreement: No more than k different values are decided.

As we have seen 1-set agreement is the uniform consensus problem. In the following, we implicitly assume $k \leq t$ (this is because, as we have seen in the introduction, k-set agreement is trivial when $k > t$).

As already mentioned, we are interested here in protocols that direct all the good processes to decide. So, we consider a stronger version of the k-set agreement problem, in which the termination property is replaced by the following property:

- Strong Termination: Every good process decides.

3   A Strongly Terminating k-Set Agreement Protocol

We first present a strongly terminating k-set agreement protocol where the good processes terminate in $\lfloor \frac{t}{k} \rfloor + 1$ rounds. The protocol is described in Figure 1. r is a global variable that defines the current round number; the processes can only read it.

A process p_i starts the protocol by invoking the function k-SET_AGREEMENT(v_i) where v_i is the value it proposes. It terminates either by crashing, by returning the default value $\bot$ at line 08, or by returning a proposed value at line 11. As we will see, only a bad process can exit at line 08 and return $\bot$. That default value cannot be proposed by a process. So, returning $\bot$ means "no decision" from the k-set agreement point of view.

3.1   Local Variables

A process p_i manages four local variables. The scope of the first two is the whole execution, while the scope of the last two is limited to each round. Their meaning is the following:

- est_i is p_i's current estimate of the decision value. Its initial value is v_i (line 01).
- $trusted_i$ represents the set of processes that p_i currently considers as being correct. Its initial value is Π (the whole set of processes). So, $i \in trusted_i$ (line 04) means that p_i considers it is correct. If $j \in trusted_i$ we say "p_i trusts p_j"; if $j \notin trusted_i$ we say "p_i suspects p_j".
- rec_from_i is a round local variable used to contain the ids of the processes that p_i does not currently suspect and from which it has received messages during that round (line 05).
- $W_i(j)$ is a set of processes identities that represents the set of the processes p_ℓ that are currently trusted by p_i and that (to p_i's knowledge) trust p_j (line 06).

3.2   Process Behavior

The aim is for a process to decide the smallest value it has seen. But, due to the send and receive omission failures possibly committed by some processes, a process cannot safely decide the smallest value it has ever seen, it can only safely decide the smallest in a subset of the values it has received during the rounds. The crucial part of the protocol consists in providing each process with correct rules that allow it to determine its "safe subset".

During each round r, these rules are implemented by the following process behavior decomposed in three parts according to the synchronous round-based computation model.

Function k-SET_AGREEMENT(v_i)
(01) $est_i \leftarrow v_i$; $trusted_i \leftarrow \Pi$; $\% \; r = 0 \%$
(02) **for** $r = 1, \ldots, \lfloor \frac{t}{k} \rfloor + 1$ **do**
(03)   **begin_round**
(04)   **if** $(i \in trusted_i)$ **then foreach** $j \in \Pi$ **do** send$(est_i, trusted_i)$ to p_j **enddo endif**;
(05)   **let** $rec\_from_i = \{j : (est_j, trust_j)$ is received from p_j during $r \wedge j \in trusted_i\}$;
(06)   **foreach** $j \in rec\_from_i$ **let** $W_i(j) = \{\ell : \ell \in rec\_from_i \wedge j \in trust_\ell\}$;
(07) $trusted_i \leftarrow rec\_from_i - \{j : |W_i(j)| < n - t\}$;
(08)   **if** $(|trusted_i| < n - t)$ **then** return $(\bot)$ **endif**;
(09) $est_i \leftarrow \min(est_j$ received during r and such that $j \in trusted_i)$
(10)   **end_round**;
(11) return (est_i)

Fig. 1. Strongly terminating k-set protocol for general omission failures, code for p_i, $t < \frac{n}{2}$

- If p_i considers it is correct ($i \in trusted_i$), it first sends to all the processes its current local state, namely, the current pair $(est_i, trusted_i)$ (line 04). Otherwise, p_i skips the sending phase.
- Then, p_i executes the receive phase (line 05). As already indicated, when it considers the messages it has received during the current round, p_i considers only the messages sent by the the processes it trusts (here, the set $trusted_i$ can be seen as a filter).
- Finally, p_i executes the local computation phase that is the core of the protocol (lines 06-09). This phase is made up of the following statements where the value $n - t$ constitutes a threshold that plays a fundamental role.
  - First, p_i determines the new value of $trusted_i$ (lines 06-07). It is equal to the current set rec_from_i from which are suppressed all the processes p_j such that $|W_i(j)| < n - t$. These processes p_j are no longer trusted by p_i because there are "not enough" processes trusted by p_i that trust them (p_j is missing "Witnesses" to remain trusted by p_i, hence the name $W_i(j)$); "not enough" means here less than $n - t$.
  - Then, p_i checks if it trusts enough processes, i.e., at least $n - t$ (line 08). If the answer is negative, as we will see in the proof, p_i knows that it has committed receive omission failures and cannot safely decide. It consequently halts, returning the default value $\bot$.
  - Finally, if it has not stopped at line 08, p_i computes its new estimate of the decision value (line 09) according to the estimate values it has received from the processes it currently trusts.

4   Proof of the Strongly Terminating Protocol

The full proof of the protocol is given in [29]. The protocol proof assumes $t < n/2$. It uses the following notations.

- Given a set of process identities $X = \{i, j, \ldots\}$, we sometimes use $p_i \in X$ for $i \in X$.
- $\mathcal{C}$ is the set of correct processes in a given execution.

- $x_i[r]$ denotes the value of p_i's local variable x at the end of round r.
  By definition $trusted_i[0] = \Pi$. When $j \in trusted_i$, we say that "p_i trusts p_j" (or "p_j is trusted by p_i").
- $Completing[r] = \{i : p_i \text{ proceeds to } r + 1 \}$. By definition $Completing[0] = \Pi$.
  (If $r = \lfloor \frac{t}{k} \rfloor + 1$, "$p_i$ proceeds to $r + 1$" means p_i executes line 11.)
- $EST[r] = \{est_i[r] : i \in Completing[r]\}$. By definition $EST[0] =$ the proposed values.
  $EST[r]$ contains the values that are present in the system at the end of round r.
- $Silent[r] = \{i : \forall j \in Completing[r] : i \notin trusted_j[r]\}$. It is important to remark that if $i \in Silent[r]$, then no process p_j (including p_i itself) takes into account est_i sent by p_i (if any) to update its local variables est_j at line 09 of the round r. ($Silent[0] = \emptyset$.)

The proof of the following relations are left to the reader: $Completing[r + 1] \subseteq Completing[r]$, $Silent[r] \subseteq Silent[r + 1]$, $\forall i \in Completing[r] : Silent[r] \subseteq \Pi - trusted_i[r]$.

4.1   Basic Lemmas

The first lemma that follows will be used to prove that a process that does not commit receive omission failure decides.

Lemma 1. *Let p_i be a process that is correct or commits only send omission failures. We have $\forall r : (1) C \subseteq trusted_i[r]$ and $(2) i \in Completing[r]$.*

The next two lemmas show that $n - t$ is a critical threshold related to the number of processes (1) for a process to become silent or (2) for the process estimates to become smaller or equal to some value. More explicitly, the first of these lemmas states that if a process p_x is not trusted by "enough" processes (i.e., trusted by less than $n - t$ processes[4]) at the end of a round $r - 1$, then that process p_x is not trusted by the processes that complete round r.

Lemma 2. $\forall r \geq 1 : \forall x : \left| \{y : y \in Completing[r - 1] \wedge x \in trusted_y[r - 1]\} \right| < n - t \Rightarrow x \in Silent[r]$.

The next lemma shows that if "enough" (i.e., at least $n - t$) processes have an estimate smaller than or equal to a value v at the end of a round $r - 1$, then no process $p_i \in Completing[r]$ has a value greater than v at the end of r.

Lemma 3. *Let v be an arbitrary value.* $\forall r \geq 1 : \left| \{x : est_x[r - 1] \leq v \wedge x \in Completing[r - 1]\} \right| \geq n - t \Rightarrow \forall i \in Completing[r] : est_i[r] \leq v$.

Finally, the next lemma states that the sequence of set values $EST[0], EST[1], \ldots$ is monotonic and never increases.

Lemma 4. $\forall r \geq 0 : EST[r + 1] \subseteq EST[r]$.

---

[4] Equivalently, trusted by at most t processes.

4.2   Central Lemma

The lemma that follows is central to prove the agreement property, namely, at most k distinct values are decided. Its formulation is early-stopping oriented. Being general, this formulation allows using the same lemma to prove both the non-early stopping version of the protocol (Theorem 3) and the early stopping protocol (Theorem 4).

Lemma 5. *Let r $(1 \leq r \leq \lfloor \frac{t}{k} \rfloor + 1)$ be a round such that (1) $C \subseteq Completing[r-1]$, and (2) $|EST[r]| > k$ (let v_m denote the kth smallest value in $EST[r]$, i.e., the greatest value among the k smallest values of $EST[r]$). Let $i \in Completing[r]$. We have $n - k\,r < |trusted_i[r]| \Rightarrow est_i[r] \leq v_m$.*

4.3   Properties of the Protocol

Theorem 1. [Validity] *A decided value is a proposed value.*

Theorem 2. [Strong Termination] *A process p_i that neither crashes nor commits receive omission failures decides.*

As a correct process does not commit receive omission failures, the following corollary is an immediate consequence of the previous theorem.

Corollary 1. [Termination] *Every correct process decides.*

Theorem 3. [Agreement] *No more than k different values are decided.*

5   A Strongly Terminating and Early Stopping k-Set Agreement Protocol

This section enriches the previous strongly terminating k-set agreement protocol to obtain an early stopping protocol, namely, a protocol where a good process decides and halts by round $\min(\lfloor \frac{f}{k} \rfloor + 2, \lfloor \frac{t}{k} \rfloor + 1)$, and a bad process executes at most $\min(\lceil \frac{f}{k} \rceil + 2, \lfloor \frac{t}{k} \rfloor + 1)$ rounds.

The protocol is described in Figure 2. To make reading and understanding easier, all the lines from the first protocol appears with the same number. The line number of each of the 10 new lines that make the protocol early stopping are prefixed by "E". We explain here only the new parts of the protocol.

5.1   Additional Local Variables

A process p_i manages three additional local variables, one (can_dec_i) whose scope is the whole computation, and two (CAN_DEC_i and REC_FROM_i) whose scope is limited to each round. Their meaning is the following.

- can_dec_i is a set of process identities that contains, to p_i's knowledge, all the processes that can decide a value without violating the agreement property. The current value of can_dec_i is part of each message sent by p_i. Its initial value is $\emptyset$.

```
Function k-SET_AGREEMENT(v_i)
(01)    est_i ← v_i; trusted_i ← Π; can_dec_i ← ∅;    % r = 0 %
(02)    for r = 1, ..., ⌊t/k⌋ + 1 do
(03)    begin_round
(04)        if (i ∈ trusted_i) then
                        foreach j ∈ Π do send (est_i, trusted_i, can_dec_i) to p_j enddo endif;
(E01)       let REC_FROM_i = {i} ∪ {j : (est_j, trust_j, c_dec_j) rec. from p_j during r};
(E02)       let CAN_DEC_i = ∪(c_dec_j : j ∈ REC_FROM_i);
(E03)       if (i ∉ trusted_i ∨ i ∈ can_dec_i) then
(E04)          if |CAN_DEC_i| > t then let EST_i = {est_j : j ∈ REC_FROM_i ∧ c_dec_j ≠ ∅};
(E05)                          return (min(EST_i))
(E06)       endif  endif;
(05)        let rec_from_i = {j : (est_j, trust_j, c_dec_j) rec. from p_j during r ∧ j ∈ trusted_i};
(06)        foreach j ∈ rec_from_i let W_i(j) = {ℓ : ℓ ∈ rec_from_i ∧ j ∈ trust_ℓ};
(07)        trusted_i ← rec_from_i − {j : |W_i(j)| < n − t};
(08)        if (|trusted_i| < n − t) then return (⊥) endif;
(09)        est_i ← min(est_j received during r and such that j ∈ trusted_i);
(E07)       can_dec_i ← ∪(c_dec_j received during r and such that j ∈ trusted_i);
(E08)       if (i ∈ trusted_i ∧ i ∉ can_dec_i)then
(E09)          if (n − k r < |trusted_i|) ∨ (can_dec_i ≠ ∅) then can_dec_i ← can_dec_i ∪ {i}
(E10)       endif  endif
(10)    end_round;
(11)    return (est_i)
```

Fig. 2. k-set early-deciding protocol for general omission failures, code for p_i, $t < \frac{n}{2}$

- REC_FROM_i is used by p_i to store its id plus the ids of all the processes from which it has received messages during the current round r (line E01). Differently from the way rec_from_i is computed (line 05), no filtering (with the set $trusted_i$) is used to compute REC_FROM_i.
- CAN_DEC_i is used to store the union of all the can_dec_j sets that p_i has received during the current round r (line E02).

5.2   Process Behavior

As already indicated, the behavior of a process p_i is modified by adding only 10 lines (E01-E10). It is important to notice that no variable used in the basic protocol is updated by these lines; the basic protocol variables are only read. This means that, when there is no early deciding/stopping at line E05, the enriched protocol behaves exactly as the basic protocol.

Let us now examine the two parts of the protocol where the new statements appear.

- Let us first consider the lines E07-E10.
  After it has updated its current estimate est_i (line 09), p_i updates similarly its set can_dec_i, to learn the processes that can early decide. As we can see, est_i and can_dec_i constitute a pair that is sent (line 04) and updated "atomically".
  Then, if p_i trusts itself ($i \in trusted_i$) and, up to now, was not allowed to early decide and stop ($i \notin can\_dec_i$), it tests a predicate to know if it can early decide. If

it can, p_i adds its identity to can_dec_i (line E09). The "early decision" predicate is made up of two parts:

- If $can\_dec_i \neq \emptyset$, then p_i learns that other processes can early decide. Consequently, as it has received and processed their estimates values (line 09), it can safely adds its identity to can_dec_i.
- If $n - k\,r < |trusted_i|$, then p_i discovers that the set of processes it trusts is "big enough" for it to conclude that it knows one of the k smallest estimate values currently present in the system. "Big enough" means here greater than $n - k\,r$. (Let us notice that threshold was used in Lemma 5 in the proof of the basic protocol.)

- Let us now consider the lines E01-E06.

  As already indicated REC_FROM_i and CAN_DEC_i are updated in the receive phase of the current round.

  To use these values to decide during the current round (at line E05), p_i must either be faulty (predicate $i \notin trusted_i$) or have previously sent its pair (est_i, can_dec_i) to the other processes (predicate $i \notin trusted_i \vee i \in can\_dec_i$ evaluated at line E03). But, when $i \in trusted_i$, $i \in can\_dec_i$ is not a sufficiently strong predicate for p_i to safely decide. This is because it is possible that p_i committed omission faults just during the current round. So, to allow p_i to early decide, we need to be sure that at least one correct process can decide (as it is correct such a process p_j can play a "pivot" role sending its (est_j, can_dec_j) pair to all the processes). Hence, the intuition for the final early decision/stopping predicate, namely $|CAN\_DEC_i| > t$ used at line E04: that additional predicate guarantees that at least one correct process can early decide and consequently has transmitted or will transmit its (est_j, can_dec_j) pair to all.

So, the early decision/stopping predicate for a process p_i spans actually two rounds r and r' ($r' > r$). This is a "two phase" predicate split as follows:

- During r (lines E08|E09): $(i \in trusted_i \wedge i \notin can\_dec_i) \wedge (n - k\,r < |trusted_i|) \vee (can\_dec_i \neq \emptyset)$, and
- During r' (lines E03|E04): $(i \notin trusted_i \vee i \in can\_dec_i) \wedge |CAN\_DEC_i| > t$.

Moreover, for a correct process p_i, the assignment $can\_dec_i \leftarrow can\_dec_i \cup \{i\}$ can be interpreted as a synchronization point separating the time instants when they are evaluated to $true$.

6 Proof of the Strongly Terminating Early Stopping Protocol

Detailed proofs of the following lemmas and theorems are given in [29].

6.1 Basic Lemmas

The next lemma extends Lemma 1 to the early stopping context.

Lemma 6. *Let r_d be the first round during which a correct process decides at line E05 (If there is no such round, let $r_d = \lfloor \frac{t}{k} \rfloor + 1$). Let p_i be a process that is correct or commits only send omission failures. $\forall r \leq r_d$: if p_i does not decide at line E05 of the round r, we have (1) $\mathcal{C} \subseteq trusted_i[r]$ and (2) $i \in Completing[r]$.*

Lemma 5 considers a round r such that $C \subseteq Completing[r-1]$ (i.e., a round executed by all the correct processes). Its proof relies on Lemma 1, but considers only the rounds $r' \leq r$. As, until a correct process decides, the Lemma 1 and the Lemma 6 are equivalent, it follows that the Lemma 1 can be replaced by Lemma 6 in the proof of Lemma 5. Let us also observe that the proofs of the Lemmas 2, 3 and 4 are still valid in the early stopping context (these proofs use the set $Completing[r]$ and do not rely on the set C). We now state and prove additional lemmas used to prove the early stopping k-set agreement protocol.

Lemma 7. *The set $EST_i[r]$ computed by p_i during round r (line E04) is not empty.*

Lemma 8. *Assuming that a process decides at line E05 during round r, let p_x be a process that proceeds to round $r+1$ (if $r = \lfloor \frac{t}{k} \rfloor + 1$, "proceed to round $r+1$" means "execute the return() statement at line 11"). We have: $x \notin trusted_x[r] \lor x \in can\_dec_x[r]$.*

Lemma 9. *Let $i \in Completing[r]$ ($1 \leq r \leq \lfloor \frac{t}{k} \rfloor + 1$). $can\_dec_i[r] \neq \emptyset \Rightarrow est_i[r]$ is one of the k smallest values in $EST[r]$.*

Lemma 10. *Assuming that a process decides at line E05 during round r, let p_x be a process that proceeds to round $r+1$ (if $r = \lfloor \frac{t}{k} \rfloor + 1$, "proceed to round $r+1$" means "execute the return() statement at line 11"). We have: $est_x[r]$ is among the k smallest values in $EST[r-1]$.*

Lemma 11. *Let $r \leq \lfloor \frac{t}{k} \rfloor$ be the first round during which a process decides at line E05. Then, (1) every process that is correct or commits only send omission failures decides at line E05 during round r or $r+1$. Moreover, (2) no process executes more than $r+1$ rounds.*

6.2   Properties of the Protocol

Theorem 4. [Agreement] *No more than k different values are decided.*

Theorem 5. [Strong Termination and Early Stopping] *(i) A process that is correct or commits only send omission failures decides and halts by round $\min(\lfloor \frac{t}{k} \rfloor + 2, \lfloor \frac{t}{k} \rfloor + 1)$. (ii) No process halts after $\min(\lceil \frac{f}{k} \rceil + 2, \lfloor \frac{t}{k} \rfloor + 1)$ rounds.*

The next corollary is an immediate consequence of the previous theorem.

Corollary 2. [Termination] *Every correct process decides.*

Theorem 6. [Validity] *A decided value is a proposed value.*

Theorem 7. [Bit Complexity] *Let b be the number of bits required to represent a proposed value. The bit complexity is upper bounded by $O(n(b+2n)f/k)$ per process.*

References

1. Aguilera M.K. and Toueg S., A Simple Bivalency Proof that t-Resilient Consensus Requires $t + 1$ Rounds. *Information Processing Letters*, 71:155-178, 1999.
2. Attiya H. and Welch J., Distributed Computing, Fundamentals, Simulation and Advanced Topics (Second edition). *Wiley Series on Parallel and Distributed Computing*, 414 pages, 2004.
3. Biran O., Moran S. and Zaks S., A Combinatorial Characterization of the Distributed 1-Solvable Tasks. *Journal of Algorithms*, 11(3): 420-440, 1990.
4. Biran O., Moran S. and Zaks S., Tight Bounds on the Round Complexity of Distributed 1-Solvable Tasks. *Theoretical Computer Science*, 145(1-2):271-290, 1995.
5. Borowsky E. and Gafni E., Generalized FLP Impossibility Results for t-Resilient Asynchronous Computations. *Proc. 25th ACM Symposium on Theory of Computation (STOC'93)*, California (USA), pp. 91-100, 1993.
6. Charron-Bost B. and Schiper A., Uniform Consensus is Harder than Consensus. *Journal of Algorithms*, 51(1):15-37, 2004.
7. Chaudhuri S., More *Choices* Allow More *Faults:* Set Consensus Problems in Totally Asynchronous Systems. *Information and Computation,* 105:132-158, 1993.
8. Chaudhuri S., Herlihy M., Lynch N. and Tuttle M., Tight Bounds for k-Set Agreement. *Journal of the ACM*, 47(5):912-943, 2000.
9. Dolev D., Reischuk R. and Strong R., Early Stopping in Byzantine Agreement. *Journal of the ACM*, 37(4):720-741, April 1990.
10. Fischer M.J., Lynch N.A., A Lower Bound on the Time to Assure Interactive Consistency. *Information Processing Letters*, 14(4):183-186, 1982.
11. Fischer M.J., Lynch N.A. and Paterson M.S., Impossibility of Distributed Consensus with One Faulty Process. *Journal of the ACM*, 32(2):374-382, 1985.
12. Gafni E., Guerraoui R. and Pochon B., >From a Static Impossibility to an Adaptive Lower Bound: The Complexity of Early Deciding Set Agreement. *Proc. 37th ACM Symposium on Theory of Computing (STOC'05)*, Baltimore (MD), pp.714-722, May 2005.
13. Guerraoui R. and Pochon B., The Complexity of Early Deciding Set Agreement: how Topology Can Help? *Proc. 4th Workshop in Geometry and Topology in Concurrency and Distributed Computing (GETCO'04)*, BRICS Notes Series, NS-04-2, pp. 26-31, Amsterdam (NL), 2004.
14. Hadzilacos V., Issues of Fault Tolerance in Concurrent Computations. *PhD Thesis, Tech Report 11-84*, Harvard University, Cambridge (MA), 1985.
15. Hadzilacos V. and Toueg S., Reliable Broadcast and Related Problems. In *Distributed Systems*, ACM Press (S. Mullender Ed.), New-York, pp. 97-145, 1993.
16. Herlihy M.P. and Penso L. D., Tight Bounds for k-Set Agreement with Limited Scope Accuracy Failure Detectors. *Distributed Computing*, 18(2): 157-166, 2005.
17. Herlihy M.P. and Shavit N., The Topological Structure of Asynchronous Computability. *Journal of the ACM*, 46(6):858-923, 1999.
18. Keidar I. and Rajsbaum S., A Simple Proof of the Uniform Consensus Synchronous Lower Bound. *Information Processing Letters*, 85:47-52, 2003.
19. Lamport L. and Fischer M., Byzantine Generals and Transaction Commit Protocols. *Unpublished manuscript*, 16 pages, April 1982.
20. Lynch N.A., Distributed Algorithms. *Morgan Kaufmann Pub.*, San Fransisco (CA), 872 pages, 1996.
21. Mostéfaoui A. and Raynal M., k-Set Agreement with Limited Accuracy Failure Detectors. *Proc. 19th ACM Symposium on Principles of Distributed Computing (PODC'00)*, ACM Press, pp. 143-152, Portland (OR), 2000.

22. Mostéfaoui A. and Raynal M., Randomized Set Agreement. *Proc. 13th ACM Symposium on Parallel Algorithms and Architectures (SPAA'01)*, ACM Press, pp. 291-297, Hersonissos (Crete), 2001.

23. Neiger G. and Toueg S., Automatically Increasing the Fault-Tolerance of Distributed Algorithms. *Journal of Algorithms*, 11:374-419, 1990.

24. Pease L., Shostak R. and Lamport L., Reaching Agreement in Presence of Faults. *Journal of the ACM*, 27(2):228-234, 1980.

25. Perry K.J. and Toueg S., Distributed Agreement in the Presence of Processor and Communication Faults. *IEEE Transactions on Software Eng.*, SE-12(3):477-482, 1986.

26. Raïpin Parvédy Ph. and Raynal M., Optimal Early Stopping Uniform Consensus in Synchronous Systems with Process Omission Failures. *Proc. 16th ACM Symposium on Parallel Algorithms and Architectures (SPAA'04)*, Barcelona (Spain), ACM Press, pp. 302-310, 2004.

27. Raïpin Parvédy Ph., Raynal M. and Travers C., Early-Stopping k-set Agreement in Synchronous Systems Prone to any Number of Process Crashes. *8th Int. Conference on Parallel Computing Technologies (PaCT'05)*, Krasnoyarsk (Russia), Springer Verlag LNCS #3606, pp. 49-58, 2005.

28. Raïpin Parvédy Ph., Raynal M. and Travers C., Decision Optimal Early-Stopping k-set Agreement in Synchronous Systems Prone to Send Omission Failures. *Proc. 11th IEEE Pacific Rim Int. Symposium on Dependable Computing (PRDC'05)*, Changsa (China), IEEE Computer Press, pp. 23-30, 2005.

29. Raïpin Parvédy Ph., Raynal M. and Travers C., Strongly Terminating Early-Stopping k-set Agreement in Synchronous Systems with General Omission Failures. *Tech Report #1711*, IRISA, Université de Rennes (France), 22 pages 2005. ftp://ftp.irisa.fr/techreports/2005/PI-1711.ps.gz

30. Raynal M., Consensus in Synchronous Systems: a Concise Guided Tour. *Proc. 9th IEEE Pacific Rim Int. Symposium on Dependable Computing (PRDC'02)*, Tsukuba (Japan), IEEE Computer Press, pp. 221-228, 2002.

31. Saks M. and Zaharoglou F., Wait-Free k-Set Agreement is Impossible: The Topology of Public Knowledge. *SIAM Journal on Computing*, 29(5):1449-1483, 2000.

32. Yang J., Neiger G. and Gafni E., Structured Derivations of Consensus Algorithms for Failure Detectors. *Proc. 17th Int. ACM Symposium on Principles of Distributed Computing (PODC'98)*, ACM Press, pp. 297-308, Puerto Vallarta (Mexico), July 1998.

On Fractional Dynamic Faults with Threshold[*]

Stefan Dobrev[1], Rastislav Královič[2], Richard Královič[2], and Nicola Santoro[3]

[1] School of Information Technology and Engineering, University of Ottawa, Ottawa,
K1N 6N5, Canada
[2] Dept. of Computer Science, Comenius University, Mlynská dolina,
84248 Bratislava, Slovakia
[3] School of Computer Science, Carleton University, Ottawa, K1S 5B6, Canada

Abstract. Unlike *localized* communication failures that occur on a fixed
(although a priori unknown) set of links, *dynamic* faults can occur on any
link. Known also as mobile or ubiquitous faults, their presence makes
many tasks difficult if not impossible to solve even in synchronous systems. Their analysis and the development of fault-tolerant protocols have
been carried out under two main models. In this paper, we introduce a
new model for dynamic faults in synchronous distributed systems. This
model includes as special cases the existing settings studied in the literature. We focus on the hardest setting of this model, called *simple
threshold*, where to be guaranteed that at least one message is delivered
in a time step, the total number of transmitted messages in that time
step must reach a threshold $T \leq c(G)$, where $c(G)$ is the edge connectivity of the network. We investigate the problem of *broadcasting* under this
model for the worst threshold $T = c(G)$ in several classes of graphs as
well as in arbitrary networks. We design solution protocols, proving that
broadcast is possible even in this harsh environment. We analyze the
time costs showing that broadcast can be completed in (low) polynomial
time for several networks including rings (with or without knowledge of
n), complete graphs (with or without chordal sense of direction), hypercubes (with or without orientation), and constant-degree networks (with
or without full topological knowledge).

1 Introduction

1.1 Dynamic Faults

In a message-passing distributed computing environment, entities communicate
by sending messages to their neighbors in the underlying communication network. However, during transmission, messages might be lost.

The presence of communication faults renders the solution of problems difficult if not impossible. In particular, in *asynchronous* settings, the mere possibility
of faults renders unsolvable almost all non trivial tasks, even if the faults are *localized* to (i.e., restricted to occur on the links of) a single entity [11]. Due to this

---

[*] Partially supported by VEGA 1/3106/06, UK/404/2006, NSERC, and TECSIS Co.

P. Flocchini and L. Gąsieniec (Eds.): SIROCCO 2006, LNCS 4056, pp. 197–211, 2006.

inherent difficulty connected with asynchrony, the focus is on *synchronous* environments, both from the point of view of theoretical investigation, and industrial application (e.g. communication protocols for wireless networks).

Since synchrony provides a perfect omission detection mechanism [2], localized faults are easily dealt with in these systems; indeed, any number of faulty links can be tolerated provided they do not disconnect the network. The immediate question is then whether synchrony allows to tolerate also *dynamic* communication faults; that is, faults that are not restricted to a fixed (but a priori unknown) set of links, but can occur between any two neighbors [17]. These types of faults, also called *mobile* or *ubiquitous*, are clearly more difficult to handle.

In this regard, the investigations have focused mostly on the basic problem of *broadcast*: an entity has some information that must communicate to all other entities in the network. Indeed, the ability or impossibility of performing this task has immediate consequence for many other tasks. Not surprisingly, a large research effort has been on the analysis of broadcasting in the presence of dynamic communication faults.

Clearly no computation, including broadcast, is possible if the amount of faults that can occur per time unit and the modality of their occurrence is unrestricted. The research quest has thus been on determining under what conditions on the faults non-trivial computations can be performed in spite of those faults. Constructively, the effort is on designing protocols that can correctly solve a problem provided some restrictions on the occurrence of faults hold.

A first large group of investigations have considered the so-called *cumulative* model; that is, there is a (known) limit L on the number[1] of messages that can be lost at each time unit. If the limit is less than the edge connectivity of the network, $L < c(G)$, then broadcast can be achieved by simply flooding and repeating transmissions for an appropriate amount of time. The research has been on determining what is the smallest amount of time in general or for specific topologies [3, 4, 5, 6, 8, 9, 10, 12, 14, 15], as well as on how to use broadcast for efficiently computing functions and achieving other tasks [7, 18, 19].

The advantage of the cumulative model is that solutions designed for it are L-tolerant; that is they tolerate up to L communication faults per time units. The disadvantage of this approach is that it neglects the fact that in real systems the number of lost messages is generally a function of the number of all message transitions. This feature leads to an anomaly of the cumulative model, where solutions that flood the network with large amounts of messages tend to work well, while their behavior in real faulty environments is often quite poor.

A setting that takes into account the interplay between amount of transmissions and number of losses is the *probabilistic* model: there is no a priori upper bound on the total number of faults per time unit, but each transmission has a (known) probability $p < 1$ to fail. The investigations in this model have focused on designing broadcasting algorithms with low time complexity and high probability of success [1, 16]. The drawback of this model is that the solutions derived for it have no deterministic guarantee of correctness.

---

[1] Since the faults are dynamic, no restriction is clearly posed on their location.

The drawbacks of these two models have been the motivation behind the introduction of the so called *fractional* model, a deterministic setting that explicitly takes into account the interaction between the number of omissions and the number of messages. In the fractional model, the amount of faults that can occur at time t is not fixed but rather a linear fraction $\lfloor \alpha\, m_t \rfloor$ of the total number m_t of messages sent at time t, where $0 \leq \alpha < 1$ is a (known) constant. The advantage of the fractional model is that solutions designed for it tolerate the loss of up to a fraction of all transmitted messages [13]. The anomaly of the fractional model is that, in this setting, transmitting a single message per communication round ensures its delivery; thus, the model leads to very counterintuitive algorithms which do not behave well in real faulty environments.

Summarizing, to obtain optimal solutions, message redundancy must be avoided in the fractional model, while massive redundancy of messages must be used in the cumulative model; in real systems, both solutions might not fare well. In many ways, the two models are opposite extremes. The lesson to be learned from their anomalies is that on one hand there is need to use redundant communication, but on the other hand brute force algorithms based on repeatedly flooding the network do not necessarily solve the problem.

In this paper we propose a deterministic model that combines the cumulative and fractional models in a way that might better reflect reality. This model is actually more general, in that it includes those models as particular, extreme cases. It also defines a spectrum of settings that avoid the anomalies of both extreme cases.

1.2   Fractional Threshold and Broadcast

The failure model we consider, and that we shall call *fractional dynamic faults with threshold* or simply *fractional threshold* model, is a combination of the fractional model with the cumulative model. Both fractional and cumulative models can be described as a game between the algorithm and an adversary: in a time step t, the algorithm tries to send m_t messages, and the adversary may destroy up to $F(m_t)$ of them. While in the cumulative model, the *dependency function* F is a constant function, in fractional model $F(m_t) = \lfloor \alpha m_t \rfloor$. The dependency function of the fractional threshold model is the maximum of those two:

$$F(m_t) = \max\{T - 1, \lfloor \alpha\, m_t \rfloor\}$$

where $T \leq c(G)$ is a constant at most equal to the edge connectivity of the graph, and α is a constant $0 \leq \alpha < 1$. The name "fractional threshold" comes from the fact that it is the fractional model with the additional requirement that the algorithm has to send at least T messages in a time step t in order to have any guarantees about the number of faults.

Note that both the cumulative and the fractional models are particular, extreme instances of this model. In fact, $\alpha = 0$ yields the *cumulative* setting: at most $T - 1$ faults occur at each time step. On the other hand, the case $T = 1$ results in the *fractional* setting. In between, it defines a spectrum of new settings never explored before, which avoid the anomalies of both extreme cases.

From this spectrum, the settings that give the maximum power to the adversary, thus making the broadcasting most difficult, are what will be called a *simple threshold* model defined by $T = c(G)$ and $\alpha = 1 - \varepsilon$ with ε infinitely close to 0. In this model, if less than $c(G)$ messages are sent in a step, none of them is guaranteed to arrive (i.e., they all may be lost); on the other hand, if at least $c(G)$ messages are transmitted, at least one message is guaranteed to be delivered.

In this paper we start the analysis of fault-tolerant computing in the fractional threshold model, focusing on the simple threshold setting. In this draconian setting the tricks from cumulative and fractional models fail: if the algorithm uses simple flooding the adversary can deliver only one message between the same pair of vertices over and over. If, on the other hand, the algorithm sends too few messages, they all may be lost.

1.3   The Results

The network is represented by a simple graph G of n vertices representing the entities and m edges representing the links. The vertices are *anonymous*, i.e. they are without distinct IDs. The communication is by means of synchronous message passing (i.e. in globally synchronized communication rounds), local computation is performed between the communication rounds and is considered instantaneous. The communication failures are dynamic omissions in the simple threshold model.

We consider the problem of *broadcasting*: At the beginning, there is a single initiator v containing the information to be disseminated. Upon algorithm termination, all entities must have learned this information. We consider *explicit* termination, i.e. when the algorithm terminates at an entity, it will not process any more messages (and, in fact, no messages should be arriving anyway).

The complexity measure of interest is *time* (i.e., number of communication rounds). We consider various levels of topological knowledge about the network (knowing network size n, being aware of the network topology, having *Sense of Direction* or having full topological knowledge).

In this paper, we focus on the hardest setting, the *simple threshold*, where to be guaranteed that at least one message is delivered in a time step, the total amount of transmitted messages in that time step must be at least $c(G)$, i.e. the edge connectivity of the network.

By definition, it is sufficient to ensure that $c(G)$ or more messages are transmitted at each time unit to guarantee that at least one of these messages is delivered. The problem however is that an entity does not know which other entities are transmitting at the same time and in general does not know which of its neighbors has already received its messages. Indeed the problem, in spite of synchrony and of the simplicity of its statement, is not simple.

We investigate the problem of *broadcasting* under this model in several classes of graphs as well as in arbitrary networks. We design solution protocols, proving that broadcast is possible even under the worst threshold $c(G)$. We analyze the time costs showing the surprising result that broadcast can be completed in (low) polynomial time for several networks including rings (with or without knowledge

Table 1. Summary of results presented in this paper

Topology	Condition	Time complexity
ring	n not necessarily known	$\Theta(n)$
complete graph	with chordal sense of direction	$O(n^2)$
complete graph	unoriented	$\Omega(n^2), O(n^3)$
hypercube	oriented	$O(n^2 \log n)$
hypercube	unoriented	$O(n^4 \log^2 n)$
arbitrary network	full topological knowledge	$O(2^{c(G)} nm)$
arbitrary network	no topological knowledge except $c(G)$, n, m	$O(2^{c(G)} m^2 n)$

of n), complete graphs (with or without chordal sense of direction), hypercubes (with or without orientation), and constant-degree networks (with or without full topological knowledge). In addition to the upper bounds, we also establish a lower bound in the case of complete graphs without sense of direction. The results are summarized in the Table 1. Due to space constraints some technical parts have been omitted.

2   Ring

The ring is a 2-connected network, i.e. $T = c(G) = 2$. Hence, at least two messages must be sent in a round to ensure that not all of them are lost.

We first present the algorithm assuming the ring size n is known, and then show how it can be extended to the case n unknown.

At any moment of time, the vertices can be either *informed* or *uninformed*. Since the information is spreading from the single initiator vertex s, informed vertices form a connected component. The initiator splits this component into the left part and the right part. Each informed vertex v can easily determine whether it is on the left part or on the right part of the informed component – this information is delivered in the message that informs the vertex v.

Each informed vertex can be further classified as either *active* or *passive*. A vertex is active if and only if it has received a message from only one of its neighbor. A passive vertex has received a message from both neighbors. This implies that, as long as the broadcast has not yet finished, there is at least one active vertex in both left and right part of the informed component (the left-most and the right-most informed vertices must be active; note, however, that also the intermediate vertices might be active).

The computation consists of $n - 1$ phases, with each phase taking four communication rounds. The goal of a phase is to ensure that at least one active vertex becomes passive.

Each phase consists of the following four steps:

1. Each active vertex sends a message to its possibly uninformed neighbor.
2. Each active vertex in the right part sends a message to its possibly uninformed neighbor. Each vertex in the left part that received a message in step 1 replies to this message.

3. Same as step 2, but left and right parts are reversed.
4. Each vertex that received a non-reply message in steps 1–3 replies to that message.

To avoid corner cases at the initiator of the broadcast, the initiator is split into two virtual vertices such that each of them starts in active state (i.e. the initiator acts as if it belongs both to the left and to the right part).

Lemma 1. *At least one reply message passes during the phase.*

Initially, there are two active (virtual) vertices (the left- and right- part of the initiator). Lemma 1 ensures that during each of the subsequent phases, at least one previously active vertex becomes passive. Since passive vertices never become active again, it follows that after at most $n - 1$ phases, there are $n - 1$ passive vertices. Once there are $n - 1$ passive vertices, the remaining two must be informed (both are neighbors of a passive vertex), i.e. $n - 1$ phases are sufficient to complete the broadcast.

Note also that the algorithm does not require distinct IDs or ring orientation (it can compute them, though, as it is initiated by a single vertex).

Theorem 1. *There is $4(n - 1)$-time fault-tolerant broadcasting algorithm for (anonymous, unoriented) rings of known size.*

If n is unknown, the above algorithm cannot be directly used, as it does not know when to terminate. This is not a serious obstacle, though. Assume that the algorithm is run without a time bound, and each discover message also contains a counter how far is the vertex from the initiator. After at most n phases there will be a vertex v that has received discover messages from both directions. From the counters in those messages v can compute the ring size n. In the second part of the algorithm v broadcasts n (and the time since the start of the second broadcast) using the algorithm for known n; when that broadcast is finished, the whole algorithm terminates. In order to make this work, we have to ensure that there is no interaction between the execution of the first broadcast and the second broadcast. That can be easily accomplished by scheduling the communication steps of the first broadcast in odd time slots and the second broadcast in even time slots.

Theorem 2. *There is an $O(n)$-time fault-tolerant broadcasting algorithm for (anonymous, unoriented) rings of unknown size.*

3   Complete Graphs

As the connectivity of complete graphs is $n - 1$, we assume that least $n - 1$ messages must be sent to ensure that at least one passes through.

3.1   Complete Graphs with Chordal Sense of Direction

Chordal Sense of Direction in a complete graphs means that vertices are numbered $0, 1, \ldots, n - 1$ and the link from a vertex u to a vertex v is labelled $v - u \bmod n$.[2]

The algorithm consists of two parts. The purpose of the first part is to make sure that at least $\lceil n/2 \rceil$ vertices are informed; the second part uses these vertices to inform the remaining ones. The algorithm is executed by informed vertices. Each message contains a time counter, so a newly informed vertex can learn the time and join the computation at the right place.

The first part of the algorithm consists of phases $0, 1, \ldots, \lceil n/2 \rceil - 2$. During phase 0 the initiator sends messages to all its neighbors. The goal of phase k is to ensure that there are at least $k + 1$ informed vertices distinct from the initiator; this ensures that after the first part, there are at least $\lceil n/2 \rceil$ informed vertices.

Consider a phase k and suppose that there are exactly k informed vertices distinct from the initiator at the beginning of phase k. Let $d = \left\lfloor \frac{n-1}{k+1} \right\rfloor$, and consider $k+1$ disjoint intervals $I_0, \ldots, I_k$ each of size d, consisting of non-initiator vertices. The phase will consist of $k + 1$ steps. The idea is that during the i-th step, the informed vertices (including initiator) try to inform an additional vertex in the interval I_i by sending messages to all vertices in I_i. If I_i does not contain any informed vertices, and at least one message is delivered, then a new vertex must be informed. The problem is, however, that only $d(k+1)$ messages are sent, which may not be sufficient to guarantee delivery. To remedy this, the i-th step will span over d rounds. In a j-th round, all informed vertices send messages to all vertices in I_i and to the j-th vertex of $I_{i \oplus 1}$ (the addition is taken modulo $k + 1$). Now, in each step there are $(k + 1)(d + 1)$ messages sent, so at least one must be delivered. Hence we can argue that, during phase k, a new vertex is informed if there is an interval I_i that does not contain any informed vertices, followed by interval $I_{i \oplus 1}$ that contains at least one non-informed vertex. However, the existence of such I_i follows readily from the fact that there are only k informed vertices distinct from initiator and $d \geq 2$.

Lemma 2. *After phase k there are at least $k+1$ informed vertices distinct from the initiator.*

Each phase k consists of $k + 1$ steps with d rounds each, therefore every phase takes $O(n)$ time steps. Since there are $O(n)$ phases, the first part of the algorithm finishes in $O(n^2)$ time.

The second part of the algorithm starts with at least $\lceil n/2 \rceil$ informed vertices and informs all remaining ones. The algorithm is as follows: consider all pairs $[i, j]$ such that $1 \leq i, j \leq n - 1$, sorted in lexicographic order. In each step, all informed vertices consider one pair and send messages to vertices i and j. Since

---

[2] Strictly speaking, the vertices do not necessarily need to know their ID, the link labels are sufficient: The initiator may assume ID 0 and each message will also carry the link label it travels on and the ID of the sender, allowing the receiver to compute its ID.

at least $2\lceil n/2 \rceil \geq n - 1$ messages are sent, at least one of them is delivered. This ensures that a new vertex is informed whenever both i and j were uninformed. In this manner, all but one vertex can be informed (at any moment the two smallest unexplored vertices form a pair that has not been considered yet).

To inform the last vertex, all $n - 1$ informed vertices send in turn messages to vertices $1, 2, \ldots, n - 1$.

Theorem 3. *There is a $O(n^2)$ time fault-tolerant broadcasting algorithm for complete graphs with chordal sense of direction.*

Proof. The first part consist of $\lceil n/2 \rceil - 2$ phases, with each phase taking $O(n)$ steps. The second part consists of $n(n-1)/2$ steps and informing the last vertex takes $n - 1$ steps.

Note that the algorithm did not exploit all properties of the chordal sense of direction, it is sufficient for the informed vertices to agree on the IDs of the vertices, and to be able to determine the ID of the vertex on the other side of a link. Therefore, we get:

Corollary 1. *There is a $O(n^2)$ time broadcasting algorithm for complete graphs with neighboring (Abelian group based) sense of direction.*

3.2   Unoriented Complete Graphs

The algorithm in the previous section strongly relied on the fact that the vertices know the IDs of the vertices on the other side of the links. In this section, we use very different techniques to develop an algorithm that works for unoriented complete graphs (i.e. the only structural information available is the knowledge that the graph is complete; of course, local orientation – being able to distinguish incident ports – is also required).

We will view the flow of messages as tokens traveling through the network (and possibly spawning new tokens). A message (token) arriving to a vertex may cause the vertex to transmit some messages (either immediately, or in some of the subsequent steps). We will view those new messages as child tokens of the parent token. This means the tokens form a tree structure, and each token can be assigned unique identifier (corresponding to a path in the tree structure). Note that each vertex can also be given unique identifier (the ID of the token that first informed it). Each token carries all information about itself and its ancestors (i.e. IDs of its ancestors, traversed vertices and traversed ports).

Each token may be of two types: green and red. The intuition is that a token is green if it is "exploring", i.e. trying to traverse a port that has never been explored by its ancestors. When every port has been explored by the token's ancestors, the broadcast is finished, and no new tokens are sent. Ideally, if a token arrives to a vertex v, it would be spawned as a green token along all links that have not yet been explored by its ancestors. However, there is usually not enough unexplored ports in v[3]. In this case red tokens are sent along some already explored links.

---

[3] Recall that at least $n - 1$ messages must be sent in every step to make sure that at least one is delivered.

The meaning is that a red token carries a "request for help" to already explored vertices that are not yet engaged in helping. This request triggers new tokens to be sent from those vertices, and eventually a situation occurs when only green tokens are sent and at least one of them is delivered.

Let T be any token. The *green ancestor* of T is the closest green ancestor of T, if T is red, and T itself, if T is green. The *red tail* of token T is the path (sequence of tokens) between the green ancestor of T and T itself. Note that all tokens on the red tail are red except the first one.

We present a fault-tolerant broadcast algorithm that satisfies the following invariants:

I1. Let T be a token that is sent over an oriented edge $\langle a, b \rangle$. If T is green, then it holds that no ancestor of T has been sent over $\langle a, b \rangle$. Conversely, if T is red, there exists some ancestor of T that has been sent over $\langle a, b \rangle$.

I2. Let T be a red token. Then the red tail of T contains at most n vertices.

I3. Let T be a red token. Then T is sent exactly one round later than the parent of T.

I4. Let T be a green token. Then T is sent at most $n + 1$ rounds later than the last green ancestor of the parent of T.

I5. Let T be a green token delivered in round t. If the broadcast is not finished yet, at least one green token is delivered in some of the rounds $t+1, \ldots t+n$.

These invariants imply that the broadcast completes in $O(n^3)$ time: the invariant *I5* ensures that the algorithm can not stop before the broadcast is finished. Consider a path from root to a leaf in the tree of tokens. Invariant *I1* ensures that the leaf is green and that there are at most $O(n^2)$ green tokens on this path. Invariant *I4* implies that there are at most $n + 1$ consecutive red tokens on the path. Hence the overall time of the broadcasting algorithm is $O(n^3)$.

The algorithm works as follows. In the first round of the algorithm, the initiator sends green tokens through all its ports. All these tokens are children of some virtual root token. In each subsequent round t, each vertex gathers all received tokens in this round and processes them in parallel using procedure PROCESS described in Algorithm 1..

If some processor should send more than one token through a port in one round, it (arbitrarily) chooses single one of them to send and discards the remaining ones.

Lemma 3. *The presented algorithm satisfies invariants I1, I2, ..., I5.*

Combining Lemma 3 with the discussion about the invariants we get

Theorem 4. *There exist a $O(n^3)$ fault-tolerant broadcasting algorithm for unoriented complete networks.*

3.3   Lower Bound for Unoriented Complete Networks

The $O(n)$ algorithm for rings is obviously asymptotically optimal. An interesting question is: How far from optimal are our algorithms for oriented and unoriented complete networks? In this section we show that

---

Algorithm 1. Complete graphs without sense of direction

1: **procedure** PROCESS(T)   // process token T
2:     Let P be the set of all ports
3:     Let A be the set of ports that have never been traversed by any ancestor of T
4:     If $A = \emptyset$, the broadcast is finished.
5:     Let S be the set of vertices acting as a source of a red token in the red tail of T.
6:     Let $B \subseteq P - A$ be the set of ports that lead to a vertex in S.
7:     Let $C = P - (A \cup B)$
   // Note that since only ports already traversed by (an ancestor of) T
   // are considered, the vertex processing T can indeed compute B and C.

8:     **for** the first round of processing T **do**
9:         Send new red tokens with parent T to all ports in C
10:        Send new green tokens with parent T to all ports in A
11:    **end for**

12:    Let l be the length of the red tail of T.
13:    **for** subsequent $n - l$ rounds of processing T **do**
14:        Send new green token with parent T to all ports in A
15:    **end for**
16: **end procedure**

Theorem 5. *Any fault-tolerant broadcasting algorithm on unoriented complete networks must spend $\Omega(n^2)$ time.*

Proof. In the course of the computation there are two kinds of ports: the ports that have never been traversed by any message in any direction are called *"free"*, the ports that are not free are called *"bound"*. The lower bound proof is based on the following simple fact:

Let p be a free port of vertex u in time t. Let v be any vertex such that no bound port of u leads to v. Then it is possible that port p leads to vertex v.

Indeed, if p would lead to v, the first t rounds of computation would be the same. Hence, the computation can be viewed as a game of two players: the algorithm chooses a set of ports through which messages are to be sent. The adversary chooses one port through which the requested message passes. If this port is free, it chooses also the vertex to which this port will be bound.

We show now that it is possible for the adversary to keep the vertex n uninformed for $\frac{(n-1)(n-2)}{2} = \Omega(n^2)$ communication rounds. The idea is that some message has to traverse through all edges between vertices $1 \ldots n-1$ before any message arrives to the vertex n.

Consider the time step $i < \frac{(n-1)(n-2)}{2}$ and assume that the vertex n is not informed yet. There exist at most $2i$ bound ports, since in each time step at most one edge, i.e two ports are bounded. This means that at least $(n-1)(n-1)-2i \geq n$ ports of vertices $1 \ldots n-1$ are free.

The following cases can occur:

1. The algorithm sends some message through some bound port. The adversary passes this message, hence the vertex n stays uninformed.
2. The algorithm sends messages only through free ports.
   (a) The algorithm does not send messages from all vertices $1 \ldots n-1$. Then there have to be at least 2 messages sent from one vertex. The adversary delivers one of these messages and binds the corresponding port to any vertex other than n. (Since there are at least two free ports, it is possible for the adversary to do so.)
   (b) The algorithm sends messages from all vertices $1 \ldots n-1$. Since at least n ports of vertices $1 \ldots n-1$ are free, at least one vertex w from $1 \ldots n-1$, has 2 free ports. The adversary delivers the message sent from w, and binds corresponding port to any vertex other than n. (Again, since there are at least two free ports, it is possible for the adversary to do so.)

Hence it is possible for the adversary to keep the vertex n uninformed for the first $\Omega(n^2)$ time steps.

Now assume a stronger computation model: each vertex immediately learns for any message it has sent whether this message has been delivered or not. It is interesting to note that our lower bound is valid also in this model. Furthermore, it is easy to see that the lower bound is tight in this model.

4   Arbitrary k-Connected Graphs

In this section we consider k-edge-connected graphs and we assume the threshold is k, i.e. at least k messages must be sent to ensure that a message is delivered.

4.1   With Full Topological Knowledge

The algorithm runs in $n - 1$ phases. Each phase has an initiator vertex u (informed) and a destination vertex v (uninformed), with the source s being the initiator of the first phase. The goal of a phase is to inform vertex v, which then becomes the initiator of the next phase; the process is repeated until all vertices are informed.

The basic idea is a generalization of the idea from rings. The ring algorithm tried to "push" the information simultaneously along the left and right part of the ring. Here, the initiator u chooses k edge-disjoint paths[4] $\mathcal{P} = \{P_1 \ldots P_k\}$ from itself to v and then pushes the information through all the paths simultaneously. Let $P_i = (u_0 = u, u_1, \ldots u_{l_i} = v)$; consider an oriented edge $e = \langle u_j, u_{j+1} \rangle$. This edge can be either *sleeping*, *active* or *passive*:

1. The edge e is *passive* if and only if a message has been received over both e and the edge opposite to e, i.e. $\langle u_{j+1}, u_j \rangle$.

---

[4] Since the graph is k-edge-connected and the vertices have full topological knowledge, the initiator can always find these paths.

2. The edge e is *active* if and only if it is not passive and a message has been received over the edge $\langle u_{j-1}, u_j \rangle$. In case $j = 0$ the edge e is active whenever it is not passive.

3. The edge e is *sleeping* if and only if it is not active nor passive.

One phase consists of several rounds, each round spanning over many communication steps. The goal of one round is to ensure that a progress over at least one edge has been made: at least one active edge becomes passive, at least one sleeping edge becomes active or the vertex v becomes informed.

The procedure ROUND() defined in Algorithm 2. is the core of the algorithm; it is performed in each round by every vertex $w \in \mathcal{P}$.

Algorithm 2. k-connected graphs

1: **procedure** ROUND(vertex w)
2:     Let A be the set of active edges incident to w at the beginning of the round
3:     **for** i:=0 to k **do**   // One subround:
4:         **for** $B \subseteq \{1 \ldots k\}$ such that $|B| = i$ **do**   // one iteration per time step
5:             Let C be the set of edges incident to w via which an activating message
6:             has been received in the current round   // not in the current time step
7:             **for** $e \in C$ **do**
8:                 send deactivating message through e // all in one time step
9:             **end for**
10:            **for** $e \in A$ such that $e \in P_z \land z \notin B$ **do**
11:                send activating message through e // all in the same time step as
    in 8
12:            **end for**
13:        **end for**
14:    **end for**
15: **end procedure**

It is easy to see that the uninformed vertices never send any messages and that at any time each vertex can determine all active edges incident to it. Synchronous communication and full topological knowledge ensure that all procedures (phases/rounds/subrounds) are started and executed simultaneously by all participating vertices.

Lemma 4. *During one round at least one active edge becomes passive, or a sleeping edge becomes active, or v is informed.*

Proof. By contradiction. Assume the contrary, we show that in such case, at the beginning of the i-th subround there will be at least i paths $\mathcal{P}' \subseteq \mathcal{P}$ such that on any path $P_j \in \mathcal{P}'$ there is an edge through which an activating message has been delivered in the current round. This would mean that in the k-th subround there are at least k deactivating messages sent and therefore at least one of them will be delivered and an active edge will become passive, a contradiction.

We prove that above statement about subrounds by induction on i. The statement trivially holds for $i = 0$, as there is nothing to prove. Assume (by induction hypothesis) that at the beginning of the i-th subround there are exactly i paths $\mathcal{P}'$ with an edge over which an activating message has been delivered in the current round(if there are more, the hypothesis is already true for $i + 1$). From the definition of an active edge and from construction it follows that unless the vertex v is informed, there is at least one active edge on each path P_j. Let us focus on the time step in the i-th subround when B contains exactly the numbers of paths from $\mathcal{P}'$ (i.e. $B = \{j | P_j \in \mathcal{P}'\}$). In this time step, at least $k - i$ activating and at least i deactivating messages are sent, therefore at least one of them must be delivered. As no activating message is sent over an edge $e \in \mathcal{P}'$ and no deactivating message is delivered (by assumption that no active edge becomes passive), an activating message must be delivered on a path not in $\mathcal{P}'$. Hence, the invariant is ensured for the subround $i + 1$, too.

Theorem 6. *There is a fault-tolerant broadcasting algorithm on k-connected graphs with full topology knowledge that uses $O(2^k nm)$ time, where n is the number of vertices and m is the number of edges in the graph.*

Proof. The correctness follows straightforwardly from construction and Lemma 4.

The time complexity of one round is 2^k, as it spends one time step for each subset of $\{1, 2, \ldots, k\}$. The number of rounds per phase is[5] $2m$, as all paths in $\mathcal{P}$ together cannot contain more than all m edges and each edge can change its state at most twice (from sleeping to active to passive). Finally, the number of phases is $n - 1$ as $n - 1$ vertices need to be informed. Multiplying we get $O(2^k mn)$.

Theorem 6 can be successfully applied to many commonly used interconnection topologies. However, better results can usually be obtained by carefully choosing the order in which the vertices should be informed, allowing for short paths in $\mathcal{P}$. One such example is oriented hypercubes (i.e. each link is marked by the dimension it lies in):

Theorem 7. *There is a fault-tolerant broadcasting algorithm for oriented d-dimensional hypercubes that uses $O(n^2 \log n)$ time, where $n = 2^d$ is the number of vertices of the hypercube.*

Proof. The basic idea is to use the algorithm for k-connected graphs, with the initiator of a phase choosing as the next vertex to inform its successor in (a fixed) Hamiltonian path of the hypercube.

The algorithm for one phase is the same as in the case of k-connected graphs with the following exception: it is possible to choose d edge-disjoint paths from vertex u to its neighbor vertex v such that each of these paths has length at most 3. This results in $\mathcal{P}$ containing only $O(d)$ edges instead of $O(n \log n)$, thus reducing the cost of one phase from $O(n^2 \log n)$ to $O(nd) = O(n \log n)$. The resulting time complexity is therefore $O(n^2 \log n)$.

---

[5] Some topology-specific optimization is possible here.

4.2 Without Topological Knowledge

Finally, we show that the broadcasting on a k-connected graph with n vertices and m edges can be performed in time $O(2^k m^2 n)$ even in the case when the only known information about the graph are the values of n, m, and k. To achieve this, we combine the ideas used for complete graphs with those using full topology knowledge. In particular, the vertices accumulate topology information (using local identifiers) in a fashion similar to the algorithm for complete graphs. The algorithm works in phases, where each phase is performed within one informed component, and uses the topology knowledge of that component. However, since there may be many phases active at the same moment, great care must be given to avoid unwanted interference. The detailed result has been omitted due to space constraints.

Applying this result to the case of d-dimensional hypercube without sense of direction yields an algorithm that uses $O(n^4 \log^2 n)$ time.

5 Conclusions

We have introduced a new model for dynamic faults in synchronous distributed systems. This model includes as special cases the existing settings studied in the literature. We have focused on the *simple threshold* setting where, to be guaranteed that at least one message is delivered in a time step, the total amount of transmitted messages in that time step must be above the threshold T. We have investigated broadcasting in rings and complete graphs, as well as arbitrary networks, and we have designed solution protocols, proving that broadcast is possible also under the worst threshold (i.e., equal to the connectivity). The perhaps surprising result is that the time costs are (low) polynomial for several networks including rings, complete graphs, hypercubes, and constant-degree networks.

This investigation is the first step in the analysis of distributed computing in spite of fractional dynamic faults with threshold.

References

1. P. Berman, K. Diks, and A. Pelc, "Reliable broadcasting in logarithmic time with Byzantine link failures". *Journal of Algorithms*, 22 (2), 199–211, 1997.
2. T. Chandra, V. Hadzilacos, and S. Toueg, "The weakest failure detector for solving consensus". *Journal of ACM*, 43(4), 685–722, 1996.
3. B.S. Chlebus, K. Diks, and A. Pelc, "Broadcasting in synchronous networks with dynamic faults". *Networks* 27, 309–318, 1996.
4. G. De Marco and A. Rescigno, "Tighter time bounds on broadcasting in torus networks in presence of dynamic faults". *Parallel Processing Letters* 10 (1), 39–50, 2000.
5. G. De Marco and U. Vaccaro, "Broadcasting in hypercubes and star graphs with dynamic faults". *Information Processing Letters* 66, 309–318, 1998.
6. S. Dobrev, "Communication-efficient broadcasting in complete networks with dynamic faults". *Theory of Computing Systems* 36(6), 695–709, 2003.

7. S. Dobrev, "Computing input multiplicity in anonymous synchronous networks with dynamic faults". *Journal of Discrete Algorithms* 2, 425–438, 2004.

8. S. Dobrev and I. Vrt'o, "Optimal broadcasting in hypercubes with dynamic faults". *Information Processing Letters* 71, 81–85, 1999.

9. S. Dobrev and I. Vrt'o, "Optimal broadcasting in even tori with dynamic faults". *Parallel Processing Letters* 12, 17–22, 2002.

10. S. Dobrev and I. Vrt'o, "Dynamic faults have small effect on broadcasting in hypercubes". *Discrete Applied Mathematics* 137(2), 155–158, 2004.

11. M. J. Fischer, N.A. Lynch, and M.S. Paterson, "Impossibility of distributed consensus with one faulty process", *Journal of the ACM* 32 (2), 1985.

12. P. Fraigniaud and C. Peyrat, "Broadcasting in a hypercube when some calls fail", *Information Processing Letters* 39, 115–119, 1991.

13. R. Královič, R. Královič, and P. Ružička, "Broadcasting with many faulty links". In *Proc. 10th Colloquium on Structural Information and Communication complexity* (SIROCCO'03), 211–222, 2003.

14. Z. Liptak and A. Nickelsen, "Broadcasting in complete networks with dynamic edge faults", In *Proc. 4th International Conference on Principles of Distributed Systems* (OPODIS 00), Paris, 123–142, 2000.

15. Tz. Ostromsky and Z. Nedev, "Broadcasting a Message in a Hypercube with Possible Link Faults". In *Parallel and Distributed Processing '91* (K. Boyanov, editor), Elsevier, 231–240, 1992.

16. A. Pelc and D. Peleg, "Feasibility and complexity of broadcasting with random transmission failures". In *Proc. 24th ACM Symposium on Principles of Distributed Computing* (PODC 05), 334–341, 2005.

17. N. Santoro and P. Widmayer, "Time is not a healer". In *Proc. 6th Ann. Symposium on Theoretical Aspects of Computer Science* (STACS 89), LNCS 349, 304–313, 1989.

18. N. Santoro and P. Widmayer, "Distributed function evaluation in the presence of transmission faults". In *Proc. International Symposium on Algorithms* (SIGAL 90), Tokyo, LNCS 450, 358–367, 1990.

19. N. Santoro and P. Widmayer, "Agreement in synchronous networks with ubiquitous faults". In *Theoretical Computer Science*, 2006, to appear; preliminary version in *Proc. 12th Colloquium on Structural Information and Communication Complexity* (SIROCCO'05), LNCS, 2005.

Discovering Network Topology in the Presence of Byzantine Faults

Mikhail Nesterenko[1,*] and Sébastien Tixeuil[2,**]

[1] Computer Science Department, Kent State University Kent, OH, 44242, USA
mikhail@cs.kent.edu
[2] LRI-CNRS UMR 8623 & INRIA Grand Large
Université Paris Sud, France
tixeuil@lri.fr

Abstract. We study the problem of Byzantine-robust topology discovery in an arbitrary asynchronous network. We formally state the weak and strong versions of the problem. The weak version requires that either each node discovers the topology of the network or at least one node detects the presence of a faulty node. The strong version requires that each node discovers the topology regardless of faults.

We focus on non-cryptographic solutions to these problems. We explore their bounds. We prove that the weak topology discovery problem is solvable only if the connectivity of the network exceeds the number of faults in the system. Similarly, we show that the strong version of the problem is solvable only if the network connectivity is more than twice the number of faults.

We present solutions to both versions of the problem. Our solutions match the established graph connectivity bounds. The programs are terminating, they do not require the individual nodes to know either the diameter or the size of the network. The message complexity of both programs is low polynomial with respect to the network size.

1 Introduction

In this paper, we investigate the problem of Byzantine-tolerant distributed topology discovery in an arbitrary network. Each node is only aware of its neighboring peers and it needs to learn the topology of the entire network.

Topology discovery is an essential problem in distributed computing (*e.g.* see [1]). It has direct applicability in practical systems. For example, link-state based routing protocols such as OSPF use topology discovery mechanisms to compute the routing tables. Recently, the problem has come to the fore with the introduction of ad hoc wireless sensor networks, such as Berkeley motes [2], where topology discovery is essential for routing decisions.

---

* This author was supported in part by DARPA contract OSU-RF#F33615-01-C-1901 and by NSF CAREER Award 0347485.
** This author was supported in part by the FNS grants FRAGILE and SR2I from ACI "Sécurité et Informatique".

P. Flocchini and L. Gąsieniec (Eds.): SIROCCO 2006, LNCS 4056, pp. 212–226, 2006.

As reliability demands on distributed systems increase, the interest in developing robust topology discovery programs grows. One of the strongest fault models is *Byzantine* [3]: the faulty node behaves arbitrarily. This model encompasses rich set of fault scenarios. Moreover, Byzantine fault tolerance has security implications, as the behavior of an intruder can be modeled as Byzantine. One approach to deal with Byzantine faults is by enabling the nodes to use cryptographic operations such as digital signatures or certificates. This limits the power of a Byzantine node as a non-faulty node can verify the validity of received topology information and authenticate the sender across multiple hops. However, this option may not be available. For example, wireless sensors may not have the capacity to manipulate digital signatures. Another way to limit the power of a Byzantine process is to assume synchrony: all processes proceed in lock-step. Indeed, if a process is required to send a message with each pulse, a Byzantine process cannot refuse to send a message without being detected. However, the synchrony assumption may be too restrictive for practical systems.

Our Contribution. In this study we explore the fundamental properties of topology discovery. We select the weakest practical programming model, establish the limits on the solutions and present the programs matching those limits.

Specifically, we consider arbitrary networks of arbitrary topology where up to fixed number of nodes k is faulty. The execution model is asynchronous. We are interested in solutions that do not use cryptographic primitives. The solutions should be terminating and the individual processes should not be aware of the network parameters such as network diameter or its total number of nodes.

We state two variants of the topology discovery problem: *weak* and *strong*. In the former — either each non-faulty node learns the topology of the network or one of them detects a fault; in the latter — each non-faulty node has to learn the topology of the network regardless of the presence of faults.

As negative results we show that any solution to the weak topology discovery problem can not ascertain the presence of an edge between two faulty nodes. Similarly, any solution to the strong variant can not determine the presence of a edge between a pair of nodes at least one of which is faulty. Moreover, the solution to the weak variant requires the network to be at least $(k+1)$-connected. In case of the strong variant the network must be at least $(2k + 1)$-connected.

The main contribution of this study are the algorithms that solve the two problems: *Detector* and *Explorer*. The algorithms match the respective lower bounds. To the best of our knowledge, these are the first asynchronous Byzantine-robust solutions to the topology discovery problem that do not use cryptographic operations. *Explorer* solves the stronger problem. However, *Detector* has better message complexity. *Detector* either determines topology or signals fault in $O(\delta n^3)$ messages where δ and n are the maximum neighborhood size and the number of nodes in the system respectively. *Explorer* finishes in $O(n^4)$ messages. We extend our algorithms to (a) discover a fixed number of routes instead of complete topology and (b) reliably propagate arbitrary information instead of topological data.

Related Work. A number of researchers employ cryptographic operations to counter Byzantine faults. Avromopolus et al [4] consider the problem of secure routing. Therein see the references to other secure routing solutions that rely on cryptography. Perrig et al [5] survey robust routing methods in ad hoc sensor networks. The techniques covered there also assume that the processes are capable of cryptographic operations.

A naive approach of solving the topology discovery problem without cryptography would be to use a Byzantine-resilient broadcast [6, 7, 8, 9]: each node advertises its neighborhood. However all existing solutions for arbitrary topology known to us require that the graph topology is *a priori* known to the nodes.

Let us survey the non-cryptography based approaches to Byzantine fault-tolerance. Most programs described in the literature [10, 11, 12, 13] assume completely connected networks and can not be easily extended to deal with arbitrary topology. Dolev [7] considers Byzantine agreement on arbitrary graphs. He states that for agreement in the presence of up to k Byzantine nodes, it is necessary and sufficient that the network is $(2k + 1)$-connected and the number of nodes in the system is at least $3k + 1$. However, his solution requires that the nodes are aware of the topology in advance. Also, this solution assumes the synchronous execution model. Recently, the problem of Byzantine-robust reliable broadcast has attracted attention [6, 8, 9]. However, in all cases the topology is assumed to be known. Bhandari and Vaidya [6] and Koo [8] assume two-dimensional grid. Pelc and Peleg [9] consider arbitrary topology but assume that each node knows the exact topology a priori. A notable class of algorithms tolerates Byzantine faults locally [14, 15, 16]. Yet, the emphasis of these algorithms is on containing the fault as close to its source as possible. This is only applicable to the problems where the information from remote nodes is unimportant such as vertex coloring, link coloring or dining philosophers. Thus, local containment approach is not applicable to topology discovery.

Masuzawa [17] considers the problem of topology discovery and update. However, Masuzawa is interested in designing a self-stabilizing solution to the problem and thus his fault model is not as general as Byzantine: he considers only transient and crash faults.

The rest of the paper is organized as follows. After stating our programming model and notation in Section 2, we formulate the topology discovery problems, as well as state the impossibility results in Section 3. We present *Detector* and *Explorer* in Sections 4 and 5 respectively. We discuss the composition of our programs and their extensions in Section 6 and conclude the paper in Section 7.

2   Notation, Definitions and Assumptions

Graphs. A distributed *system* (or *program*) consists of a set of processes and a *neighbor* relation between them. This relation is the system *topology*. The topology forms a graph G. Denote n and e to be the number of nodes[1] and edges in G respectively. Two processes are *neighbors* if there is an edge in G connecting

---

[1] We use terms *process* and *node* interchangeably.

them. A set P of neighbors of process p is *neighborhood* of p. In the sequel we use small letters to denote singleton variables and capital letters to denote sets. In particular, we use a small letter for a process and a matching capital one for this process' neighborhood. Since the topology is symmetric, if $q \in P$ then $p \in Q$. Denote δ to be the maximum number of nodes in a neighborhood.

A *node-cut* of a graph is the set of nodes U such that $G \setminus U$ is disconnected or trivial. A *node-connectivity* (or just *connectivity*) of a graph is the minimum cardinality of a node-cut of this graph. In this paper we make use of the following fact about graph connectivity that follows from Menger's theorem (see [18]): if a graph is k-connected (where k is some constant) then for every two vertices u and v there exists at least k internally node-disjoint paths connecting u and v in this graph.

Program Model. A process contains a set of variables. When it is clear from the context, we refer to a variable *var* of process p as *var.p*. Every variable ranges over a fixed domain of values. For each variable, certain values are *initial*. Each pair of neighbor processes share a pair of special variables called *channels*. We denote *Ch.b.c* the channel from process b to process c. Process b is the *sender* and c is the *receiver*. The value for a channel variable is chosen from the domain of (potentially infinite) sequences of messages.

A *state* of the program is the assignment of a value to every variable of each process from its corresponding domain. A state is *initial* if every variable has initial value. Each process contains a set of actions. An action has the form $\langle name \rangle : \langle guard \rangle \longrightarrow \langle command \rangle$. A *guard* is a boolean predicate over the variables of the process. A *command* is sequence of assignment and branching statements. A guard may be a receive-statement that accesses the incoming channel. A command may contain a send-statement that modifies the outgoing channel. A parameter is used to define a set of actions as one parameterized action. For example, let j be a parameter ranging over values 2, 5 and 9; then a parameterized action $ac.j$ defines the set of actions $ac.(j = 2)\ []\ ac.(j = 5)\ []\ ac.(j = 9)$. Either guard or command can contain quantified constructs [19] of the form: $(\langle quantifier \rangle \langle bound\ variables \rangle : \langle range \rangle : \langle term \rangle)$, where *range* and *term* are boolean constructs.

Semantics. An action of a process of the program is *enabled* in a certain state if its guard evaluates to **true**. An action containing receive-statement is enabled when appropriate message is at the head of the incoming channel. The execution of the command of an action updates variables of the process. The execution of an action containing receive-statement removes the received message from the head of the incoming channel and inserts the value the message contains into the specified variables. The execution of send-statement appends the specified message to the tail of the outgoing message.

A *computation* of the program is a maximal fair sequence of states of the program such that the first state s_0 is initial and for each state s_i the state s_{i+1} is obtained by executing the command of an action whose state is enabled in s_i. That is, we assume that the action execution is *atomic*. The maximality of a computation means that the computation is either infinite or it terminates in a state where none of the actions are enabled. The fairness means that if an action is enabled in all but finitely many states of an infinite computation then this action is executed infinitely often. That is, we assume *weak fairness* of action execution. Notice that we define the receive statement to appear as a standalone guard of an action. This means, that if a message of the appropriate type is at the head of the incoming channel, the receive action is enabled. Due to weak fairness assumption, this leads to *fair message receipt* assumption: each message in the channel is eventually received. Observe that our definition of a computation considers *asynchronous* computations.

To reason about program behavior we define boolean predicates on program states. A program *invariant* is a predicate that is **true** in every initial state of the program and if the predicate holds before the execution of the program action, it also holds afterwards. Notice that by this definition a program invariant holds in each state of every program computation.

Faults. Throughout a computation, a process may be either Byzantine (faulty) or non-faulty. A Byzantine process contains an action that assigns to each local variable an arbitrary value from its domain. This action is always enabled. Observe that this allows a faulty node to send arbitrary messages. We assume, however, that messages sent by such node conform to the format specified by the algorithm: each message carries the specified number of values, and the values are drawn from appropriate domains. This assumption is not difficult to implement as message syntax checking logic can be incorporated in receive-action of each process. We assume *oral record* [3] of message transmission: the receiver can always correctly identify the message sender. The channels are reliable: the messages are delivered in FIFO order and without loss or corruption. Throughout the paper we assume that the maximum number of faults in the system is bounded by some constant k.

Graph Exploration. The processes discover the topology of the system by exchanging messages. Each message contains the identifier of the process and its neighborhood. Process p *explored* process q if p received a message with (q, Q). When it is clear from the context, we omit the mention of p. An *explored* subgraph of a graph contains only explored processes. A Byzantine process may potentially circulate information about the processes that do not exist in the system altogether. A process is *fake* if it does not exist in the system, a process is *real* otherwise.

3   Topology Discovery Problem: Statement and Solution Bounds

Problem Statement

Definition 1 (Weak Topology Discovery Problem). A program is a solution to the weak topology discovery problem if each of the program's computation satisfies the following properties: *termination* — either all non-faulty processes determine the system topology or at least one process detects a fault; *safety* — for each non-faulty process, the determined topology is a subset of the actual system topology; *validity* — the fault is detected only if there are faulty processes in the system.

Definition 2 (Strong Topology Discovery Problem). A program is a solution to the strong topology discovery problem if each of the program's computations satisfies the following properties: *termination* — all non-faulty processes determine the system topology; *safety* — the determined topology is a subset of the actual system topology.

According to the safety property of both problem definitions each non-faulty process is only required to discover a subset of the actual system topology. However, the desired objective is for each node to discover as much of it as possible. The following definitions capture this idea. A solution to a topology discovery problem is *complete* if every non-faulty process always discovers the complete topology of the system. A solution to the problem is *node-complete* if every non-faulty process discovers all nodes of the system. A solution is *adjacent-edge complete* if every non-faulty node discovers each edge adjacent to at least one non-faulty node. A solution is *two-adjacent-edge complete* if every non-faulty node discovers each edge adjacent to two non-faulty nodes.

Solution Bounds. The proofs for the theorems stated in this section are to be found elsewhere [20].

Theorem 1. There does not exist a complete solution to the weak topology discovery problem.

Theorem 2. There exists no node- and adjacent-edge complete solution to the weak topology problem if the connectivity of the graph is lower or equal to the total number of faults k.

Observe that for $(k+1)$-connected graphs an adjacent-edge complete solution is also node complete.

Theorem 3. There does not exist an adjacent-edge complete solution to the strong topology discovery problem.

Theorem 4. There exists no node- and two-adjacent-edge complete solution to the strong topology problem if the connectivity of the graph is less than or equal to twice the total number of faults k.

4 Detector

Outline. *Detector* solves the weak topology discovery problem for system graphs whose connectivity exceeds the number of faulty nodes k. The algorithm leverages the connectivity of the graph. For each pair of nodes, the graph guarantees the presence of at least one path that does not include a faulty node. The topology data travels along every path of the graph. Hence, the process that collects information about another process can find the potential inconsistency between the information that proceeds along the path containing faulty nodes and the path containing only non-faulty ones.

Care is taken to detect the fake nodes whose information is introduced by faulty processes. Since the processes do not know the size of the system, a faulty process may potentially introduce an infinite number of fake nodes. However, the graph connectivity assumption is used to detect fake nodes. As faulty processes are the only source of information about fake nodes, all the paths from the real nodes to the fake ones have to contain a faulty node. Yet, the graph connectivity is assumed to be greater than k. If a fake node is ever introduced, one of the non-faulty processes eventually detects a graph with too few paths leading to the fake node.

Detailed Description. The program is shown in Figure 1. Each process p stores the identifiers of its immediate neighbors. They are kept in set P. Each process keeps the upper bound k on the number of faulty processes. Process p maintains the following variables. Boolean variable *detect* indicates if p discovers a fault

process p
const
      P: set of neighbor identifiers of p
      k: integer, upper bound on the number of faulty processes
parameter
      $q : P$
var
      detect : boolean, initially **false**, signals fault
      start : boolean, initially **true**, controls sending of p's neighborhood info
      TOP : set of tuples, initially $\{(p, P)\}$, (process ids, neighbor id set)
                        received by p

                  $*[$
init: $start \longrightarrow$
                      $start :=$ **false**,
                      $(\forall j : j \in P : $**send** (p, P) **to** $j)$
              $[\![$
accept: **receive** (r, R) **from** $q \longrightarrow$
                    if $(\exists s, S : (s, S) \in TOP : s = r \wedge S \neq R) \vee$
                      (**path_number**$(TOP \cup \{(r, R)\}) < k + 1)$
                    then
                      detect := **true**
                  else
                      if $(\nexists s, S : (s, S) \in TOP : s = r)$ **then**
                        $TOP := TOP \cup \{(r, R)\}$,
                        $(\forall j : j \in P : $**send** (r, R) **to** $j)$
            $]$

Fig. 1. Process of *Detector*

in the system. Boolean variable *start* guards the execution of the action that sends p's neighborhood information to its neighbors. Set TOP (for topology) stores the subgraph explored by p; TOP contains tuples of the form: (*process identifier, its neighborhood*). In the initial state, TOP contains (p, P).

Function **path_number** evaluates the topology of the subgraph stored in TOP. Recall that a node u is unexplored by p if for every tuple $(s, S) \in TOP$, s is not the same as u. That is u may appear in S only. We construct graph G' by adding an edge to every pair of unexplored processes present in TOP. We calculate the value of **path_number** as follows. If the information of TOP is inconsistent, that is:

$$(\exists u, v, U, V : ((u, U) \in TOP) \wedge ((v, V) \in TOP) :$$
$$(u \in V) \wedge (v \notin U))$$

then **path_number** returns 0. If there is exactly one explored node in TOP, **path_number** returns $k+1$. Otherwise the function returns the minimum number of internally node disjoint paths between two explored nodes in G'. In the correctness proof for this program we show that unless there is a fake node, the **path_number** of G' is no smaller than the connectivity of G.

Processes exchange messages of the form (*process identifier, its neighborhood id set*). A process contains two actions: *init* and *accept*. Action *init* starts the propagation of p's neighborhood throughout the system. Action *accept* receives the neighborhood data of some process, records it, checks against other data already available for p and possibly further disseminates the data. If the data received from neighbor q about a process r contradicts what p already holds about r in TOP or if the newly arrived information implies that G is less than $(k + 1)$-connected p indicates that it detected a fault by setting *detect* to **true**. Alternatively, if p did not previously have the information about r, p updates TOP and sends the received information to all its neighbors.

Theorem 5. *Detector* is an adjacent-edge complete solution to the weak topology discovery problem in case the connectivity of system topology graph exceeds the number of faults.

A correctness proof of the theorem can be found elsewhere [21].

Efficiency Evaluation. Since we consider an asynchronous model, the number of messages a Byzantine process can send in a computation is infinite. To evaluate the efficiency of *Detector* we assume that each process is familiar with the upper bound on the number of processes in the system and this upper bound is in $O(n)$. A non-faulty process then detects a fault if the number of processes it explores exceeds this bound or if it receives more than one identical message from the same neighbor. We assume that the process stops and does not send or receive any more messages if it detects a fault.

In this case we can estimate the number of messages that are received by non-faulty processes before one of them detects a fault or before the computation terminates. To make the estimation fair, the assume that the unit is $log(n)$ bits.

Since it takes that many bits to assign unique process identifiers to n processes, we assume that one identifier is exactly one unit of information. A message in *Detector* carries up to $\delta + 1$ identifiers, where δ is the maximum number of nodes in the neighborhood of a process. Observe that a process can receive at most n messages from each incoming channel. Thus, the total number of messages that can be sent by *Detector* is $2en$, where e is the number of edges in the graph. The message complexity of the program is in $O(2en\delta)$. If e is proportional to n^2, then the complexity of the program is in $O(\delta n^3)$.

5   Explorer

Outline. The main idea of *Explorer* is for each process to collect information about some node's neighborhood such that the information goes along more than twice as many paths as the maximum number of Byzantine nodes. While the paths are node-disjoint, the information is correct if it comes across the majority of the paths. In this case the recipient is in possession of confirmed information. It turns out that the topology information does not have to come directly from the source. Instead it can come from processes with confirmed information. The detailed description of *Explorer* follows.

To simplify the presentation, we describe and prove correct the version of *Explorer* that tolerates only one Byzantine fault. We describe how this version can be extended to tolerate multiple faults in the end of the section.

Description. Since we first describe the 1-fault tolerant version of *Explorer* we assume that the graph is 3-connected. The program is shown in Figure 2. Similar to *Detector*, each process p in *Explorer*, stores the ids of its immediate neighbors. Process p maintains the variable *start*, whose function is to guard the execution of the action that initiates the propagation of p's own neighborhood. Unlike *Detector*, however, p maintains two sets that store the topology information of the network: $uTOP$ and $cTOP$. Set $uTOP$ stores the topology data that is unconfirmed; $cTOP$ stores confirmed topology data. Set $uTOP$ contains the tuples of neighborhood information that p received from other nodes. Besides the process id and the set of its neighbor ids, each such tuple contains a set of process identifiers, that relayed the information. We call it *visited set*. The tuples in $cTOP$ do not require visited set.

Processes exchange messages where, along with the neighbor identifiers for a certain process, a visited set is propagated. A process contains two actions: *init* and *accept*. The purpose of *init* is similar to that in the process of *Detector*. Action *accept* receives the neighborhood information of some process r, its neighborhood R which was relayed by nodes in set S. The information is received from p's neighbor — q.

First, *accept* checks if the information about r is already confirmed. If so, the only manipulation is to record the received information in $uTOP$. Actually, this update of $uTOP$ is not necessary for the correct operation of the program, but it makes the its proof of correctness easier to follow.

process p
const
      P, set of neighbor identifiers of p
parameter
      $q : P$
var
      $start$: boolean, initially **true**, controls sending of p's neighbor ids
      $cTOP$: set of tuples, initially $\{(p, P)\}$,
            (process id, neighbor id set) confirmed topology info
      $uTOP$: set of tuples, initially $\varnothing$,
            (process id, neighbor id set, visited id set)
            unconfirmed topology info

      $*[$

$init$: $start \longrightarrow$
               $start := $ **false**,
               $(\forall j : j \in P : $ **send** $(p, P, \varnothing)$ **to** $j)$

        $[\![$

$accept$:   **receive** (r, R, S) **from** $q \longrightarrow$
           if $(\forall t, T : (t, T) \in cTOP : t \neq r)$ **then**
               if $(\forall t, T, U : (t, T, U) \in uTOP : t \neq r \vee T \neq R)$ **then**
                   $(\forall j : j \in P : $ **send** $(r, R, S \cup \{q\})$ **to** $j)$
               elsif $(\exists t, T, U : (t, T, U) \in uTOP : $
                   $t = r \wedge R = T \wedge ((U \cap (S \cup \{q\}))) \subset \{r\}))$
               then
                   $cTOP := cTOP \cup \{(r, R)\}$,
                   $(\forall j : j \in P : $ **send** $(r, R, \varnothing)$ **to** $j)$
            $uTOP := uTOP \cup \{(r, R, S \cup \{q\})\}$

       $]$

Fig. 2. Process of *Explorer*

If the received information does not concern already confirmed process, *accept* checks if this information differs from what is already recorded in $uTOP$ either in r or in R. In either case the information is broadcast to all neighbors of p. Before broadcasting p appends the sender — q to the visited set S.

If the information about r and R has already been received and recorded in $uTOP$, *accept* checks if the previously recorded information came along an internally node disjoint path. If so, the information about r is added to $cTOP$. In this case, this information is also broadcast to all p's neighbors. Note, however, that p is now sure of the information it received. Hence, the visited set of nodes in the broadcast message is empty.

Theorem 6. *Explorer* is a two-adjacent-edge complete solution to the strong topology discovery problem in case of one fault and the system topology graph is at least 3-connected.

A correctness proof of the theorem be found elsewhere [21].

Modification to Handle $k > 1$ faults. Observe that *Explorer* confirms the topology information about a node's neighborhood, when it receives two messages carrying it over internally node disjoint paths. Thus, the program can handle a single Byzantine fault. The explorer can handle $k > 1$ faults, if it waits until it receives $k + 1$ messages before it confirms the topology info. All the messages have to travel along internally node disjoint paths. For the correctness of the algorithm, the topology graph has to be $(2k + 1)$-connected.

Proposition 1. *Explorer* is a two-adjacent-edge complete solution to the strong topology discovery problem in case of k faults and the system topology graph is at least $(2k + 1)$-connected.

Efficiency evaluation. Unlike *Detector*, *Explorer* does not quit when a fault is discovered. Thus, the number of messages a faulty node may send is arbitrary large. However, we can estimate the message complexity of *Explorer* in the absence of faults. Each message carries a process identifier, a neighborhood of this process and a visited set. The number of the identifiers in a neighborhood is no more than δ, and the number of identifiers in the visited set can be as large as n. Hence the message size is bounded by $\delta + n + 1$ which is in $O(n)$.

Notice, that for the neighborhood A of each process a, every process broadcasts a message twice: when it first receives the information, and when it confirms it. Thus, the total number of sent messages is $4e \cdot n$ and the overall message complexity of *Explorer* if no faults are detected is in $O(n^4)$.

6   Composition and Extensions

Composing *Detector* and *Explorer*. Observe that *Detector* has better message complexity than *Explorer* if the neighborhood size is bounded. Hence, if the incidence of faults is low, it is advantageous to run *Detector* and invoke *Explorer* only if a fault is detected. We assume that the processes can distinguish between message types of *Explorer* and *Detector*. In the combined program, a process running *Detector* switches to *Explorer* if it discovers a fault. Other processes follow suit, when they receive their first *Explorer* messages. They ignore *Detector* messages henceforth. A Byzantine process may potentially send an *Explorer* message as well, which leads to the whole system switching to *Explorer*. Observe that if there are no faults, the system will not invoke *Explorer*. Thus, the complexity of the combined program in the absence of faults is the same as that of *Detector*. Notice that even though *Detector* alone only needs $(k+1)$-connectivity of the system topology, the combined program requires $(2k + 1)$-connectivity.

Message Termination. We have shown that *Detector* and *Explorer* comply with the functional termination properties of the topology discovery problem. That is, all processes eventually discover topology. However, the performance aspect of termination, viz. message termination, is also of interest. Usually an algorithm is said to be message terminating if all its computations contain a finite number of sent messages [22].

However, a Byzantine process may send messages indefinitely. To capture this, we weaken the definition of message termination. We consider a Byzantine-tolerant program *message terminating* if the system eventually arrives at a state where: (a) all channels are empty except for the outgoing channels of a faulty process; (b) all actions in non-faulty processes are disabled except for possibly the receive-actions of the incoming channels from Byzantine processes, these receive-actions do not update the variables of the process. That is, in a terminating program, each non-faulty process starts to eventually discard messages it receives from its Byzantine neighbors.

Making *Detector* terminating is fairly straightforward. As one process detects a fault, the process floods the announcement throughout the system. Since the topology graph for *Detector* is assumed $(k+1)$-connected, every process receives such announcement. As the process learns of the detection, it stops processing or forwarding of the messages. Notice that the initiation of the flood by a Byzantine node itself, only accelerates the termination of *Detector* as the other processes quickly learn of the faulty node's existence.

The addition of termination to *Explorer* is more involved. To ensure termination, restrictions have to be placed on message processing and forwarding. However, the restrictions should be delicate as they may compromise the liveness properties of the program.

By the design of *Explorer*, each process may send at most one message about its own neighborhood to its neighbors. Hence, the subsequent messages can be ignored. However, a faulty process may send messages about neighborhoods of other processes. These processes may be real or fake. We discuss these cases separately.

Note that each process in *Explorer* can eventually obtain an estimate of the identities of the processes in the system and disregard fake process information. Indeed, a path to a fake node can only lead through faulty processes. Thus, if a process discovers that there may be at most k internally node disjoint paths between itself and a certain node, this node is fake. Therefore, the process may cease to process messages about the fake node's neighborhood. Notice, that since the system is $(2k+1)$-connected, messages about real nodes will always be processed. Therefore, the liveness properties of *Explorer* are not affected.

As to the real processes, they can be either Byzantine or non-faulty. Recall that each non-faulty process of *Explorer* eventually confirms neighborhoods of all other non-faulty processes. After the neighborhood of a process is confirmed, further messages about it are ignored.

The last case is a Byzantine process u sending a message to its correct neighbor v about the neighborhood of another Byzantine process w. By the design of *Explorer*, v relays the message about w provided that the neighborhood information about w differs from what previously received about w. As we discussed above, eventually v estimates the identities of all real processes in the system. Therefore, there is a finite number of possible different neighborhoods of w that u can create. Hence, eventually they will be exhausted, and v starts ignoring further messages form u about w.

Thus, *Explorer* can be made terminating as well.

Other Extensions. Observe that *Explorer* is designed to disseminate the information about the complete topology to all processes in the system. However, it may be desirable to just establish the routes from all processes in the system to one or a fixed number of distinguished ones. To accomplish this *Explorer* needs to be modified as follows. No, neighborhood information is propagated. Instead of the visited set, each message carries the propagation path of the message. That is the order of the relays is significant.

Only the distinguished processes initiate the message propagation. The other processes only relay the messages. Just as in the original *Explorer*, a process confirms a path to another process only if it receives $2k + 1$ internally node disjoint paths from the source or from other confirming nodes. Again, like in *Explorer*, such process rebroadcasts the message, but empties the propagation path. In the outcome of this program, for every distinguished process, each non-faulty process will contain paths to at least $2k + 1$ processes that lead to this distinguished node. Out of these paths, at least $k + 1$ ultimately lead to the distinguished node.

In *Explorer*, for each process the propagation of its neighborhood information is independent of the other neighborhoods. Thus, instead of topology, *Explorer* can be used for efficient fault-tolerant propagation of arbitrary information from the processes to the rest of the network.

7   Conclusion

In conclusion, we would like to outline a couple of interesting avenues of further research.

The existence of Byzantine-robust topology discovery solutions opens the question of theoretical limits of efficiency of such programs. The obvious lower bound on message complexity can be derived as follows. Every process must transmit its neighborhood to the rest of the nodes in the system. Transmitting information to every node requires at least n messages, so the overall message complexity is at least δn^2. If k processes are Byzantine, they may not relay the messages of other nodes. Thus, to ensure that other nodes learn about its neighborhood, each process has to send at least $k+1$ messages. Thus, the complexity of any Byzantine-robust solution to the topology discovery problem is at least in $\Omega(\delta n^2 k)$.

Observe that *Explorer* and *Detector* may not explicitly identify faulty nodes or the inconsistent view of the their immediate neighborhoods. We believe that this can be accomplished using the technique used by Dolev [7]. In case there are $3k+1$ non-faulty processes, they may exchange the topologies they collected to discover the inconsistencies. This approach, may potentially expedite termination of *Explorer* at the expense of greater message complexity: if a certain Byzantine node is discovered, the other processes may ignore its further messages.

References

1. Spinelli, J.M., Gallager, R.G.: Event-driven topology broadcast without sequence numbers. IEEE trans. on commun. **COM-37, 5** (1989) 468–474
2. Hill, J., Culler, D.: Mica: A wireless platform for deeply embedded networks. IEEE Micro **22** (2002) 12–24

3. Lamport, L., Shostak, R., Pease, M.: The byzantine generals problem. ACM Transactions on Programming Languages and Systems **4** (1982) 382–401

4. Avramopoulos, I.C., Kobayashi, H., Wang, R., Krishnamurthy, A.: Highly secure and efficient routing. In: Proceedings of INFOCOM: The Conference on Computer Communications, joint conference of the IEEE Computer and Communications Societies, Hong Kong (2004)

5. Perrig, A., Stankovic, J., Wagner, D.: Security in wireless sensor networks. Communications of the ACM **47** (2004) 53–57

6. Bhandari, V., Vaidya, N.H.: On reliable broadcast in a radio network. In: Proceedings of the Twenty-Fourth Annual ACM SIGACT-SIGOPS Symposium on Principles of Distributed Computing (PODC 2005), Las Vegas, Nevada (2005) to appear

7. Dolev, D.: The Byzantine generals strike again. Journal of Algorithms **3** (1982) 14–30

8. Koo, C.Y.: Broadcast in radio networks tolerating byzantine adversarial behavior. In: PODC '04: Proceedings of the twenty-third annual ACM symposium on Principles of distributed computing, New York, NY, USA, ACM Press (2004) 275–282

9. Pelc, A., Peleg, D.: Broadcasting with locally bounded byzantine faults. Information Processing Letters **93** (2005) 109–115

10. Attiya, H., Welch, J.: Distributed Computing: Fundamentals, Simulations, and Advanced Topics. McGraw-Hill Publishing Company, New York (1998) 6.

11. Malkhi, D., Reiter, M., Rodeh, O., Sella, Y.: Efficient update diffusion in byzantine environments. In: The 20th IEEE Symposium on Reliable Distributed Systems (SRDS '01), Washington - Brussels - Tokyo, IEEE (2001) 90–98

12. Malkhi, D., Mansour, Y., Reiter, M.K.: Diffusion without false rumors: on propagating updates in a Byzantine environment. Theoretical Computer Science **299** (2003) 289–306

13. Minsky, Y., Schneider, F.B.: Tolerating malicious gossip. Distributed Computing **16** (2003) 49–68

14. Masuzawa, T., Tixeuil, S.: A self-stabilizing link-coloring protocol resilient to unbounded byzantine faults in arbitrary networks. Technical Report 1396, Laboratoire de Recherche en Informatique (2005)

15. Nesterenko, M., Arora, A.: Tolerance to unbounded byzantine faults. In: Proceedings of 21st IEEE Symposium on Reliable Distributed Systems. (2002) 22–29

16. Sakurai, Y., Ooshita, F., Masuzawa, T.: A self-stabilizing link-coloring protocol resilient to byzantine faults in tree networks. In: Proceedings of the 2004 International Conference on Principles of Distributed Systems (OPODIS'2004). Lecture Notes in Computer Science, Springer-Verlag (2004)

17. Masuzawa, T.: A fault-tolerant and self-stabilizing protocol for the topology problem. In: Proceedings of the Second Workshop on Self-Stabilizing Systems. (1995) 1.1–1.15

18. Yellen, J., Gross, J.L.: Graph Theory & Its Applications. CRC Press (1998) ISBN: 0–849–33982–0.

19. Dijkstra, E.W., Scholten, C.S.: Predicate Calculus and Program Semantics. Springer-Verlag, Berlin (1990)

20. Nesterenko, M., Tixeuil, S.: Bounds on topology discovery in the presence of byzantine faults. Technical Report TR-KSU-CS-2006-01, Dept. of Computer Science, Kent State University (2006) http://www.cs.kent.edu/techreps/TR-KSU-CS-2006-01.pdf.

21. Nesterenko, M., Tixeuil, S.: Discovering network topology in the presence of byzantine faults. Technical Report TR-KSU-CS-2005-01, Dept. of Computer Science, Kent State University (2005) http://www.cs.kent.edu/techreps/TR-KSU-CS-2005-01.pdf.
22. Dijkstra, E., Scholten, C.: Termination detection for diffusing computations. Information Processing Letters **11** (1980) 1–4

Minimum Energy Broadcast and Disk Cover in Grid Wireless Networks[*]

(Extended Abstract)

Tiziana Calamoneri[2], Andrea E.F. Clementi[1], Miriam Di Ianni[1],
Massimo Lauria[2], Angelo Monti[2], and Riccardo Silvestri[2]

[1] Dipartimento di Matematica, Università degli Studi di Roma"Tor Vergata"
{clementi, diianni}@mat.uniroma2.it
[2] Dipartimento di Informatica, Università degli Studi di Roma "La Sapienza"
{calamo, lauria, monti, silver}@di.uniroma1.it

Abstract. The *Minimum Energy Broadcast* problem consists in finding the minimum-energy range assignment for a given set S of n stations of an ad hoc wireless network that allows a source station to perform broadcast operations over S.

We prove a nearly tight asymptotical bound on the optimal cost for the Minimum Energy Broadcast problem on square grids. We emphasize that finding tight bounds for this problem restriction is far to be easy: it involves the *Gauss's Circle* problem and the *Apollonian Circle Packing*. We also derive near-tight bounds for the *Bounded-Hop* version of this problem. Our results imply that the best-known heuristic, the MST-based one, for the Minimum Energy Broadcast problem is far to achieve optimal solutions (even) on very regular, well-spread instances: its worst-case approximation ratio is about π and it yields $\Omega(\sqrt{n})$ hops.

As a by product, we get nearly tight bounds for the *Minimum Disk Cover* problem and for its restriction in which the allowed disks must have *non-constant* radius.

Finally, we emphasize that our upper bounds are obtained via polynomial time constructions.

1   Introduction

An *ad-hoc* wireless network consists of a set S of radio stations connected by wireless links. We assume that stations are located on the Euclidean plane. A transmission range is assigned to every station: a *range assignment* $r : S \to R$ determines a directed *communication graph* $G(S, E)$ where edge $(i, j) \in E$ if and only if $\texttt{dist}(i, j) \leq r(i)$ where $\texttt{dist}(i, j)$ is the Euclidean distance between i and j. In other words, $(i, j) \in E$ if and only if j belongs to the *disk* of radius $r(i)$ centered at i. The transmission range of a station depends on the energy power supplied to the station. In particular, the power P_s required by a station s to transmit data to another station t must satisfy the inequality

$$\frac{P_s}{\texttt{dist}(s, t)^\alpha} \geq 1$$

---

[*] Research partially supported by the EC Project *AEOLUS*.

P. Flocchini and L. Gąsieniec (Eds.): SIROCCO 2006, LNCS 4056, pp. 227–239, 2006.
© Springer-Verlag Berlin Heidelberg 2006

where $\alpha \geq 1$ is the *distance-power gradient*. In the empty space, $\alpha = 2$ (see [20]): this is the case considered in this paper.

Stations of an ad-hoc network cooperate in order to provide specific network connectivity properties by adapting their transmission ranges. A *Broadcast Range Assignment* (for short *Broadcast*) is a range assignment that yields a communication graph G containing a directed spanning tree rooted at a given source station s. A fundamental problem in the design of ad-hoc wireless networks is the *Minimum Energy Broadcast* Problem (for short *Minimum Broadcast*): it consists in finding a Broadcast of minimal *overall energy power* [7, 10, 18]. A range assignment r can be represented by the corresponding family $\mathcal{D} = \{D_1, \ldots, D_\ell\}$ of disks, and its overall energy power (i.e. $\mathsf{cost}(\mathcal{D})$) is defined as

$$\mathsf{cost}(\mathcal{D}) = \sum_{i=1}^{\ell} r_i^2 \text{ where } r_i \text{ is the radius of } D_i \qquad (1)$$

The Minimum Broadcast problem is known to be NP-hard [5] and the best-known approximation algorithm is the MST-based heuristic [1, 10]. The MST-based heuristic computes the minimum spanning tree of the complete graph induced by S, then, it assigns a direction to the edges from the source s to the leaves; finally, it assigns to each node i a range equal to the length of the longest edge outgoing from i. This heuristic is efficient and easy to implement, so, its worst-case approximation analysis has been the subject of several works over the last five years. In particular, the first *constant* upper bound ($\simeq 40$) on the approximation ratio was determined in [5]. A rather sophisticated analysis, recently introduced in [1], yields the *tight* upper bound 6. The tightness follows from the lower bound proved in [4, 10] by considering *unlike* input configurations. The worst-case analysis is often not sufficient to evaluate the practical interest of a heuristic. It might be the case that the MST-based heuristic provides *nearly* optimal solutions *for most* of natural and practically-relevant instances. Recently, experimental studies have been presented on this issue [11, 6, 10].

1.1   Our Results

Minimum Broadcast Problem. In this paper, we address the above issue by adopting an analytical approach: we consider Minimum Broadcast and some other related problems on *square grids*. Square grids have been often considered in wireless networks since they well-model some *well-spread*, practically relevant ad-hoc network topologies [8, 19, 20]. One can see that the MST-based heuristic, on a square *grid* of n points (without loss of generality, adjacent points are placed at unit distance), returns, *in the worst-case*, a solution of cost $n - 1$. On the other hand, what is the optimal cost on the square grids? One may think that determining this cost is an easy task for so simple instances. On the contrary, this is far to be true: as we will see later, this analysis involves the well-known mathematical *Gauss' Circle problem* [15, 17] and the *Apollonian Circle Packing* [13, 21]. Our first contribution is the following result.

Theorem [Broadcast]. *If $\mathcal{B}^*$ is any optimal Broadcast for the square grid $\mathcal{G}$ of n points, then*

$$\frac{n}{\pi} - O(\sqrt{n}) \leq \mathsf{cost}(\mathcal{B}^*) \leq 1.01013\frac{n}{\pi} + O(\sqrt{n})$$

The upper bound is achieved via a polynomial time construction.

The above upper bound implies that the MST-based heuristic yields, in the worst-case, a solution cost which is about π times larger than the optimum.

Minimum Cover Problem. Any Broadcast yields a *(disk) cover* of the grid and a communication graph that contains a spanning tree. A cover $\mathcal{C}$ of a set S of points is a set of disks $\mathcal{C} = \{D_1, \ldots, D_\ell\}$ of radius at least 1, centered at some points of S, that covers all points in S. The cost of $\mathcal{C}$ is defined as $\mathsf{cost}(\mathcal{C})$ (see Eq. 1). The *Minimum Cover* problem consists in finding a cover for S of minimum cost. Observe that this is a variant of the well-known NP-hard *Minimum Geometric Disk Cover* [9, 16].

In general, a cover does not suffice to provide a feasible solution for the Minimum Broadcast problem. A natural question here is whether (or when) the minimum cover cost is asymptotically equivalent to the minimum broadcast cost. This question is formally addressed by determining the cost of a minimum cover for square grids.

Theorem [Cover]. *If $\mathcal{C}^*$ is any optimal cover of the square grid $\mathcal{G}$ of n points, then*

$$n/5 \leq \mathsf{cost}(\mathcal{C}^*) \leq n/5 + O(\sqrt{n})$$

The upper bound is achieved via a polynomial time construction.

From the above theorems, it turns out that the cost of the cover is significantly lower than the cost of the broadcast. However, next theorem shows that this is not the case when we require that the disks are sufficiently large.

Theorem [Large Disk Cover]. *Let $f(n) = \omega(1)$. The cost of any cover of $\mathcal{G}$ with disks of radius at least $f(n)$ is at least $\frac{n}{\pi} - o(n)$. The upper bound is achieved via a polynomial time construction.*

We emphasize that there are important network scenarios in which the *installing* cost (i.e. the cost of installing an omni-directional transmitter at a given location) is rather high and it must be "amortized" by a *relevant* use of the antenna. In such cases, it is convenient to assign positive range to a station only if such a range (so, disk) is large enough.

Bounded-Hop Broadcast. An important version of the Minimum Broadcast problem is the one in which feasible solutions must guarantee a *bounded number of hops*: The number of links (i.e. *hops*) in the path from the source to *any* other node must be not larger than a fixed bound. This problem version is relevant since the number of hops is closely related to the delay transmission time. The hop restriction finds another application in the context of *reliability*: Assume that, in a communication network, link faults happen with probability p and that all faults occur independently. Then, the probability that a multi-hop transmission fails exponentially increases with the number of hops. For further motivations in studying bounded hops communication see [2, 12, 14, 22].

A main question here is the following: Does broadcasting with a bounded number of hops require a *significantly* larger cost than broadcasting with an unbounded number of hops? Intuitively speaking, one might figure out that the right answer is the positive one since the cost is proportional to the area of the solution disks and bounded-hop solutions require larger disks. Observe also that the use of *large* disks yields *large* disk overlapping. Rather surprisingly, this is not the case: we derive a broadcast for grids that uses only a constant (i.e. not depending on n) number of disks and thus yields a *constant* number of hops. This solution has a cost which is very close to that of the unbounded-hops version.

Theorem [Broadcast with few Hops]. *A positive constant c exists such that it is possible to construct in polynomial time a broadcast $\mathcal{B}$ for $\mathcal{G}$ with (only) c disks (of radius $\Omega(\sqrt{n})$) and such that*

$$\mathsf{cost}(\mathcal{B}) < 1.1171\frac{n}{\pi} + O(\sqrt{n}).$$

By comparing the above theorem with Theorem [Large Disks Cover], we can state that covering and broadcasting over grids have almost asymptotically-equivalent cost when the solution disks have *non-constant* radius (remind that any broadcast is also a cover). We also remark that the MST-based heuristic *always* returns a solution for the grid that has an *unbounded* (i.e. $\Omega(\sqrt{n})$) number of hops. So, our almost optimal polynomial-time construction yields bounded-hop solutions whose structure significantly departs from that of the MST-based solutions.

Square grids are thus the first family of well-spread, natural instances that perfectly capture the "hardness" of solving the Minimum Broadcast problem via the MST-based heuristic. It is our opinion that the set of results presented in this paper provides strong theoretical arguments that open new possibilities in the design of an efficient heuristic that significantly improves over the MST-based one (at least) in the case of *well-spread* and *uniform-random* instances.

1.2   Preliminaries

We consider a Cartesian coordinates system and a square grid $\mathcal{G}$ of side length $m - 1$ with its bottom left vertex in the origin. $\mathcal{G}$ contains $n = m^2$ points at integer coordinates; the coordinates of point P of the grid will be denoted as x_p and y_p. A $\mathcal{G}$-*disk* D is a disk centered at any point of the grid and having at least one point of the grid on its boundary. We also denote as D the set of points of grid $\mathcal{G}$ covered by D.

2   The Minimum Cover Problem on the Grid

In this section we study two versions of the disk cover problem of the grid $\mathcal{G}$. In the first version, we consider coverings by disks of arbitrary radius, while, in the second one, disks are required to have a minimal non constant radius. For both versions, we need to evaluate the number $N(r)$ of points of the *infinite* grid covered by a $\mathcal{G}$-disk of radius r. This problem, known as *Gauss' Circle problem*, has been extensively studied [15, 17] in order to derive the best exponent $\delta < 1$ such that $N(r) \leq \pi r^2 + cr^\delta$ for some constant c. However, all these studies are not useful to provide a good bound on c: instead, we need an upper bound on $N(r)$ with a small constant c while the exponent δ can be 1. The proof of next lemma is given in the full version of the paper [3].

Lemma 1. *For any radius $r \geq 1$, it holds that $N(r) < \pi r^2 + (\pi\sqrt{2} - 2)r + \frac{1}{5}\sqrt{r} + \frac{\pi}{2}$. Moreover, for $r > \sqrt{10}$, it holds that $N(r) < \pi r^2 + 2\sqrt{2}r - 5$.*

The above lemma is now exploited to prove asymptotically tight lower and upper bounds on the minimum cost of a cover of grid $\mathcal{G}$.

Theorem 1. *If C^* is any minimum cover of the square grid $\mathcal{G}$ of n points, then*

$$n/5 \leq \mathsf{cost}(C^*) \leq n/5 + O(\sqrt{n})$$

The upper bound is achieved via a polynomial time construction.

Proof. We first observe that, for any $r > 0$, it holds that

$$N(r) \leq 5r^2. \tag{2}$$

Indeed, $N(1) = 5$, $N(\sqrt{2}) = 9$, and Lemma 1 implies that $N(r) \leq 5r^2$, for any $r \geq 2$. Let $D_1, D_2, \ldots D_t$ be the $\mathcal{G}$-disks of an optimal cover and let cost^* be its cost. Let r_i be the radius of D_i, $1 \leq i \leq t$. Since D_i covers $N(r_i)$ points, Inequality (2) implies that

$$n \leq \sum_{i=1}^{t} N(r_i) \leq \sum_{i=1}^{t} 5r_i^2 = 5 \cdot \mathsf{cost}^*$$

and so $\mathsf{cost}^* \geq \frac{n}{5}$.

A cover of $\mathcal{G}$ with cost $\frac{n}{5} + O(\sqrt{n})$ is shown in Figure 1 for $m = 11$. It is easy to see that the number of grey $\mathcal{G}$-disks (i.e. disks not completely contained in $\mathcal{G}$) is $O(\sqrt{n})$, and the number of white $\mathcal{G}$-disks (i.e. disks completely contained in $\mathcal{G}$) is not greater than $\frac{n}{5}$. Since all $\mathcal{G}$-disks have unit radius, then the cost $\frac{n}{5} + O(\sqrt{n})$ follows. It is easy to check that the above construction can be computed in time polynomial in n. $\qquad\square$

The cover resulting by the construction in Theorem 1 uses only $\mathcal{G}$-disks of unit radius. Next theorem investigates the cost of covers using only $\mathcal{G}$-disks of large, non constant radius.

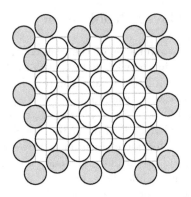

Fig. 1. An asymptotically optimum disk cover for $\mathcal{G}$ with $m = 11$

Theorem 2. *Let $f(n) = \omega(1)$. The cost of any cover of $\mathcal{G}$ with $\mathcal{G}$-disks of radius at least $f(n)$ is at least $\frac{n}{\pi} - o(n)$.*

Proof. Let $D_1, D_2, \ldots D_t$ be the $\mathcal{G}$-disks of a cover of $\mathcal{G}$ and let cost be its cost. Let r_i be the radius of D_i, $1 \le i \le t$. As D_i covers $N(r_i)$ points, Lemma 1 implies that

$$n < \sum_{i=1}^{t} N(r_i) < \sum_{i=1}^{t} \left(\pi r_i^2 + (\pi\sqrt{2})r_i + \frac{1}{5}\sqrt{r_i} + \frac{\pi}{2} \right) < \tag{3}$$

$$< \sum_{i=1}^{t} \left(\pi r_i^2 + 2\pi r_i \right) = \pi\mathsf{cost} + 2\pi \sum_{i=1}^{t} r_i$$

By hypothesis $r_i \ge f(n)$, hence we get

$$\mathsf{cost} = \sum_{i=1}^{t} r_i^2 \ge f(n) \sum_{i=1}^{t} r_i$$

and thus

$$\sum_{i=1}^{t} r_i \le \frac{\mathsf{cost}}{f(n)}$$

From the above inequality and from Inequality 3, we get $n < \pi\mathsf{cost} + 2\pi \frac{\mathsf{cost}}{f(n)}$ and, finally,

$$\mathsf{cost} > n \left(\frac{f(n)}{\pi f(n) + 2\pi} \right) = \frac{n}{\pi} \left(1 - \frac{2}{f(n) + 2} \right) = \frac{n}{\pi} - o(n). \qquad \square$$

As we shall see in the next section, the lower bound of this theorem is almost tight.

3   The Minimum Broadcast Problem on the Grid

The aim of this section is to prove lower and upper bounds on the cost of an optimal broadcast. In particular, in order to prove the lower bound, we introduce the following definitions. A *chain* $\mathcal{H}$ is a sequence of $\mathcal{G}$-disks $D_1, D_2, \ldots, D_k$, $k \geq 1$, such that D_{i+1} is centered at some point contained in D_i for $1 \leq i < k$. We also say that a chain $\mathcal{H}$ *activates* a disk D if (i) D does not belong to $\mathcal{H}$, (ii) the center of D is contained in D_k, and (iii) D does not contain the center of D_1. Furthermore, we define

$$\mathcal{U}(\mathcal{H}) = \bigcup_{i=1}^{k} D_i$$

where the union refers to points of the infinite grid contained in disks D_i.

For any $r \geq 1$, consider any disk D of radius r; we define

$$M(r) = \min\{|\mathcal{U}(\mathcal{H}) \cap D| \text{ such that } \mathcal{H} \text{ activates } D\}.$$

Notice that $M(r)$ does not depend on the choice of D and that any disk of a broadcast tree not containing the source is activated by a chain of disks belonging to the tree. The cardinality of the intersection between the disk and the chain is at least $M(r)$, where r is the radius of the disk. In order to evaluate the broadcast cost, we need a lower bound on $M(r)$. The proof of next lemma is given in the full version of the paper [3].

Lemma 2. *For any $r \geq 1$, it holds that $M(r) \geq 2\sqrt{2}r - 5$.*

Theorem 3. *The cost of any broadcast of $\mathcal{G}$ is at least $\frac{n}{\pi} - O(\sqrt{n})$.*

Proof. Let $D_1, D_2, \ldots D_t$ be the $\mathcal{G}$-disks of an optimal broadcast of $\mathcal{G}$ and let cost^* be its cost. Let r_i be the radius of D_i, $1 \leq i \leq t$. If there exists a disk D_i with radius $r_i \geq \sqrt{\frac{n}{\pi}}$, the thesis holds. Hence, we assume that $r_i < \sqrt{\frac{n}{\pi}}$, $1 \leq i \leq t$. In order to exploit Lemma 1, we partition the set $\{D_1, D_2, \ldots D_t\}$ into two sets: X and its complement $\overline{X}$, where

$$X = \{D_i \mid r_i > \sqrt{10}\}$$

From Lemma 1, it follows that

$$\sum_{i=1}^{t} N(r_i) = \sum_{D_i \in X} N(r_i) + \sum_{D_i \in \overline{X}} N(r_i) \leq \sum_{D_i \in X} (\pi r_i^2 + 2\sqrt{2}r_i - 5) + \sum_{D_i \in \overline{X}} N(r_i)$$

$$= \pi \cdot \mathsf{cost}^* + 2\sqrt{2} \sum_{D_i \in X} r_i - 5|X| + \sum_{D_i \in \overline{X}} \left(N(r_i) - \pi r_i^2 \right) \qquad (4)$$

As a consequence, we have that

$$\pi \cdot \mathsf{cost}^* \geq \sum_{i=1}^{t} N(r_i) - 2\sqrt{2} \sum_{D_i \in X} r_i + 5|X| - \sum_{D_i \in \overline{X}} \left(N(r_i) - \pi r_i^2 \right) \qquad (5)$$

Now, we derive a lower bound on $\sum_{i=1}^{t} N(r_i)$. Observe that the communication graph yielded by the optimal broadcast contains a directed spanning tree T rooted at the source node. We partition $\{D_1, D_2, \ldots D_t\}$ into two sets Y and $\overline{Y}$, where Y is the set of $\mathcal{G}$-disks that cover the source point. We observe that every $\mathcal{G}$-disk $D_i \in \overline{Y}$ is activated by a chain of $\mathcal{G}$-disks whose centers induce a directed path in T. This implies that the number of intersection points between the activating chain and D_i is at least $M(r_i)$. Now we prove the following inequality:

$$\sum_{i=1}^{t} N(r_i) \geq n + \sum_{D_i \in \overline{Y}} M(r_i)$$

We consider a numbering of the T disks such that the disks on a root→leaf path have strictly increasing numbers. Let

$$E = \{(p, i) \mid \exists i : 1 \leq i \leq t \wedge p \in D_i\} \quad \text{and}$$

$$F = \{(p, j) \mid (p, j) \in E \wedge j = \min\{k \mid (p, k) \in E\}\}$$

In other words, $(p, j) \in F$ if and only if D_j is the "first" disk that covers p. Clearly, it holds that $|E| = \sum_i N(r_i)$, $F \subseteq E$, and $|F| \geq n$. Now, for every $i \in \overline{Y}$, let $\mathcal{H}_i$ be the chain that activates D_i. Define $E_i = \{(p, i) \mid p \in \mathcal{U}(\mathcal{H}_i) \cap D_i\}$. The following properties hold: (a) $E_i \subseteq E - F$; (b) if $i \neq j$ then $E_i \cap E_j = \emptyset$; (c) $|E_i| \geq M(r_i)$. As for (a), clearly $E_i \subseteq E$. Furthermore, if $(p, i) \in E_i$ then $p \in \mathcal{U}(\mathcal{H}_i) \cap D_i$; thus, there exists a disk $D_j \in \mathcal{H}_i$ such that $p \in D_j$ and $j < i$. This implies that $\min\{k \mid (p, k) \in E\} \leq j < i$ and so $(p, i) \notin F$. The proofs of (b) and (c) are immediate from the definitions of E_i and $M(\cdot)$. Finally, it holds that

$$\sum_{i=1}^{t} N(r_i) = |E| = |F| + (|E| - |F|) \geq n + \sum_{i \in \overline{Y}} |E_i| \geq n + \sum_{i \in \overline{Y}} M(r_i).$$

Lemma 2 implies that

$$\sum_{D_i \in \overline{Y}} M(r_i) = \sum_{D_i \in \overline{Y} \cap X} M(r_i) + \sum_{D_i \in \overline{Y} \cap \overline{X}} M(r_i) \geq$$

$$\geq 2\sqrt{2} \sum_{D_i \in \overline{Y} \cap X} r_i - 5|\overline{Y} \cap X| + \sum_{D_i \in \overline{Y} \cap \overline{X}} M(r_i)$$

From the above inequality, Inequality (5), and simple calculations, we get:

$$\pi \cdot \text{cost}^* \geq n - 2\sqrt{2} \sum_{D_i \in \overline{Y} \cap X} r_i + 5|X| - 5|\overline{Y} \cap X| + \sum_{D_i \in \overline{Y} \cap \overline{X}} M(r_i) - \sum_{D_i \in X} \left(N(r_i) - \pi r_i^2\right)$$

and

$$\text{cost}^* > \frac{n}{\pi} - \frac{2\sqrt{2}}{\pi} \sum_{D_i \in \overline{Y} \cap X} r_i + \tag{6}$$

$$+\frac{1}{\pi} \sum_{D_i \in \overline{Y} \cap X} \left(M(r_i) - N(r_i) + \pi r_i{}^2\right) - \frac{1}{\pi} \sum_{D_i \in Y \cap \overline{X}} \left(N(r_i) - \pi r_i{}^2\right)$$

Now we bound $\sum_{D_i \in Y \cap X} r_i$. Consider the sets

$$B_k = \{D_j \in Y \mid 2^{k-1} \le r_j < 2^k\}, \ 1 \le k \le l$$

where $l = \lceil \log r_{max} \rceil + 1$ and $r_{max} = \max\{r_j \mid D_j \in Y\}$. It holds that

$$\sum_{D_i \in Y \cap X} r_i \le \sum_{D_i \in Y} r_i = \sum_{k=1}^{l} \sum_{D_i \in B_k} r_i \le \sum_{k=1}^{l} \frac{1}{2^{k-1}} \sum_{D_i \in B_k} r_i{}^2 \qquad (7)$$

Replace the $\mathcal{G}$-disks in $B_1 \cup B_2 \cup \ldots B_k$ by a $\mathcal{G}$-disk with radius (2^{k+1}) and centered in the source point. This operation produces a new broadcast with cost

$$\text{cost}^* - \sum_{D_i \in B_1 \cup B_2 \cup \ldots B_k} r_i{}^2 + (2 \cdot 2^k)^2$$

Hence, from the optimality of the previous broadcast it must be

$$\sum_{D_i \in B_1 \cup B_2 \cup \ldots B_k} r_i{}^2 \le (2 \cdot 2^k)^2$$

From the above inequality and from Inequality (7) we have

$$\sum_{D_i \in Y \cap X} r_i \le \sum_{k=1}^{l} \frac{2^{2k+2}}{2^{k-1}} = \sum_{k=1}^{l} 2^{k+3} < 2^{l+4} < 2^6 r_{max} = O(\sqrt{n}) \qquad (8)$$

where the last step follows from the initial assumption that broadcast $\mathcal{G}$-disks have radii less than $\sqrt{\frac{n}{\pi}}$. It is possible to exhaustively prove that $M(r) - N(r) + \pi r^2 > 0$ when $r \le \sqrt{10}$, i.e., $r \in \{1, \sqrt{2}, 2, \sqrt{5}, \sqrt{8}, 3, \sqrt{10}\}$. Hence,

$$\sum_{D_i \in \overline{Y} \cap X} \left(M(r_i) - N(r_i) + \pi r_i{}^2\right) > 0 \qquad (9)$$

Moreover, the number of $\mathcal{G}$-disks in $Y \cap \overline{X}$ is bounded by constant $N(\sqrt{10})$. Thus,

$$\sum_{D_i \in Y \cap \overline{X}} \left(N(r_i) - \pi r_i{}^2\right) = O(1) \qquad (10)$$

Finally, by combining Inequality (6) with bounds (8), (9) and (10) we get the thesis. $\qquad \square$

The construction of optimal Broadcasts for the grid is somewhat connected with the famous problem known as *Apollonian Circle Packing* [13, 21]. More precisely, we observe that if it were possible to evaluate the cost of the Apollonian Circle Packing of the grid then it would be possible to obtain the optimal bound on the Broadcast cost. We strongly believe that this is the *only* way to obtain such an optimal bound. The former problem is known to be a hard mathematical problem. In order to get a near-tight bound, we here adopt a simpler construction.

Theorem 4. *Given any source $s \in \mathcal{G}$, it is possible to construct, in polynomial time, a Broadcast for $\mathcal{G}$ of cost $1.01013\frac{n}{\pi} + O(\sqrt{n})$.*

Proof. In order to provide a Broadcast of cost $1.01013\frac{n}{\pi} + O(\sqrt{n})$, we assume that $m - 1$ is a multiple of 6. If this is not the case, we can add $O(m)$ new unit radius $\mathcal{G}$-disks to our construction in order to broadcast to the remaining points.

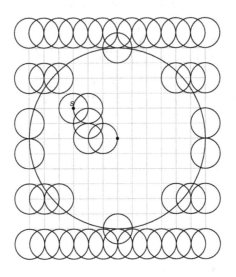

Fig. 2. An almost optimal Broadcast for the grid where $m = 19$

Consider the Broadcast shown in figure 2. Its cost can be computed by summing up the following three contributions.

- A chain of $\mathcal{G}$-disks of radius 1 from the source point to the middle point of $\mathcal{G}$. The cost of this chain is $O(m)$.
- A big $\mathcal{G}$-disk of radius $r = \frac{m-1}{2}$ centered in the middle point of $\mathcal{G}$. This disk has cost $r^2 = \frac{n}{4} - \Theta(m)$.
- A set of $\mathcal{G}$-disks of radius 1 that broadcast to all nodes of $\mathcal{G}$ out of the big $\mathcal{G}$-disk. In order to compute the cost of this set, assume that the origin of the Cartesian plane lies in the middle point of $\mathcal{G}$ and compute only the cost of the $\mathcal{G}$-disks in the first quadrant, multiplied by 4. Furthermore, observe that the contribution of the first quadrant consists of $\frac{m-1}{6}$ horizontal chains of unit-radius $\mathcal{G}$-disks whose length depends on their y-coordinates. So the cost of this contribution is:

$$C = 4 \sum_{i=0}^{\frac{r}{3}} \left(r - \left\lfloor \sqrt{r^2 - (3i)^2} \right\rfloor \right) < \frac{4}{3}r^2 - 4\sum_{i=0}^{\frac{r}{3}-1} \left(\sqrt{r^2 - (3i)^2} - 1 \right) <$$

$$< \frac{4}{3}r^2 + \frac{4}{3}r - 4 - 4\int_0^{\frac{r}{3}} \sqrt{r^2 - (3x)^2}dx < \frac{4}{3}r^2 + \frac{4}{3}r - 4 - \frac{4}{3}\int_0^r \sqrt{r^2 - x^2}dx <$$

$$< \frac{4}{3}r^2 + \frac{4}{3}r - 4 - \frac{4}{3}\left[\frac{r^2}{2}\arcsin\frac{x}{r} + \frac{x}{2}\sqrt{r^2 - x^2}\right]_0^r =$$

$$= \left(\frac{4-\pi}{3}\right)\frac{n}{4} + O(m)$$

Finally, the cost of this Broadcast is $\frac{n}{4} + \left(\frac{4-\pi}{3}\right)\frac{n}{4} + O(m) = 1.01013\frac{n}{\pi} + O(\sqrt{n})$. The construction of this solution can be clearly performed in time polynomial in n.   □

Even when the $\mathcal{G}$-disks must be very large, we are able to provide a Broadcast whose cost is very close to the lower bound, as shown in the following result. We remark that its proof makes use of a construction that approximates the Apollonian Circle Packing of the grid. The proof of next lemma is given in the full version of the paper [3].

Lemma 3. *Let $0 < c < 1$ be a constant. For any source $s \in \mathcal{G}$, it is possible to construct, in polynomial time, a broadcast $\mathcal{B}$ for $\mathcal{G}$ with disks of radius at least $c\sqrt{n}$ and such that*

$$\mathsf{cost}(\mathcal{B}) = f(c)\frac{n}{\pi} + O(\sqrt{n})$$

where

$$f(c) < \pi\left(0.35483 + 24.6814c^{2-\log_{1+\sqrt{2}}3} - 0.5551c + 0.5c^2\right)$$

The following upper bound is an easy consequence of the previous lemma.

Theorem 5. *For any source point, there exists a (polynomial-time computable) Broadcast $\mathcal{B}$ for $\mathcal{G}$ that uses disks with radius at least $\frac{\sqrt{n}}{10^6}$ and such that*

$$\mathsf{cost}(\mathcal{B}) < 1.1171\frac{n}{\pi} + O(\sqrt{n})$$

As a consequence, $\mathcal{B}$ consists of a constant number of disks.

Observe that Theorem 3 implies that the upper bound of Theorem 5 is almost tight.

4   Future Research

Our asymptotical bounds on the Broadcast Problem on grids are not tight: achieving tight bounds here is an interesting theoretical open problem. However, as mentioned in the Introduction, we believe that our results open new promising directions in the design of new, good heuristics for a wide and practically relevant class of input configurations: well-spread, regular instances and uniform random instances [8, 19]. This is, in our opinion, the most relevant challenge in this topic. Efficient implementation, performance analysis and tests of some heuristics inspired by our constructive upper bounds are the subject of our present research activity.

References

1. C. Ambuehl. An optimal bound for the MST algorithm to compute energy efficient broadcast trees in wireless networks. In Proc. of *32th ICALP*, 1139–1150, 2005.

2. A. Balakrishnan and K. Altinkemer. Using a hop-constrained model to generate alternative communication network design. *ORSA Journal of Computing*, 4, 147–159, 1992.

3. T. Calamoneri, A. Clementi, M. Di Ianni, M. Lauria, A. Monti, and R. Silvestri. Minimum Energy Broadcast and Disk Cover in Grid Wireless Networks (Full Version). Available at http://www.dsi.uniroma1.it/ calamo/papers.html.

4. G. Călinescu, X.Y. Li, O. Frieder, and P.J. Wan. Minimum-Energy Broadcast Routing in Static Ad Hoc Wireless Networks. In Proc. of *20th IEEE INFOCOM*, 1162–1171, April 2001.

5. A. Clementi, P. Crescenzi, P. Penna, G. Rossi and P. Vocca. On the Complexity of Computing Minimum Energy Consumption Broadcast Subgraphs. In Proc. of 18th *STACS*, LNCS 2010, 121–131, February 2001.

6. A. Clementi, G. Huiban, P. Penna, G. Rossi, Y. C. Verhoeven. On the Approximation Ratio of the MST-based Heuristic for the Energy-Efficient Broadcast Problem in Static Ad-Hoc Radio Networks. In Proc. of *3rd IEEE Intern. Workshop on Wireless, Mobile and Ad Hoc Networks (WMAN'03)*, 2003.

7. A. Clementi, G. Huiban, P. Penna, G. Rossi, and Y. Verhoeven. Some Recent Theoretical Advances and Open Questions on Energy Consumption in Static Ad-Hoc in Wireless Networks. In Proc. of *3rd Int. Workshop ARACNE*, Carleton Scientific, 23–38, 2002.

8. A. Clementi, P. Penna, and R. Silvestri. On the Power Assignment Problem in Radio Networks. *Mobile Networks and Applications (MONET)*, 9, 125–140, 2004.

9. P. Crescenzi and V. Kann, A Compendium of NP Optimization Problems. http://www.nada.kth.se/ viggo/wwwcompendium/

10. A. Ephremides, G.D. Nguyen, and J.E. Wieselthier. On the Construction of Energy-Efficient Broadcast and Multicast Trees in Wireless Networks. In Proc. of *19th IEEE INFOCOM*, 585–594, 2000.

11. M. Flammini, A. Navarra, and S. Perennes. The Real Approximation Factor of the MST Heuristic for the Minimum Energy Broadcast. In Proc. of *WEA*, 22–31, 2005.

12. L. Gouveia. Using the Miller-Tucker-Zemlin constraints to formulate a minimal spanning tree problem with hop-constraint. *European Journal of Operational Research*, 95, 170–190, 2001.

13. R.L. Graham, J.C. Lagarias, C.L. Mallows, A. R. Wilks, and C.H. Yan. Apollonian Circle Packings: Number Theory. *J. Number Theory*, 100, 1-45, 2003.

14. M. Haenggi. Twelve Reasons not to Route over Many Short Hops, in Proc. of *IEEE Vehicular Technology Conference (VTC'04 Fall)*, (5), 3130- 3134, 2004.

15. D. Hilbert and S. Cohn-Vossen. *Geometry and the Immagination*, Chelsea, 33–35, 1999.

16. D.S. Hochbaum and W. Maass. Approximation Schemes for Covering and Packing problems in Image Processing and VLSI. *Journal of ACM*, 32, 130–136, 1985.

17. M.N. Huxley. Exponential sums and lattice points. *Proc. London Math. Soc.*, 60, 471–502, 1990.

18. L. M. Kirousis, E. Kranakis, and D. Krizanc, and A. Pelc. Power Consumption in Packet Radio Networks. *Theoretical Computer Science*, 243, 289–305, 2000.

19. E. Kranakis, D. Krizanc, and A. Pelc. Fault-tolerant broadcasting in radio networks. *Journal of Algorithms*, 39, 47–67, 2001.

20. K. Pahlavan and A. Levesque. *Wireless Information Networks*. Wiley-Interscience, 1995.
21. B. Söderberg. Apollonian Tiling, the Lorentz Group, and Regular Trees. *Physical Review A*, 46, 1859-1866, 1992.
22. S. Voss, The steiner tree problem with hop constraint. *Annals of Operations Research*, 86, 321-345, 1999.

3-D Minimum Energy Broadcasting*

Alfredo Navarra

Computer Science Department, University of L'Aquila
Via Vetoio I-67100 L'Aquila, Italy
navarra@di.univaq.it

Abstract. The Minimum Energy Broadcast Routing problem was extensively studied during the last years. Given a sample space where wireless devices are distributed, the aim is to perform the broadcast pattern of communication from a given source while minimizing the total energy consumption. While many papers deal with the 2-dimensional case where the sample space is given by a flat area, few results are known about the more interesting and practical 3-dimensional case. In this paper we study such a case and we present a tighter analysis of the minimum spanning tree heuristic in order to considerably decrease its approximation factor from the known 26 to roughly 18.8. This decreases the gap with the known lower bound of 12 given by the so called kissing number.

1 Introduction

The study of a basic pattern of communication such as the Broadcast is of main interest in the context of Wireless Ad Hoc Networks. The broadcast can be in fact used to setup the network or to rapidly spread useful information. The wireless environment allows to all the devices in the range of a transmitter to receive the message. The range of a transmission basically depends by the environment in which the devices are distributed. According to the mostly used power attenuation model [1], for some constants $\alpha, \beta \in \mathbb{R}^+$, when a station s transmits with power P_s, a station r can receive its message if and only if

$$\frac{P_s}{\|s, r\|^\alpha} > \beta,$$

where $\|s, r\|$ is the Euclidean distance between s and r. Clearly in environments with obstacles the needed power α increases. Due to the nonlinear power attenuation, multi-hop transmission of messages through intermediate devices may result in energy saving. Thus, a naturally arising issue is that of supporting the broadcast with a minimum total energy consumption. The problem is called *Minimum Energy Broadcast Routing* (*MEBR*) and it is NP-hard, while if $\alpha = 1$ or $d = 1$ it is solvable in polynomial time [2, 3]. One of the most extensively studied cases concerns the 2-dimensional Euclidean space with $\alpha = 2$. Several papers

---

* Work supported by the European project COST Action 293, "Graphs and Algorithms in Communication Networks" (GRAAL).

P. Flocchini and L. Gąsieniec (Eds.): SIROCCO 2006, LNCS 4056, pp. 240–252, 2006.

progressively reduced the estimate of the approximation ratio of the fundamental Minimum Spanning Tree (MST) heuristic from 40 to 6 [4, 5, 6, 7, 8, 9, 10]. In [6] it was proven that for any considered dimension $d > 1$, the critical case to study is when $\alpha = d$ while for $\alpha > d$ any result can be easily extended to any power between d and α. Note that for $\alpha < d$ the ratios cannot be bounded by any function of α and d [4]. The MST and other heuristics have been presented in [1, 11] also for the multicasting variation of the problem. As already noted, the performance of the MST heuristic has been investigated by several authors and in the 2-dimensional Euclidean space, for $\alpha = 2$, the performed approximation ratio is 6 [5], and it is optimal [10]. Such a value coincides with the so called *kissing number* that was proven to be a lower bound for the approximation ratio of the MST heuristic for any dimension $d > 2$ and power $\alpha \geq d$ [4]. More precisely, the kissing number is the maximum number of d-spheres (or hyperspheres) of a given radius r that can simultaneously touch a d-sphere of radius r in the d-dimensional Euclidean space [12]. In the 3-dimensional Euclidean space the kissing number is 12 (see Figure 1) but the best known approximation ratio so far is 26 [6].

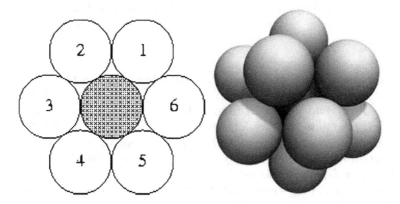

Fig. 1. The kissing number in the 2- and in the 3-dimensional case. It is given by 6 circles and 12 spheres respectively, simultaneously touching a central one.

In this paper we are interested in investigating more carefully this 3-dimensional case. We reduce the gap between upper and lower bound by decreasing the upper bound to roughly 18.8 (the exact obtained ratio is 18.802). Note that the 3-dimensional space better models practical environments since, in real life scenarios, radio stations are distributed over a 3-dimensional Euclidean space. Again the presence of obstacles can be overcome by the increasing of the power of transmission α. The main analysis is based on the study presented in [9] where a 6.33-approximation ratio of the MST heuristic for the 2-dimensional case was proven.

The paper is organised as follows. In the next section, we introduce the $MEBR$ problem with notations and the necessary definitions. In Section 3, we describe the technique that was used in [6] to prove the mentioned upper bound of 26 and we explain how to modify it in order to obtain a tighter bound for the 3-dimensional case. In Section 4, we present our main contribution that leads to the 18.8-approximation ratio. Finally, in Section 5, we give some conclusive remarks and discuss some open questions.

2   Minimum Energy Broadcast Routing

Let us first provide a formal definition of the Minimum Energy Broadcast Routing problem. Given a set of points S in a d-dimensional Euclidean space that represents the set of radio stations, an integer $\alpha \geq 1$ and a constant $\beta \in \mathbb{R}^+$, let $G_\alpha(S)$ be the complete weighted graph obtained as follows. The nodes of $G_\alpha(S)$ represent the points of S and the weight of each edge $\{x, y\}$ is the power consumption needed for a correct communication between x and y, that is $\beta \cdot \|(x, y)\|^\alpha$. For any subset of stations $Q \subseteq S$, let $G_\alpha(Q)$ be the subgraph of $G_\alpha(S)$ induced by Q.

A range assignment for S is a function $r : S \rightarrow \mathbb{R}^+$ such that the range $r(x)$ of a station x denotes the maximal distance from x at which signals can be correctly received. The total cost of a range assignment is then

$$cost(r) = \sum_{x \in S} \beta \cdot r(x)^\alpha.$$

A range assignment r for S yields a directed communication graph $G^r = (S, A)$ such that, for each $(x, y) \in S^2$, the directed edge (x, y) belongs to A if and only if y is at distance at most $r(x)$ from x. In other words, (x, y) belongs to A if and only if the emission power of x is at least equal to the weight of $\{x, y\}$ in $G_\alpha(S)$. In order to perform the required $MEBR$ from a given source $s \in S$, G^r must contain a directed spanning tree rooted at s and must have a minimum cost, from now on denoted as $m_\alpha^*(S, s)$.

One fundamental algorithm, called the MST heuristic [1], is based on the idea of tuning ranges so as to include a spanning tree of minimum cost. Roughly speaking, the heuristic computes the directed minimum spanning tree from the given source to the leaves. Such a computation is made over the complete weighted graph obtained from the set of nodes in which weights are the power of α of the distances of the endpoints of the edges. For each node, then, the heuristic assigns a power of transmission equal to the weight of the longest outgoing edge.

More precisely, let $T_\alpha(S)$ be a minimum spanning tree of $G_\alpha(S)$ and $MST(G_\alpha(S))$ its cost. Considering $T_\alpha(S)$ rooted at the source station s, the heuristic directs the edges of $T_\alpha(S)$ toward the leaves and sets the range $r(x)$ of every internal station x of $T_\alpha(S)$ with k children $x_1, \ldots, x_k$ in such a way that $r(x) = \beta \cdot max_{i=1,\ldots,k} \|x, x_i\|^\alpha$. In other words, r is the range assignment of minimum cost inducing the directed tree derived from $T_\alpha(S)$ and is such that $cost(r) \leq MST(G_\alpha(S))$. Therefore, in order to bound the approximation ratio of the heuristic, it is sufficient to bound the ratio between the cost $MST(G_\alpha(S))$ of a minimum spanning tree of $G_\alpha(S)$ and the optimal cost $m_\alpha^*(S, s)$.

Starting from the definition of minimum spanning tree given in [13], in [6] an interesting way to evaluate the cost of the heuristic is provided. For any subset of stations $Q \subseteq S$, let $G_\alpha(Q, r)$ be the graph obtained by considering only the edges of $G_\alpha(Q)$ of length at most r (that clearly have cost at most βr^α) and let $CC(Q, r)$ be the set of the connected components of $G_\alpha(Q, r)$. Let $n(Q, r) = |CC(Q, r)|$ be the number of connected components in $G_\alpha(Q, r)$ and $r_{max}(Q)$ be the minimum r such that $G_\alpha(Q, r)$ is connected (i.e. $n(Q, r_{max}) = 1$).

Corollary 1. *[6] For any subset of stations $Q \subseteq S$,*

$$MST(G_\alpha(Q)) = \alpha\beta \int_0^{r_{max}(Q)} (n(Q, r) - 1)r^{\alpha-1}\partial r.$$

For any set of stations Q let $e(Q) = \min_{x \in Q} \max_{y \in Q} \|x, y\|$ be the eccentricity of Q. Hence, there exists a station $x \in Q$ such that $\|x, y\| \leq e(Q)$ for every other $y \in Q$. Once chosen such a station x, let $c(Q)$ be the sphere of radius $e(Q)$ centered at x. The following general lemma is useful in the estimation of the approximation ratio of the MST heuristic.

Lemma 1. *[6] If $MST(G_\alpha(Q)) \leq \rho\beta e(Q)^\alpha$ for any subset of stations $Q \subseteq S$, then the MST heuristic is a ρ-approximation algorithm for the $MEBR$ problem.*

In the following we will concentrate on the $MEBR$ problem with $\alpha = 3$ in the 3-dimensional case. Thus, the cost of each edge of the weighted complete graph $G_3(S)$ representing the input network is proportional to the cube of the distance between its endpoints. For ease of notation, for any set of stations Q we will denote $G_3(Q)$ simply as $G(Q)$. Moreover, for the sake of simplicity, without loss of generality we assume $\beta = 1$ and $e(Q) = 1$, as all the results provided under this assumption can be directly extended to the general case [6].

3   Description of the Approach

In this section we firstly describe the general technique presented in [6]. Such a technique leads to the $(3^d - 1)$-approximation ratio of the MST heuristic for the $MEBR$ problem for any $d > 1$ and any $\alpha \geq d$. In our specific case, that is $d = 3$, $\alpha = 3$, the obtained approximation is 26. Secondly, by following the ideas in [9], we describe how to modify the previous technique hence leading to a new and tighter estimation of the upper bound, that is, of roughly 18.8.

For the general case the technique was based on a growing process (from now on called *basic*) in which d-spheres of equal radii centered in the stations of the subset Q are synchronously grown (see for instance Figure 2). The process starts by setting the radius $r = 0$ and ends when $r = \frac{r_{max}(Q)}{2} \leq \frac{1}{2}$, that is, when $G(Q, 2r)$ becomes connected. This is accomplished by increasing at any infinitesimal step the current radii, all equal to a given r, by ∂r.

Starting from the equality established in Corollary 1 on the cost $MST(G(Q))$ of any minimum spanning tree of $G(Q)$, the idea was to provide suitable lower

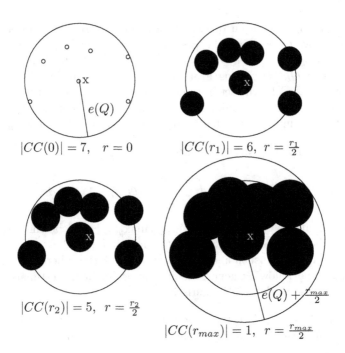

$|CC(0)| = 7, \quad r = 0$

$|CC(r_1)| = 6, \quad r = \frac{r_1}{2}$

$|CC(r_2)| = 5, \quad r = \frac{r_2}{2}$

$|CC(r_{max})| = 1, \quad r = \frac{r_{max}}{2}$

Fig. 2. The growing process of circles around the radio stations of the set Q in the 2-dimensional case

and upper bounds on the overall volume covered by the union of all the d-spheres at the end of the described process. In [6] the bound $MST(G(Q)) \leq 3^d - 1$ was proven, that by Lemma 1 implies the 26-approximability of the MST heuristic in the 3-dimensional case. Note that the lower bound is instead 12 and it is given by the kissing number [4, 12].

We now show how to improve the 26-approximation ratio by means of a new technique. The new analysis is based on the method presented in [9] where the 2-dimensional case was considered. The idea is to slightly change the shapes that are grown around stations at each infinitesimal step of the previously described basic growing process. More precisely, being in the 3-dimensional case we consider $c(Q)$ as the spherical place inside which the radio stations are thrown uniformly at random. While before each station was wrapped by a sphere, now things remain the same inside $c(Q)$, but the volume is thinned when growing outside $c(Q)$. Informally speaking, this allows to maintain the lower bound on the covered volume at the end of the growing process. On the other hand, the upper bound decreases since all the volume can be now included in a smaller sphere with respect to [6], thus improving the bound on the cost of the returned solution.

For the sake of clarity from now on we often drop Q from the notation, thus for instance writing G, $G(r)$, $CC(r)$, $n(r)$ and r_{max} instead of $G(Q)$, $G(Q, r)$, $CC(Q, r)$, $n(Q, r)$ and $r_{max}(Q)$, respectively.

In order to better explain the new reshaping technique we describe it in two phases. For any given radius r, the shape of radius r associated to a given

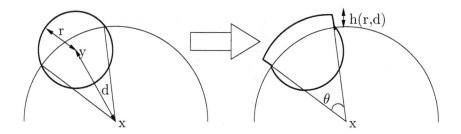

Fig. 3. Section of the new associated growing shape to each radio station

station y inside $c(Q)$ having distance d from the central station x is such that its intersection with $c(Q)$ coincides with the circular intersection of $c(Q)$ with a sphere of radius r centered at y. In other words, the intersection with $c(Q)$ of the new shape coincides with the *basic shape* given by the sphere of [6]. Outside $c(Q)$, the remaining portion of the sphere of radius r, if any, is reshaped as a kind of cylinder of suitable height $h(r, d)$ wrapping the outside spherical surface of $c(Q)$. In Figure 3 it is showed a cut section of the sphere $c(Q)$ centered at x and of the new shape. The height $h(r, d)$ is evaluated in such a way that its volume coincides with the volume of the corresponding portion of the basic shape outside $c(Q)$. This implies that the total volume remains $\frac{4}{3}\pi r^3$. With $\theta(r, d)$ we identify a conic angle obtained by connecting the center x with the circular intersection of the shape with $c(Q)$ (see Figure 3).

At each infinitesimal step in which the radius r grows by ∂r, given any function g depending on r, we denote by $\partial g(r) = g(r+\partial r) - g(r)$ the infinitesimal variation of $g(r)$.

At each infinitesimal step, while the growth of the spherical part inside $c(Q)$ is the same as in the basic case, the angle $\theta(r, d)$ of the outside part augments by a given quantity $\partial \theta(r, d)$. This is done according to the intersection of the increased sphere of radius $r+\partial r$ with $c(Q)$. About the height $h(r, d)$, it augments by $\partial h(r, d)$ in such a way that the total volume added to the shape is $4\pi r^2 \partial r$ as in the basic case.

Clearly, two shapes corresponding to a given radius r overlap if and only if the corresponding centers are at distance at most $2r$, as in the basic case. Starting from the observation that the shapes never meet at the circular intersections with the spherical surface of $c(Q)$,[1] it is possible to slightly enlarge the outside part of each shape.

This introduces the second phase of our shape modification by which enlarging the shape outside $c(Q)$ decreases its height. This must be done by increasing the angles $\theta(r, d)$ without violating the constraint that two shapes never meet outside $c(Q)$ before they meet inside. This allows to decrease the maximum

---

[1] The only exception is given when such intersections are subtended by the biggest section of the current sphere that they represent. To better explain this concept, in the 2-dimensional Euclidean space, this happen when the intersections are the endpoints of the diameter of the corresponding growing circle.

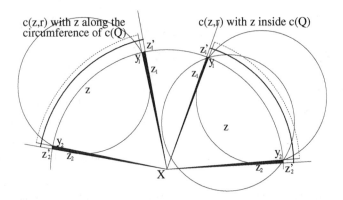

Fig. 4. Section of the new shape given by the increase of the angle θ by the black portions, yielding the new angle θ' and the decrease of the height from the dotted lines to the bold ones

height of the outside part of the shapes, thus yielding a further improvement on the approximation ratio. In other words the new shape will be larger but lower and it is defined as follows. Consider any point z inside $c(Q)$. Let $c(z, r)$ be the sphere of radius r centered at z and let $I(z, r)$ be the circular intersection of $c(z, r)$ with the spherical surface of $c(Q)$. Consider the sphere $c(z', r')$ centered on the border of $c(Q)$ and having the same intersection with the spherical surface of $c(Q)$, i.e., $I(z, r) \equiv I(z', r')$. The conic angle associated to z is now defined by the vertex x and the cone tangent to $c(z', r')$ (see the cut section of the conic angle in the right of Figure 4). Note that, in the case in which z lies on the spherical surface of $c(Q)$, $c(z, r)$ and $c(z', r')$ coincide (see the cut section of the new conic angle on the left of Figure 4). Indeed their angle does not, since, as already described, it is given by the tangent cone to the internal spherical shape and not, as before, by the cone wrapping the intersection with the surface of $c(Q)$.

When two new shapes are centered along the spherical surface of $c(Q)$ at distance $2r$, by construction, they meet outside at the same moment they meet inside, that is, when the radius grows till r. If we move one or both the corresponding centers more inside $c(Q)$ and leaving their distance at $2r$, the corresponding reshaped outside volumes remain disjoint.

4   18.8-Approximation Analysis of the MST Heuristic

In this section we formalise what was previously described. We provide a set of lemmata that describe a corresponding set of properties of the defined new shape that are useful in order to prove the 18.8-approximation claimed in the concluding theorem. The new shape must guarantee some properties that were true by means of the standard sphere. One of those properties is that two shapes growing according to a given radius r, touch each other only when the corresponding centers are at distance at most $2r$. Note that this is the fundamental property without which Corollary 1 cannot be applied for the estimation of the cost of the MST heuristic.

Lemma 2. *Given any subset of stations $Q \subseteq S$, for any $r < \frac{r_{max}}{2}$, two shapes overlap if and only if the corresponding points are at distance at most $2r$.*

Proof. If two shapes meet inside $c(Q)$, the property easily holds since the shape has the same behavior of spheres. In order to prove the claim we have to show that two shape never meet outside if they do not meet inside also. By construction, the external part of a shape is more extended (in terms of occupancy of the outer spherical surface) when the center resides along the spherical surface of $c(Q)$. In such a case, if two shapes touch each other, they do exactly at their intersection with the spherical surface of $c(Q)$ (see Figure 4). If one them has the center more inside, its growing outside part is less extended hence it cannot touch any other outer part of another shape.   □

The following two lemmata consider more carefully the structure of the new shape by considering the conic angle and the outside growing height, respectively. About the angle, it is proven that the more a station, whose associated shape grows also outside, is closer to x, the more its angle grows at each infinitesimal step.

Lemma 3. *Given any subset of stations $Q \subseteq S$, for any $r < \frac{r_{max}}{2}$ and any $d_1 \leq 1$ and $d_2 \leq 1$ such that $1 - r \leq d_1 \leq d_2$, $\partial\theta(r, d_1) \geq \partial\theta(r, d_2)$.*

The following lemma, instead, proves that the further a station is from x, the more its height outside $c(Q)$ grows. Moreover it gives also a very useful lower bound to the infinitesimal growth of the height and its maximum extension. The new shape, in fact, grows in height, outside $c(Q)$ as at least $\frac{3}{5}$ the growth of the basic shape at any infinitesimal step. This guarantees that the growth of such a shape is quite uniform during the whole process hence it is still suitable for bounding the MST heuristic cost. Moreover the maximal height outside $c(Q)$ is bounded by $.3527$ hence decreasing the maximal extension of the basic shape that was of $.5$.

Lemma 4. *Given any subset of stations $Q \subseteq S$, for any $r < \frac{r_{max}}{2}$ and any $d_1 \leq 1$ and $d_2 \leq 1$ such that $1 - r \leq d_1 \leq d_2$, $h(r, d_1) \leq h(r, d_2)$. Moreover for any $d \leq 1$, $h(r, d) \leq .3527...$ and $\partial h(r, d) \geq \frac{3}{5}\partial r$.*

Note that Lemma 3 and Lemma 4 follow directly from the corresponding lemmata of the 2-dimensional case [9]. In order to better understand this, it is sufficient to consider the new shape as the 2-dimensional one rotated along the line passing through its center and the center of $c(Q)$ (see Figure 5). In this way it is clear that what was true about the angle of the 2-dimensional case is now straightforward for the new conic angle θ (Lemma 3). And the same happens for the height h that remains unchanged (Lemma 4).

With the last lemma we ensure that the new shape guarantees an infinitesimal growth, for each connected component equal to at least the same growth of one

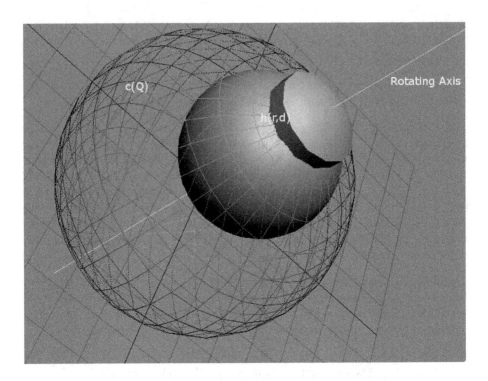

Fig. 5. The new shape obtained by means of a rotation of the 2-dimensional one along the line passing through its center and the center of $c(Q)$

sphere for each component. This was straightforward in the general case of d-spheres while it is quite complicated both in the 2- and the 3-dimensional case for the modified shapes.

Lemma 5. *The infinitesimal growth of the volume $v(P, r)$ of the region $P(r)$ covered by the shapes of a connected component $P \in CC(2r)$ of $G(2r)$ is $\partial v(P, r) \geq 4\pi r^2 \partial r$.*

Proof. (Sketch) If P contains just one station, then, by construction, the claim clearly holds. In fact, if the growth of the shape associated to such a station does not concern outside $c(Q)$ then it coincides with a growing sphere. Since the spherical surface is given by $4\pi r^2$, the infinitesimal growth is $\partial v(P, r) = 4\pi r^2 \partial r$. In the case the growth of the shape associated to the considered station goes outside $c(Q)$, then, by construction, the new shape is made in such a way that inside $c(Q)$ things do not change. Outside, the volume is maintained equal to the spherical case at every infinitesimal step, hence its growing too. When P contains more than one station, intuitively things can just go better, i.e., the growth of the union of the associated shapes is at least the growth of one sphere. This is given by the fact that both inside and outside $c(Q)$ when two shapes join in one connected component, their physical extension contains the shape corresponding to just one station. This suggest that at any infinitesimal step, its growth is bigger than the sphere.   □

From all the above lemmata we can finally obtain the following theorem.

Theorem 1. *In the 3-dimensional Euclidean space the MST heuristic is a 18.8-approximation algorithm for the MEBR problem.*

Proof. It is enough to prove that for any subset of stations $Q \subseteq S$, $MST(G(Q)) < 18.8$. The claim then follows by Lemma 1. Exploiting Lemma 5, we can easily provide a lower bound for the total region of the space covered by the union of all the shapes related to Q of radius $\frac{r_{max}}{2}$, that is $v(Q, \frac{r_{max}}{2})$, the covered volume at the end of the described growing process. In fact, recalling that by Corollary 1 $MST(G(Q)) = 3 \int_0^{r_{max}(Q)} (n(Q,r) - 1)r^2 \partial r$,

$$v\left(Q, \frac{r_{max}}{2}\right) = \int_0^{\frac{r_{max}}{2}} \sum_{P \in CC(2r)} \partial v(P,r) \partial r \geq \int_0^{\frac{r_{max}}{2}} n(2r)4\pi r^2 \partial r =$$

$$= \frac{1}{8}4\pi \int_0^{r_{max}} n(r)r^2 \partial r = \frac{1}{2}\pi \int_0^{r_{max}} (n(r) - 1)r^2 \partial r + \frac{1}{2}\pi \int_0^{r_{max}} r^2 \partial r =$$

$$= \frac{\pi}{6}MST(G) + \frac{\pi}{6}r_{max}^3.$$

Moreover, by Lemma 4, $v(Q, \frac{r_{max}}{2})$ is included in a sphere of radius $1 + h(\frac{r_{max}}{2}, 1)$ centered at the station x. Therefore, $v(Q, \frac{r_{max}}{2}) \leq \frac{4}{3}\pi(1 + h(\frac{r_{max}}{2}, 1))^3$, so that

$$\frac{\pi}{6}MST(G) + \frac{\pi}{6}r_{max}^3 \leq v\left(Q, \frac{r_{max}}{2}\right) \leq \frac{4}{3}\pi\left(1 + h\left(\frac{r_{max}}{2}, 1\right)\right)^3,$$

hence,

$$MST(G) \leq 8\left(1 + h\left(\frac{r_{max}}{2}, 1\right)\right)^3 - r_{max}^3.$$

Standard maximization argument obtained for r_{max} ranging from 0 to 1 shows that the quantity $8(1 + h(\frac{r_{max}}{2}, 1))^3 - r_{max}^3$ is maximised for $r_{max} = 1$, and since by Lemma 4, $h(r, d) \leq .3527...$, it finally results

$$MST(G) \leq 8\left(1 + h\left(\frac{1}{2}, 1\right)\right)^3 - 1 < 18.802.$$ □

5   Conclusion

In this paper we have investigated the Minimum Energy Broadcast Routing problem in the 3-dimensional Euclidean space. We have improved the previous known upper bound on the approximation ratio of the *MST* heuristic from 26 to 18.8, considerably decreasing the gap with the lower bound of 12 [4]. It is worth noting that, according to the considered method, such a new bound is not tight in terms of the associated volume outside $c(Q)$ as it was in the 2-dimensional case. Let us consider, in fact, the instance of the lower bound obtained by thirteen stations distributed like the centers of the spheres of the kissing number, i.e., everyone at distance at least $r_{max} = 1$ from each other inside $c(Q)$. The resulting associated volume of the new shapes does not fulfil neither $c(Q)$ as it was for the 2-dimensional case, nor the external volume in between the two spheres of radii 1 and $1 + h_{max} \approx 1.3527$ respectively, see Figure 6. Assuming the lower bound of 12 as the real bound of the MST heuristic in the 3-dimensional Euclidean space, the loss of 6.8 with respect to it must be found then in those "holes" inside and outside $c(Q)$, that is, the shaded volumes of Figure 6.

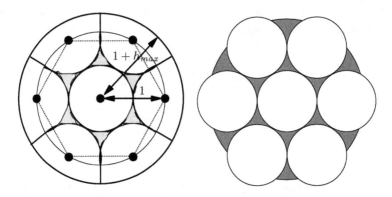

Fig. 6. On the right, a cut section of the lower bound case with the associated shapes. Shaded areas represent the mentioned holes inside the sphere $c(Q)$ of radius 1. On the left, a squeesed representation of what happens outside $c(Q)$. Again the shaded surfaces represent the mentioned holes outside $c(Q)$.

An interesting issue for a future work is of trying to apply the arguments of [5] in this 3-dimensional case and check whether they lead to anything better than the obtained 18.8 bound. The 3-D Delaunay triangulation is something known [14, 15] but it is not clear if the 2-dimensional arguments of [5] can be directly extended to the 3-dimensional case.

Another interesting case in the 3-dimensional environment is given for $2 \leq \alpha < d$. Since it can happen in practical application that the presence of obstacles can be both in contrast and in favor of communications, it depends on the desired directions. In the former case the given solution for the free 3-dimensional case is still valid since it is enough to suitably increase the value of α. In the latter, things become harder. In this case, in fact, it is not clear what the best solution may be. Moreover, the 18.8-approximation ratio does not hold for values of α smaller than d.

As last remark, from the experimental point of view, no results are known concerning the 3-dimensional case. All the experimental papers and the proposed heuristics start to investigate the 2-dimensional case (see for instance [16, 17, 18]). Is there any property not already captured that may lead to a better heuristic in the 3-dimensional case? In [7,19], for instance, nice approaches to better understand the behavior of the MST heuristic in the 2-dimensional case are provided. The experiments have shown how good is the heuristic when applied on practical instances, like the high-density ones. It may be of deep interest to investigate in this direction for the 3-dimensional case as well.

References

1. Wieselthier, J.E., Nguyen, G.D., Ephremides, A.: On the construction of energy-efficient broadcast and multicast trees in wireless networks. In: Proceedings of the 19th Annual Joint Conference of the IEEE Computer and Communications Societies (INFOCOM), IEEE Computer Society (2000) 585–594
2. Caragiannis, I., Kaklamanis, C., Kanellopoulos, P.: New results for energy-efficient broadcasting in wireless networks. In: Proceedings of the 13th International Symposium on Algorithms and Computation (ISAAC), Springer-Verlag (2002) 332–343
3. Clementi, A.E.F., Ianni, M.D., Silvestri, R.: The minimum broadcast range assignment problem on linear multi-hop wireless networks. Theoretical Computer Science **299**(1-3) (2003) 751–761
4. Clementi, A., Crescenzi, P., Penna, P., Rossi, G., Vocca, P.: On the complexity of computing minimum energy consumption broadcast subgraph. In: Proceedings of the 18th Annual Symposium on Theoretical Aspects of Computer Science (STACS). Volume 2010 of Lecture Notes in Computer Science., Springer-Verlag (2001) 121–131
5. Ambuehl, C.: An optimal bound for the mst algorithm to compute energy efficient broadcast trees in wireless networks. In: Proceedings of the 32nd International Colloquium on Automata, Languages and Programming (ICALP). Volume 3580 of Lecture Notes in Computer Science., Springer Verlag (2005) 1139–1150
6. Flammini, M., Klasing, R., Navarra, A., Perennes, S.: Improved approximation results for the Minimum Energy Broadcasting Problem. In: Proceedings of ACM Joint Workshop on Foundations of Mobile Computing (DIALM-POMC). (2004) 85–91. To appear on the associated Special Issue of *Algorithmica*.
7. Flammini, M., Navarra, A., Perennes, S.: The "Real" approximation factor of the MST heuristic for the Minimum Energy Broadcasting. In: Proceedings of the 4th International Workshop on Efficient and Experimental Algorithms (WEA). Volume 3503 of Lecture Notes in Computer Science., Springer Verlag (2005) 22–31. To appear on the associated Special Issue of *Journal of Experimental Algorithmics*.
8. Klasing, R., Navarra, A., Papadopoulos, A., Perennes, S.: Adaptive Broadcast Consumption (ABC), a new heuristic and new bounds for the minimum energy broadcast routing problem. In: Proceedings of the 3rd IFIP-TC6 International Networking Conference. Volume 3042 of Lecture Notes in Computer Science., Springer Verlag (2004) 866–877
9. Navarra, A.: Tighter bounds for the Minimum Energy Broadcasting problem. In: Proceedings of the 3rd International Symposium on Modeling and Optimization in Mobile, Ad Hoc and Wireless Networks (WiOpt). (2005) 313–322
10. Wan, P.J., Calinescu, G., Li, X., Frieder, O.: Minimum energy broadcasting in static ad hoc wireless networks. Wireless Networks **8**(6) (2002) 607–617
11. Liang, W.: Constructing minimum-energy broadcast trees in wireless ad hoc networks. In: Proceedings of the 3rd ACM international symposium on Mobile ad hoc networking and computing (MOBIHOC). (2002) 112–122

12. Conway, J.H., Sloane, N.J.A.: "The Kissing Number Problem" and "Bounds on Kissing Numbers". Ch. 2.1 and Ch. 13 in: Sphere Packings, Lattices, and Groups. Springer-Verlag, New York (3rd edition, 1998)

13. Frieze, A.M., McDiarmid, C.J.H.: On Random Minimum Length Spanning Trees. Combinatorica **9** (1989) 363–374

14. Attali, D., Boissonnat, J.D.: A linear bound on the complexity of the delaunay triangulation of points on polyhedral surfaces. In: Proceedings of the 7th ACM symposium on Solid modeling and applications (SMA). (2002) 139–146

15. Fang, T.P., Piegl, L.A.: Delaunay triangulation in three dimensions. IEEE Computer Graphics and Applications **15**(5) (1995) 62–69

16. Athanassopoulos, S., Caragiannis, I., Kaklamanis, C., Kanellopoulos, P.: Experimental Comparison of Algorithms for Energy-Efficient Multicasting in Ad Hoc Networks. In: Proceedings of the 3rd International Conference on Ad-Hoc Networks and Wireless (ADHOC-NOW). Volume 3158 of Lecture Notes in Computer Science., Springer Verlag (2004) 183–196

17. Penna, P., Ventre, C.: Energy-efficient broadcasting in ad-hoc networks: combining msts with shortest-path trees. In: Proceedings of the 1st ACM International Workshop on Performance Evaluation of Wireless, Ad Hoc, Sensor and Ubiquitous Networks (PE-WASUN). (2004) 61–68

18. Yuan, D.: Computing Optimal or Near-Optimal Trees for Minimum-Energy Broadcasting in Wireless Networks. In: Proceedings of the 3rd International Symposium on Modeling and Optimization in Mobile, Ad Hoc and Wireless Networks (WiOpt). (2005) 323–331

19. Clementi, A., Huiban, G., Penna, P., Rossi, G., Verhoeven, Y.C.: On the approximation ratio of the mst-based heuristic for the energy-efficient broadcast problem in static ad-hoc radio networks. In: Proceedings of the 3rd IEEE IPDPS Workshop on Wireless, Mobile and Ad Hoc Networks (WMAN). (2003) 222

Average-Time Complexity of Gossiping
in Radio Networks

Bogdan S. Chlebus[1,*], Dariusz R. Kowalski[2], and Mariusz A. Rokicki[1,*]

[1] Department of Computer Science and Eng., UCDHSC, Denver, CO 80217, USA
[2] Department of Computer Science, University of Liverpool, Liverpool L69 7ZF, UK

Abstract. Radio networks model wireless synchronous communication with only one wave frequency used for transmissions. In the problem of many-to-all (M2A) communication, some nodes hold input rumors, and the goal is to have all nodes learn all the rumors. We study the average time complexity of distributed many-to-all communication by deterministic protocols in directed networks under two scenarios: of combined messages, in which all input rumors can be sent in one packet, and of separate messages, in which every rumor requires a separate packet to be transmitted. Let n denote the size of a network and k be the number of nodes activated with rumors; the case when $k = n$ is called gossiping. We give a gossiping protocol for combined messages that works in the average time $\mathcal{O}(n/\log n)$, which is shown to be optimal. For the general M2A communication problem, we show that it can be performed in the average time $\mathcal{O}(\min\{k\log(n/k), n/\log n\})$ with combined messages, and that $\Omega(k/\log n + \log n)$ is a lower bound. We give a gossiping protocol for separate messages that works in the average time $\mathcal{O}(n\log n)$, which is shown to be optimal. For the general M2A communication problem, we develop a protocol for separate messages with the average time $\mathcal{O}(k\log(n/k)\log n)$, and show that $\Omega(k\log n)$ is a lower bound.

1 Introduction

Packet radio networks are a class of wireless networks in which only one wave frequency is used for communication. The restricted bandwidth results in a conflict when different messages arrive simultaneously at a node. The main challenge, in developing communication protocols for such networks, is in resolving local conflicts for access to the limited bandwidth.

The networks we consider are directed, which captures a scenario in which a possibility of a direct transmission from node x to node y does not necessarily make it possible for node y to transmit directly to node x. Networks are ad-hoc, in that protocols do not rely on the knowledge of the topology; the only information about the network that may be a part of code of a protocol is the size n, which is the number of nodes. We consider deterministic distributed communication protocols. Initially, some k among the nodes are simultaneously activated with input data; these data are called rumors. The communication task is to make

---

* The work of this author is supported by the NSF Grant 0310503.

all the nodes in the network learn all the input rumors. This communication task can be called many-to-all communication (M2A). The special case in which $k = n$ is called gossiping. The underlying network is assumed to be strongly connected, so that gossiping is always possible to achieve.

Nodes exchange packets carrying rumors. A node sends at most one packet per round. In the model of *combined messages*, a packet can carry all the rumors. In such a setting, it is natural to have a node send all the rumors learned so far in any transmitted packet. In the model of *separate messages*, a packet can carry only one rumor. With such a restriction, a protocol needs to rely on a mechanism to prioritize rumors so that a node sends a rumor of the highest current priority at a round.

The time of an execution of a protocol is defined to be the first round when the communication goal has been achieved. Such a completion of the communication task is not required to be known by the nodes. The complexity measure we investigate is the average time as a function of the size n of the network. To find the average time for size n, first compute conceptually the durations of executions of the protocol on all strongly connected networks of size n, and next take the average of the times accrued for these networks.

Our results. We give upper and lower bounds on the average-case complexity of gossiping and many-to-all communication. Protocols are distributed and designed for both the models of combined and separate messages. Let n denote the size of a network and k the number of nodes activated with rumors. The summary of the contributions is as follows.

I. Gossiping with combined messages can be performed in the average time $\mathcal{O}(n/\log n)$, which is shown to be optimal.

II. We show that M2A communication can be performed in the average time $\mathcal{O}(\min\{k\log(n/k), n/\log n\})$ with combined messages and that $\Omega(k/\log n + \log n)$ is a lower bound.

III. Gossiping with separate messages can be performed in the average time $\mathcal{O}(n\log n)$, which is shown to be optimal.

IV. M2A communication can be achieved in the average time $\mathcal{O}(k\log(n/k)\log n)$ with separate messages. We show a lower bound $\Omega(k\log n)$.

Previous work. Broadcasting in radio networks with topology modeled by random graphs was considered by Elsässer and Gąsieniec [12], who showed that $\mathcal{O}(\log n)$ expected time was optimum for distributed protocols. This result can be interpreted as giving the optimum average-time complexity of broadcasting. The authors of this paper do not know of any other results related to the average-time complexity of communication in radio networks.

A many-to-many communication problem in radio networks, similar to what we consider, was studied by Gąsieniec, Kranakis, Pelc and Xin [13]. The problem is defined as follows: There is a set S of k nodes initialized with rumors and every one among these nodes needs to get to know all the rumors. Networks are undirected, each node knows the topology of the network but does not know the set S, the maximum distance d among any pair of nodes in S is an additional parameter. A protocol solving this problem of time complexity $\mathcal{O}(d \log^2 n + k \log^3 n)$ was given in [13].

Related work. The model of multi-hop radio networks was introduced by Chlamtac and Kutten [4] who considered sequential algorithms to find an efficient broadcast protocol for a given input network. The first distributed randomized broadcast protocols of sub-quadratic expected-time performance were given by Bar-Yehuda, Goldreich and Itai [2]. The first distributed deterministic explicit broadcast protocol with sub-quadratic time performance was given by Chlebus, Gąsieniec, Gibbons, Pelc and Rytter [5]. Alon, Bar-Noy, Linial and Peleg [1] showed that there exists a bipartite graph of n nodes for which any broadcasting protocol requires time $\Omega(\log^2 n)$. The fastest known deterministic distributed broadcasting protocol was given by Czumaj and Rytter [10], who showed that it works in time $\mathcal{O}(n \log^2 D)$, where D is the diameter of the network.

Gossiping was initially studied for the model of combined messages. The first distributed protocol of sub-quadratic time complexity was given by Chrobak, Gąsieniec and Rytter [8]. The fastest known distributed deterministic protocol works in time $\mathcal{O}(n^{4/3} \log^4 n)$; it was given by Gąsieniec, Radzik and Xin [14]. The best randomized protocol operates in the expected time $\mathcal{O}(n \log^2 n)$, it was given by Czumaj and Rytter [10].

Oblivious gossiping was first studied by Chlebus, Gąsieniec, Lingas, and Pagourtzis [6]. The paper gave a deterministic gossiping protocol that works in time $\mathcal{O}(n^{3/2})$ on undirected networks; this was shown to be optimal by Kowalski and Pelc [15]. Randomized oblivious gossiping protocols working in the expected time $\mathcal{O}(n \log^2 n)$ on undirected networks and $\mathcal{O}(\min\{m, D\Delta\} \log^2 n)$ on directed networks, where Δ is the maximum node in-degree, were presented in [6].

The model of separate messages was first considered by Bar-Yehuda, Israeli and Itai [3] and Clementi, Monti and Silvestri [9]. Christersson, Gąsieniec, and Lingas [7] considered gossiping in undirected networks; they gave an adaptive deterministic gossiping protocol with time complexity $\mathcal{O}(n^{3/2} \log n)$ and a randomized protocol of the expected time complexity $\mathcal{O}(n \log^2 n)$.

2 Technical Preliminaries

A radio network is modeled as a graph $G = (V, E)$, in which the set of vertices V represents the physical nodes of the network, and the set of edges E represents the possibilities of direct transmissions among the nodes. If node x of a radio network can send a message directly to y, then node y is *reachable* from x. For any ordered pair $\langle u, v \rangle$ of nodes in the network, edge $u \to v$ is in the graph G if and only if node v is reachable from node u. The *size* of the network is defined to be the number of nodes $|V|$, which we usually denote by n.

We assume full synchrony in that all nodes are equipped with local clocks that are clicking at the same rate and indicate the same round numbers. Protocols we consider are for a scenario when all the nodes activated with inputs start simultaneously at round zero. When some two nodes v and v' transmit simultaneously at a given round, and are both in-neighbors of node x in the reachability graph of the network, then a *conflict* occurs at x. A conflict at x results in all the messages arriving at x interfering with one another so that each is received as garbled. A message is said to be *heard* when it is received as fully readable in its correct form. Radio networks have the following properties:

(a) If a node performs a transmission, then it transmits a single message.
(b) The message transmitted by a node is delivered in that round to all the reachable nodes.
(c) A node can hear a message delivered at a round, if exactly one among its in-neighbors transmitted in this round.

Communication problems. Initially some nodes hold their input data called *rumors*. When node i is initialized with a rumor, then this rumor is denoted by r_i. The goal of communication protocols is to disseminate such input rumors.

In the problem of *gossiping*, each node v is a source for its private input rumor r_v, and the goal is to have all the nodes learn all the rumors. Gossiping may be called all-to-all communication problem.

A generalization of gossiping called *many-to-all* problem, or simply *M2A*, is about a scenario in which only some of the nodes have input rumors. Such nodes are called *activated*. The goal is to have all the nodes in the network get to know all the rumors of the activated nodes. All nodes in the network participate in forwarding messages in the course of an execution of an M2A protocol.

To have a communication problem in radio networks meaningful, we need to assume that the topology of the underlying graph makes the communication task at hand possible to perform. In the case of gossiping and M2A, the graph is assumed to be strongly connected.

Communication protocols. Correctness of an M2A or gossiping protocol means that the communication goal is eventually achieved on any strongly connected network. Nodes running a protocol are not required to reach eventually a state representing the completion of a task. This is assumed in order to decouple termination from complexity considerations. The *time complexity* of a protocol at hand, for a given strongly connected network, is defined to be the first round when the communication goal has been achieved.

Nodes of a network of size n are identified by their unique names. We assume that *names* give a one-to-one correspondence between the nodes and integers in the range $[0, n - 1]$. While designing communication protocols, we assume that the size of the network is known.

A simple protocol called ROUND-ROBIN operates as follows. In round i, the unique node with name k such that $i \equiv k \pmod{n}$ is scheduled to perform a transmission. There are variants of this protocol depending on the size of packets. In the model of combined messages, a node scheduled to transmit at the current round transmits a message with all the rumors it has learned so far. In the model of combined messages, the protocol is augmented by a selection rule to choose a rumor to transmit from among those that have been learned by the given round. Usually the selection is made by resorting to a queuing mechanism.

```
for k := 0 to ℓ_n do
    if v is in 𝒢_k then transmit
call ROUND-ROBIN
```

Fig. 1. Protocol GOSSIP-COMBINED-MESSAGES; the code for node v

Average complexity. We consider the average time complexity of gossiping and M2A communication on strongly-connected networks. This is the same as the expected time complexity when the probabilistic space has all strongly connected networks on n nodes as elementary events, each occurring with the same probability. A random directed network is strongly connected with the probability exponentially close to 1. This fact allows to obtain expected time estimates while working with arbitrary random directed networks. These estimates are the same when conditioned on the networks being strongly connected, provided the time estimates are polynomial. An explicit termination in polynomial time could be obtained for all protocols we develop, since there are polynomial-time worst-case time estimates for these protocols, valid for strongly connected networks.

We do not want M2A protocols to have their performance biased towards specific sets of activated nodes. Therefore we work with the average complexity of M2A protocols defined in an adversarial manner as follows. Suppose there is an adversary who is given a protocol $\mathcal{P}$ for n nodes together with a number $k \leq n$. The adversary chooses a set K of k specific names of nodes to be activated; the goal of the adversary is to show a scenario maximizing the complexity of the protocol. The average complexity of the protocol $\mathcal{P}$, for n-node networks, is defined to be the average complexity of protocol $\mathcal{P}$ measured when exactly the nodes in K are activated with rumors.

3   Gossiping with Combined Messages

We show that gossiping can be performed with the average time $cn/\lg n$, for any fixed $c > 1$, and that the average time always has to be at least $cn/\lg n$, for any fixed $c < 1/2$. (The logarithm of x to the base 2 is denoted by $\lg x$.)

Gossiping protocol for combined messages. Let $\ell_n = \lceil n/b \lg n \rceil$, where $b = \frac{1}{2}(1 + \frac{1}{c})$. Observe that the inequalities $1 > b > 1/c$ hold. Define group $\mathcal{G}_k$, for $0 \leq k \leq \ell_n$, to consist of nodes i, for $0 \leq i < n$, with the property that the congruence $i \equiv k \pmod{\ell_n}$ holds. The size of a group is about $b \lg n$. The sizes of two groups differ by at most 1. We consider an oblivious protocol GOSSIP-COMBINED-MESSAGES, which is given in Figure 1. A transmission by a node contains all the rumors that the node has already learnt in the execution.

Theorem 1. *For any $c > 1$, the average number of rounds to complete gossiping by protocol GOSSIP-COMBINED-MESSAGES on a network of n nodes is smaller than $cn/\lg n$, for a sufficiently large n.*

Proof. Take a node y and group $\mathcal{G}_k$. Let x be in $\mathcal{G}_k$. The node y can hear x at the kth round of the first phase when the following two events hold:

(i) x is an in-neighbor of y; and

(ii) no other node in $\mathcal{G}_k$ is an in-neighbor of y.

It follows that the probability of the event that y hears $x \in \mathcal{G}_k$ during the kth round of the first phase of the protocol is $2^{-|\mathcal{G}_k|} = 2^{-b \lg n} = n^{-b}$. Let x and v be two nodes. Node y is called a *relay* for the pair $\langle x, v \rangle$ when the following holds:

(i) y is an in-neighbor of v; and

(ii) y heard x in the first phase.

Observe that

$$\Pr[y \text{ is a relay for } \langle x, v \rangle] = \frac{1}{2n^b} ,$$

because the events "y heard x in the first phase" and "y is an in-neighbor of v" are independent.

Consider the first t nodes, that is, the nodes i with $0 \le i < t$. These nodes are scheduled to perform a transmission among the first t rounds of the protocol ROUND-ROBIN in the second phase. Node i may make node v learn the rumor r_x of x if i is a relay for the pair $\langle x, v \rangle$. For all such nodes i making the first t transmission during ROUND-ROBIN and different from x and v, the events "i is a relay for the pair (x, v)" are independent.

If v has not learnt r_x in the first t rounds of the second phase, then no i such that $0 \le i < t$ is a relay for the pair $\langle x, v \rangle$. The latter event holds with the probability $(1 - \frac{1}{2n^b})^t$ by independence of the events of being a relay node. It follows that v does not learn r_x in the first t rounds of ROUND-ROBIN with the probability of at most $(1 - \frac{1}{2n^b})^t$.

We use the inequality

$$\left(1 - \frac{1}{s}\right)^s < \exp\left(-1 + \frac{1}{2s}\right), \tag{1}$$

which holds for real $s > 1$. It yields the following estimate:

$$\left(1 - \frac{1}{2n^b}\right)^t < \exp\left(\left(-1 + \frac{1}{4n^b}\right)\frac{t}{2n^b}\right) = \exp\left(-\frac{t}{2n^b}\right)\exp\left(\frac{t}{8n^{2b}}\right). \tag{2}$$

Let $d = \min\{2b, 1\}$. Take $t = n^a$ where $b < a < d$. Now the right-hand side of (2) becomes

$$\exp(-n^{a-b}/2)\exp(n^{a-d}/8) = \exp(-n^{a-b}/2)(1 + o(1)). \tag{3}$$

Consider the event that for any pair of nodes $\langle x, v \rangle$ there is a relay node during the first t rounds of the second phase. The event does not hold with the probability of at most $n^2 \exp(-n^{a-b}/2)(1 + o(1))$ by the estimate (3). If this event holds, then gossiping is completed by round $\frac{n}{b \lg n} + n^a$, which is smaller than $\frac{cn}{\lg n}$ for a sufficiently large n. Otherwise the time of gossiping can be estimated by $\frac{n}{b \lg n} + n^2$. These two estimates contribute to the expected value of the time of protocol $\mathcal{A}$ to complete gossiping, which together is smaller than $\frac{cn}{\lg n}$, for all sufficiently large n.

Lower bound for gossiping with combined messages. We show that any gossiping protocol for the model of combined messages has the average time complexity $\Omega(n/\log n)$. This implies that protocol GOSSIP-COMBINED-MESSAGES is asymptotically optimal.

Theorem 2. *For any $c < 1/2$ and gossiping protocol $\mathcal{A}$ for the model of combined messages, the average number of rounds to complete gossiping by $\mathcal{A}$ on a network of n nodes is larger than $cn/\lg n$, for a sufficiently large n.*

Proof. Let X be the random variable defined on the domain of all directed graphs of n nodes. For such a graph G, run $\mathcal{A}$ on G and let s be the first round when the gossiping has been completed. Define $X(G) = s$. Let an execution of $\mathcal{A}$ be given as a sequence $\langle T_0, T_1, T_2, \ldots \rangle$ of transmissions.

We estimate the probability of the event $X > s$, for integer $s > 0$. Take event $\mathrm{H}(v, s)$, for node v and round s, which holds when no node has heard from node v by round s. Observe that

$$\Pr[X > s] \geq \Pr[\mathrm{H}(v, s)]. \tag{4}$$

We want to estimate the probability that $\mathrm{H}(v, s)$ holds.

We start with choosing v. If some node v does not belong to any of the first s transmissions, then such v yields the best possible estimate $\Pr[\mathrm{H}(v, s)] = 1$, which also implies that $\mathbb{E}X > s$.

Assume that every node belongs to at least one among the first s transmissions of protocol $\mathcal{A}$. Next we restrict our attention only to these transmissions. We claim that there is a node, say, v with the property that every transmission T_i that v belongs to, for $i \leq s$, is of a size at least $|T_i| \geq n/s$. This is because otherwise, even if every node belonged to only one transmission, the total number of nodes in the initial segment of s transmissions of $\mathcal{A}$ were smaller than n, which would contradict the assumption that these transmissions include all the nodes.

A node x hears from v at round $i \leq s$, provided $v \in T_i$, when the following two events hold:

(i) v is an in-neighbor of x in T_i, and
(ii) no other node $y \neq x$ in T_i is an in-neighbor of x.

This implies that the estimate

$$\Pr[x \text{ hears from } v \text{ at round } i \mid v \in T_i] \leq 2^{-n/s}$$

holds. Node v could belong to a number of transmissions T_i for $i \leq s$, so we use the estimate

$$\Pr[x \text{ hears from } v \text{ in the first } s \text{ rounds}] \leq s2^{-n/s}.$$

Node x was arbitrary, and we need to be concerned with all the nodes. We use the estimate

$$\Pr[\text{some node hears from } v \text{ in the first } s \text{ rounds}] \leq ns2^{-n/s}.$$

The event $H(v, s)$, that no node hears v during the first s transmissions, holds with a probability of at most

$$\Pr[H(v, s)] \geq 1 - ns2^{-n/s}. \tag{5}$$

To estimate the expected value $\mathbb{E}X$ of X, we use the formula

$$\mathbb{E}X = \sum_{k=0}^{\infty} \Pr[X > k],$$

which holds true for any random variable X with non-negative integer values. Combining this with the estimates (4) and (5), we obtain the inequality

$$\mathbb{E}X \geq \sum_{k=1}^{t} \left(1 - nk \cdot 2^{-n/k}\right), \tag{6}$$

for any integer $t > 0$.

Let k_0 be the largest value of k for which the expression $1 - nk \cdot 2^{-n/k}$ is positive. We take the upper bound t on the range of summation in (6) to be close to k_0.

Next we estimate the magnitude of k_0 as a function of n. Observe that k_0 is the largest k for which the inequality

$$nk \leq 2^{n/k} \tag{7}$$

holds. Take the binary logarithm $\lg$ of both sides of (7) to obtain the equivalent inequality $\lg n + \lg k \leq \frac{n}{k}$, which implies $k_0 = \frac{n}{2 \lg n}(1 + o(1))$. We use the value $t = n/(2 \lg n)$ in the estimate (6) to obtain

$$\mathbb{E}X \geq \sum_{k=1}^{n/(2 \lg n)} \left(1 - nk \cdot 2^{-n/k}\right) = \frac{n}{2 \lg n} - n \sum_{k=1}^{n/(2 \lg n)} k \cdot 2^{-n/k}. \tag{8}$$

The function $f(k) = k2^{-n/k}$ is increasing as $k \to \infty$. The value $f(t) = f(n/(2 \lg n))$ is the largest term in the sum on the right-hand side of (8). Observe that

$$f(t) = \frac{n}{2 \lg n} \cdot 2^{-2 \lg n} = \frac{n}{2 \lg n} \cdot n^{-2} = \frac{1}{2n \lg n}$$

and hence the estimate

$$\sum_{k=1}^{n/(2 \lg n)} k \cdot 2^{-n/k} \leq \frac{n}{2 \lg n} \cdot \frac{1}{2n \lg n} = \frac{1}{4 \lg^2 n}$$

holds. Therefore (8) can be bounded from below as follows:

$$\mathbb{E}X \geq \frac{n}{2 \lg n} - \frac{n}{4 \lg^2 n} = \frac{n}{2 \lg n}\left(1 - \frac{1}{2 \lg n}\right),$$

which completes the proof of Theorem 2.

```
for i := lg k downto 0 do
    call SELECTOR-SUBROUTINE(2^i)
    continue ROUND-ROBIN for 10 lg n rounds
```

Fig. 2. Procedure M2A-COMBINED(k)

4 M2A Communication with Combined Messages

Suppose k nodes among n in the network are activated with rumors. We give a protocol with average time complexity $\mathcal{O}(\min\{k \log(n/k), n/\log n\})$. We assume that k is a power of 2.

Two schedules of transmissions $\mathcal{P}_1$ and $\mathcal{P}_2$ are said to be *interleaved*, when the consecutive actions as specified by $\mathcal{P}_1$ are performed in even-numbered rounds, while $\mathcal{P}_2$ determines the actions for the odd-numbered rounds. Infinite schedules of transmissions are called *protocols*, while finite schedules are called *procedures* in this paper. When a procedure $\mathcal{P}_1$ is interleaved with a protocol $\mathcal{P}_2$, then eventually $\mathcal{P}_1$ ends. At this point we make the protocol $\mathcal{P}_2$ take over completely, such that its actions are performed in all the following rounds; this is explicitly marked in the pseudocode of our protocols by the instruction continue $\mathcal{P}_2$. Another mode of using a protocol $\mathcal{P}$ specifies that the schedule of $\mathcal{P}$ is repeatedly executed for an interval of x rounds, then it is frozen. This is indicated in the pseudocode by the instruction continue $\mathcal{P}$ for x rounds.

We use families of sets called (n, j)-selectors in [8]. They are defined as follows. A set Y *selects* element v from a set X when $X \cap Y = \{v\}$. A family $\mathcal{F}$ of subsets of $[n] = [0, n-1]$ is an (n, j)-*selector* when, for any set $X \subseteq [n]$ of size ℓ, at least $|X|/2$ elements in X can be selected by sets in $\mathcal{F}$. The size of $\mathcal{F}$ is called its *length*. We refer to any used selector $\mathcal{F}$ as a sequence $\mathcal{F} = \langle F_1, F_2, \ldots \rangle$ in an arbitrary fixed order.

Selectors are used to determine schedules of transmissions. Given positive integer number ℓ and a (n, ℓ)-selector $\mathcal{F}$, we define SELECTOR-SUBROUTINE(ℓ) as follows. Node v transmits in round i if $v \in F_i$; rounds are counted from the call of this subroutine. We use $(n, 2^i)$-selectors of length $\Theta(2^i \log(n/2^i))$, which were proved to exist in [11].

A M2A procedure, representing the case when k may be a part of code, is given in Figure 2. Protocol M2A-COMBINED-MESSAGES is given in Figure 3. Next we analyze the average complexity of the protocol.

A node v is said to be a *unique transmitter* at a round, when v is the only node transmitting at that round. We say that *broadcast of r_v was successful/completed*, or that *node v broadcast successfully*, when every node has received r_v.

Lemma 1. *Suppose that node v transmits its rumor r_v as the unique transmitter, and after this* ROUND-ROBIN *is executed. Then the broadcast of r_v is completed in at most $10 \lg n$ following rounds of* ROUND-ROBIN *with the probability of at least $1 - 1/n^3$.*

---

for $j := 0$ **to** $\lg n$ **do**
    call M2A-COMBINED(2^j) interleaved with GOSSIP-COMBINED-MESSAGES
    continue GOSSIP-COMBINED-MESSAGES

Fig. 3. Protocol M2A-COMBINED-MESSAGES

Theorem 3. *Protocol* M2A-COMBINED-MESSAGES, *on networks of* n *nodes with any* k *activated nodes, works in average time* $\mathcal{O}(\min\{k\log(n/k), n/\log n\})$.

Proof. First we show that the protocol completes M2A by the end of the loop for $j = \lceil \lg k \rceil$ with the probability of at least $1 - 4k/n^3$. The protocol runs M2A-COMBINED(2^j) which involves SELECTOR-SUBROUTINE(2^j). Since there are $k \leq 2^j$ activated nodes, during SELECTOR-SUBROUTINE(2^j) at most 2^{j-1} activated nodes did not transmit as unique transmitters. Being a unique transmitter results in a successful broadcast during the next ROUND-ROBIN part, with probability at least $1 - 1/n^3$ by Lemma 1.

During SELECTOR-SUBROUTINE(2^{j-1}), at most 2^{j-2} activated nodes did not transmit as unique transmitters, since there are at most 2^{j-1} participating nodes with probability at least $1 - 2^j/n^3$. Those which transmitted as unique transmitters have a successful broadcast during the next ROUND-ROBIN rounds with probability at least $1 - 2^j/n^3 - 2^{j-1}/n^3$.

In general, in an execution of SELECTOR-SUBROUTINE(2^i) within M2A-COMBINED(2^j), there are at most 2^i activated nodes for which broadcast was not successful during previous iterations with probability at least $1 - \sum_{a=i+1}^{j} 2^a/n^3$. Conditioned on this event, during SELECTOR-SUBROUTINE(2^i) at most 2^{i-1} of activated nodes did not transmit as unique transmitters. It follows that after the ith iteration of the loop, at most 2^{i-1} rumors have not been broadcast successfully with probability at least $1 - \sum_{a=i}^{j} 2^a/n^3$.

Considering only M2A-COMBINED(2^j), it completes M2A for k activated nodes in time $\sum_{i=0}^{j} \mathcal{O}(2^i \log(n/2^i) + \log n) \leq \mathcal{O}(k\log(n/k))$ with probability at least $1 - \sum_{a=0}^{j} 2^a/n^3 \geq 1 - 4k/n^3$. Including also previous executions of M2A-COMBINED$(2^{j'})$ for $j' < j$ produces time estimate $\sum_{j' \leq j} \mathcal{O}(2^{j'} \log(n/2^{j'})) = \mathcal{O}(k\log(n/k))$.

Since $\mathcal{O}(n^2)$ is the worst-case time bound, the average time of M2A-COMBINED-MESSAGES is $\mathcal{O}(k\log(n/k)) + \mathcal{O}(n^2) \cdot 4k/n^3 = \mathcal{O}(k\log(n/k))$.

Theorem 4. *The average cost of any M2A protocol, for the model of combined messages, executed on network of* n *nodes with some* k *of them activated is* $\Omega(k/\log n + \log n)$.

Proof. Let $\mathcal{A}$ be a M2A protocol. Fix a set K of activated nodes, where $|K| = k$. Let $\langle T_0, T_1, \ldots \rangle$ be the sequence in which T_i denotes the set of nodes transmitting at round i in the execution of $\mathcal{A}$. There are two kinds of rounds i:

Case 1: Rounds i in which T_i includes at most $4\lg n$ nodes in K that transmit for the first time.

Even $k/(4 \lg n)$ such rounds are not sufficient to exhaust all the elements in K.

Case 2: Rounds i in which there are more than $4 \lg n$ nodes from K transmitting for the first time.

We show that with a large probability in any round, up to round $s = k/(4 \lg n)$, there is no successful transmission between any pair of nodes. Take node v. Let $a = |T_i| > 4 \lg n$. The probability that v receives a rumor for a node in T_i at round i is $(a/2)(1/2)^{a-1} > 1/n^3$, for sufficiently large n. It follows that the probability of existence of a node that receives a rumor at round i is smaller than $1/n^2$. The probability that some node receives a rumor by round s is smaller than s/n^2. The expected value of the number of rounds by completion of the communication task is at least $s \cdot (1 - s/n^2) > s/2$ for $n > 2$.

The complexity of our protocol is close to the lower bound by a factor of $\mathcal{O}(\log n \log(n/k))$.

5   Gossiping with Separate Messages

We consider now gossiping in the case when input rumors are so large that it takes a separate packet to carry one rumor. We show that gossiping can be performed with the average time $\mathcal{O}(n \log n)$, and that the average time has to be $\Omega(n \log n)$.

Gossiping protocol for separate messages. Every node v maintains a priority queue $\texttt{Queue}_v$ in the private memory. The queue is used to store rumors that v still needs to transmit. There is a set $\texttt{Received}_v$ to store all the rumors learned so far. A newly received message with a rumor that is not stored in $\texttt{Received}_v$ is added to both $\texttt{Received}_v$ and $\texttt{Queue}_v$. The protocol working according to these rules is called GOSSIP-SEPARATE-MESSAGES; it is given in Figure 4.

Let the nodes be ordered cyclically by their names in $[n] = [0, n-1]$, so that i is followed by number $(i + 1) \bmod n$. This ordering governs which nodes transmit in any ROUND-ROBIN type of protocol, like GOSSIP-SEPARATE-MESSAGES in particular.

The priority queue $\texttt{Queue}_v$ has its own queuing discipline. Rumors are ordered cyclically, starting from the own input rumor r_v. This rumor is followed by rumors with larger indices according to their order, that is, r_{v+1}, r_{v+2}, until r_{n-1}, which is then followed by r_0, r_1, through the final r_{v-1}.

Theorem 5. *The average number of rounds to complete gossiping by protocol* GOSSIP-SEPARATE-MESSAGES *on a network of n nodes is $\mathcal{O}(n \log n)$.*

Proof. Every node transmits every rumor exactly once. The worst-case time complexity of this gossiping protocol is n^2. The full cycle of n rounds makes an *epoch*. During the first epoch, every node v transmits its input rumor r_v.

Take some rumor r and consider an event $\mathcal{E}_a(r)$ which holds when r has been transmitted by $a \lg n$ different nodes. The probability of the event that some

```
initialize Received_v := Queue_v := {r_v};
for round i := 0 to ∞ do
    if v ≡ i (mod n) then
        if Queue_v nonempty then
            transmit the first rumor r in Queue_v
                and remove r from Queue_v
    else
        attempt to receive a message;
        if rumor r received then
            if r is not in Received_v then
                insert r into Queue_v and add to Received_v
```

Fig. 4. Protocol GOSSIP-SEPARATE-MESSAGES; the code for node v

node y has not heard r, conditioned on $\mathcal{E}_a(r)$, is n^{-a}. The probability that some node has not heard r, conditioned on $\mathcal{E}_a(r)$, is at most $n \cdot n^{-a} = n^{1-a}$. The probability that some node has not heard some rumor, conditioned on the events $\mathcal{E}_a(r)$ for all rumors r, is at most n^{2-a}. In the following application we will use $a = 4$ to obtain the probability $n^{2-4} = n^{-2}$.

Consider the following event $\mathcal{B}$: every rumor was transmitted at least $3 \lg n$ times during the first $b \lg n$ epochs, for some fixed integer b to be determined later. Take a node v and the $b \lg n$ nodes preceding v in the cyclic ordering. If any of these nodes receives rumor r_v in the first epoch from v, then it transmits r_v in the first $b \lg n$ epochs. When v transmits in the first epoch, then every other node receives r_v with probability $1/2$ independently over all the nodes. The expected value of the number of these nodes that receive r_v in the first epoch is $\mu = \frac{b}{2} \lg n$. Take δ determined by the equality $(1-\delta)\frac{b}{2} \lg n = 3 \lg n$, that is, $\delta = 1 - \frac{6}{b}$. Then by the Chernoff bound, the probability that less than $3 \lg n$ nodes receives rumor r_v in the first epoch is at most

$$\exp\left\{-\left(1 - \frac{6}{b}\right)^2 \frac{b}{4} \lg n\right\} \le n^{-(1-\frac{6}{b})^2 \frac{b}{4} \lg e} .$$

Take integer $b > 6$ for which the inequality $(1 - \frac{6}{b})^2 \frac{b}{4} \lg e \ge 3$ holds. This b is sufficient to guarantee that event $\mathcal{B}$ does not to hold with the probability of at most n^{-2}.

Conditional on $\mathcal{B}$, the expected time of gossiping is at most $bn \lg n + n^{-2} \cdot n^3 = n(1 + b \lg n)$, because the worst-case time complexity is n^3. Since event $\mathcal{B}$ does not hold with probability at most n^{-2}, the unconditional expected time complexity is at most $n(1 + b \lg n) + n^{-2} \cdot n^3 = n(2 + b \lg n)$, for a similar reason.

The average number of rounds to complete gossiping on a network of n nodes is $\Omega(n \log n)$; this is a corollary of a more general lower bound for M2A communication shown in Section 6.

```
for i := lg k downto 0 do
    for j := 1 to m(n, 2^i) do
        (a) if v ∈ F_j(n, 2^i) then transmit rumor r_v
            else attempt to hear a message
                ( this is jth round of SELECTOR-SUBROUTINE(2^i) )
        (r) continue ROUND-ROBIN-STACK in next 10 lg n rounds
```

Fig. 5. Procedure M2A-SEPARATE(k); the code for node v

6   M2A Communication with Separate Messages

We give a protocol with average time $\mathcal{O}(k \log(n/k) \log n)$. Let k be a power of 2.

SELECTOR-SUBROUTINE(2^i) is similar to the one described for the protocol with combined messages, in that it uses $(n, 2^i)$-selector. There are two main differences in how they are used. The first difference is that after *each* round of SELECTOR-SUBROUTINE(2^i) we continue with ROUND-ROBIN-STACK for $10 \lg n$ rounds, while in the case of combined messages we put $10 \lg n$ of ROUND-ROBIN rounds after every used $(n, 2^i)$-selector. The second difference is that specific rumor needs to be selected for each transmission by a node.

An M2A procedure, representing the case when k may be a part of code, is given in Figure 5. An auxiliary protocol ROUND-ROBIN-STACK used in procedure M2A-SEPARATE(k) is defined as follows. A node maintains a stack of rumors different from its original one. A rumor heard by the node is pushed on its stack. A rumor to transmit is obtained by popping the stack; when the stack is empty, then the node pauses. The stack is initialized to be empty, and is made empty just before ROUND-ROBIN-STACK is to be continued for $10 \lg n$ rounds, see Figure 5.

Protocol M2A-SEPARATE-MESSAGES is given in Figure 6. Next we analyze the average complexity and optimality of the protocol.

Theorem 6. *Protocol* M2A-SEPARATE-MESSAGES, *on a network of n nodes with k nodes initially activated, has the average time $\mathcal{O}(k \log(n/k) \log n)$.*

Proof. First, M2A task is completed by the end of M2A-SEPARATE(2^j), where $j = \lceil \lg k \rceil$, with probability at least $1 - 4k/n^3$. Consider M2A-SEPARATE(2^j). It follows that during SELECTOR-SUBROUTINE(2^j) of M2A-SEPARATE(2^j) at most 2^{j-1} activated nodes do not transmit as unique transmitters in rounds (a). Those who transmit as unique transmitters in some rounds (a) have also successful broadcasts in the following $10 \lg n$ rounds of ROUND-ROBIN-STACK in code line (r), with probability at least $1 - 1/n^3$ each, by Lemma 1.

Consider SELECTOR-SUBROUTINE(2^{j-1}), which is the second subroutine of M2A-SEPARATE(2^j). During this part at most 2^{j-2} activated nodes did not transmit as unique transmitters in rounds (a). Conditioned on this event, certain rumors are completed during $10 \lg n$ following rounds of ROUND-ROBIN-STACK in part (r) of the loop, with probability at least $1 - 1/n^3$ each, again by Lemma 1. We continue analyzing subroutines of M2A-SEPARATE(2^j) which are based on

---

```
for j = 0 to lg n do
    call M2A-SEPARATE(2^j)
call GOSSIP-SEPARATE-MESSAGES
```

Fig. 6. Protocol M2A-SEPARATE-MESSAGES

$(n, 2^i)$-selectors for $i = \lg(k/4), \lg(k/8), \ldots, 1, 0$. Quantitatively, by the beginning of SELECTOR-SUBROUTINE(2^i) at most 2^i selected nodes have not broadcasted successfully, with the probability of at least $1 - \sum_{a=i+1}^{j} 2^a/n^3$. Conditioned on this event, during SELECTOR-SUBROUTINE(2^i) at most 2^{i-1} activated nodes did not transmit as unique transmitters in rounds (a), while those which have transmitted as unique transmitters in rounds (a) complete broadcast during next $10 \lg n$ rounds in line (r) of the code, with the probability of at least $1 - 2^i/n^3$. Consequently, by the beginning of SELECTOR-SUBROUTINE(2^{i-1}) at most 2^{i-1} of activated nodes have not complete broadcast, with probability at least $1 - \sum_{a=i}^{j} 2^a/n^3$.

M2A-SEPARATE(2^j) takes $\sum_{i=0}^{\lg k} \mathcal{O}(2^i \log(n/2^i) \lg n) = \mathcal{O}(k \log(n/k) \log n)$ rounds, and during this procedure M2A task is completed with probability at least $1 - \sum_{a=0}^{j} 2^a/n^3 \geq 1 - 4k/n^3$.

The number of rounds in M2A-SEPARATE-MESSAGES by the end of execution of M2A-SEPARATE(2^j) is $\sum_{j'=0}^{j} \mathcal{O}(2^{j'} \log(n/2^{j'}) \log n) = \mathcal{O}(k \log(n/k) \log n)$. The worst-case $\mathcal{O}(n^3)$ can occur with probability at most $4k/n^3$. This justifies the estimate $\mathcal{O}(k \log(n/k) \log n) + \mathcal{O}(n^3) \cdot 4k/n^3 = \mathcal{O}(k \log(n/k) \log n)$ to be an upper bound on the average time.

We also show a lower bound for M2A communication with separate messages.

Theorem 7. *For any M2A protocol for the model of separate messages, the average number of rounds to complete gossiping on a network of n nodes with k nodes initially activated is $\Omega(k \log n)$.*

Corollary 1. *For any gossiping protocol for the model of separate messages, the average number of rounds to complete gossiping on a network of n nodes is $\Omega(n \log n)$.*

Our M2A protocol is within a factor of at most $\mathcal{O}(\log(n/k))$ close to optimality. In the case of $k = \Omega(n)$, which includes gossiping, the protocol is asymptotically optimal.

References

1. N. Alon, A. Bar-Noy, N. Linial, and D. Peleg, A lower bound for radio broadcast, *Journal of Computer and System Sciences*, 43 (1991) 290 - 298.
2. R. Bar-Yehuda, O. Goldreich, and A. Itai, On the time complexity of broadcast in radio networks: An exponential gap between determinism and randomization, *J. Computer and System Sciences*, 45 (1992) 104 - 126.

3. R. Bar-Yehuda, A. Israeli, and A. Itai, Multiple communication in multi-hop radio networks, *SIAM J. on Computing*, 22 (1993) 875 - 887.
4. I. Chlamtac, and S. Kutten, On broadcasting in radio networks - problem analysis and protocol design, *IEEE Transactions on Communication*, 33 (1985) 1240 - 1246.
5. B.S. Chlebus, L. Gąsieniec, A.M. Gibbons, A. Pelc, and W. Rytter, Deterministic broadcasting in ad hoc radio networks, *Distributed Computing*, 15 (2002) 27 - 38.
6. B.S. Chlebus, L. Gąsieniec, A. Lingas, and A. Pagourtzis, Oblivious gossiping in ad-hoc radio networks, in *Proc., 5th International Workshop on Discrete Algorithms and Methods for Mobile Computing and Communications (DIALM)*, 2001, pp. 44 - 51.
7. M. Christersson, L. Gąsieniec, and A. Lingas, Gossiping with bounded size messages in ad-hoc radio networks, in *Proc., 29th International Colloquium on Automata, Languages and Programming (ICALP)*, 2002, pp. 377 - 389.
8. M. Chrobak, L. Gąsieniec, and W. Rytter, Fast broadcasting and gossiping in radio networks, *Journal of Algorithms*, 43 (2002) 177 - 189.
9. A.E.F. Clementi, A. Monti, and R. Silvestri, Distributed broadcasting in radio networks of unknown topology, *Theoretical Computer Science*, 302 (2003) 337 - 364.
10. A. Czumaj, and W. Rytter, Broadcasting algorithms in radio networks with unknown topology, in *Proc., 44th IEEE Symposium on Foundations of Computer Science (FOCS)*, 2003, pp. 492 - 501.
11. A. De Bonis, L. Gąsieniec, U. Vaccaro, Generalized framework for selectors with applications in optimal group testing, in *Proc., 30th International Colloquium on Automata, Languages and Programming (ICALP)*, 2003, pp. 81 - 96.
12. R. Elsässer, and L. Gąsieniec, Radio communication in random graphs, in *Proc., 17th ACM Symposium on Parallelism in Algorithms and Architectures (SPAA)*, 2005, pp. 309 - 315.
13. L. Gąsieniec, E. Kranakis, A. Pelc, and Q.Xin, Deterministic M2M multicast in radio networks, in *Proc., 31st International Colloquium on Automata, Languages and Programming (ICALP)*, 2004, pp. 670 - 682.
14. L. Gąsieniec, T. Radzik, and Q. Xin, Faster deterministic gossiping in directed ad-hoc radio networks, in *Proc., 9th Scandinavian Workshop on Algorithm Theory (SWAT)*, 2004, pp. 397 - 407.
15. D.R. Kowalski, and A. Pelc, Time of radio broadcasting: adaptiveness vs. obliviousness and randomization vs. determinism, in *Proc., 10th International Colloquium on Structural Information and Communication Complexity (SIROCCO)*, 2003, pp. 195 - 210.

L(h,1,1)-Labeling of Outerplanar Graphs[*]

Tiziana Calamoneri[1], Emanuele G. Fusco[1],
Richard B. Tan[2,3], and Paola Vocca[4]

[1] Dipartimento di Informatica
Università di Roma "La Sapienza", via Salaria, 113-00198 Rome, Italy
{calamoneri, fusco}@di.uniroma1.it
[2] Institute of Information and Computing Sciences
Utrecht University, Padualaan 14, 3584 CH Utrecht, The Netherlands
rbtan@cs.uu.nl
[3] Department of Computer Science
University of Sciences & Arts of Oklahoma
Chickasha, OK 73018, U.S.A.
[4] Dipartimento di Matematica "Ennio de Giorgi"
Università diegli Studi di Lecce, via Provinciale Lecce-Arnesano, P.O. Box 193,73100
Lecce, Italy
paola.vocca@unile.it

Abstract. An $L(h, 1, 1)$-labeling of a graph is an assignment of labels from the set of integers $\{0, \cdots, \lambda\}$ to the vertices of the graph such that adjacent vertices are assigned integers of at least distance $h \geq 1$ apart and all vertices of distance three or less must be assigned different labels. The aim of the $L(h, 1, 1)$-labeling problem is to minimize λ, denoted by $\lambda_{h,1,1}$ and called *span* of the $L(h, 1, 1)$-labeling.

As outerplanar graphs have bounded treewidth, the $L(1, 1, 1)$-labeling problem on outerplanar graphs can be exactly solved in $O(n^3)$, but the multiplicative factor depends on the maximum degree Δ and is too big to be of practical use. In this paper we give a linear time approximation algorithm for computing the more general $L(h, 1, 1)$-labeling for outerplanar graphs that is within additive constants of the optimum values.

1 Introduction

In multi-hop radio networks, one of the problems that have been studied extensively is the radio-frequency assignment problem. Each station and its neighbors are assigned frequencies so as to avoid signal collisions. This is equivalent to a graph coloring problem, where vertices are stations and edges represent interferences between the stations.

The type of graph coloring problem varies depending on the kinds of frequency collisions that are to be avoided. If the only requirement is to avoid direct collisions between two neighbors, then this coincides with the normal graph coloring problem with its associated chromatic number χ. We call this $L(1)$-labeling

---

[*] This work was partially supported by the Università di Roma "La Sapienza", Italy.

P. Flocchini and L. Gąsieniec (Eds.): SIROCCO 2006, LNCS 4056, pp. 268–279, 2006.

problem of a graph G. Should it be desired that each station and all of its neighbors have distinct frequencies, we have the $L(1,1)$-labeling problem. This is also known as the distance-two coloring of a graph or coloring of the square of the graph, and has been well-studied.

In [7], Griggs and Yeh introduced a variation of a graph coloring problem which they called λ-coloring problem. In this problem, each vertex is assigned a color from the set of integers $\{0, \cdots, \lambda\}$ in such a way that adjacent vertices must be assigned colors of at least two apart and vertices of distance two must have distinct colors. This is also known as the $L(2,1)$-labeling problem. The motivation of this type of coloring problem comes from radio frequency adjacent-band interference problem, where adjacent frequencies may leak across the frequency bands. Subsequently, the problem has been extended to $L(h,k)$-labeling, where adjacent vertices must be assigned colors of distance at least $h \geq 0$ apart and vertices of distance two must be assigned colors at least $k \geq 0$ apart. The $L(h,k)$-labeling problem has been studied on many different graphs. Of particular interest is the class of planar graphs and its subclass the outerplanar graphs. See [5] for a comprehensive survey.

In practice, the distances in some wireless networks can be quite close (for example, the cellular network). Thus it may be necessary that not only stations of distance two apart must have distinct frequencies, but perhaps distance three or more. This motivates the study of $L(h,1,1)$-labeling problem, where adjacent nodes must have frequencies at least $h \geq 1$ bands apart and all nodes of distance two or three must also have distinct frequencies.

In this paper we only focus on $L(h,1,1)$-labeling of outerplanar graphs. More precisely, we start from $L(1,1,1)$-labeling of outerplanar graphs, i.e. the distance three coloring, where colors are distinct for vertices that are within distance three of each other, then we extend it to $L(h,1,1)$-labeling of outerplanar graphs for any $h \geq 1$.

1.1   Our Results

For an outerplanar graph G of maximum degree Δ we present lower bounds of $3\Delta - 3$ for the maximum number of colors that are needed to perform the $L(1,1,1)$-labeling. We show that by using a simple greedy approach $4\Delta - 2$ colors are necessary in the worst case for $L(1,1,1)$-labeling an outerplanar graph. Then we give a linear time approximation algorithm to $L(1,1,1)$-label an outerplanar graph using no more than $3\Delta+9$ colors for $\Delta \geq 6$ and extend it to $L(h,1,1)$-label an outerplanar graph using no more than $3\Delta + 2h + 7$ colors for $\Delta \geq h + 5$.

1.2   Related Results

The distance-d coloring problem, $L(1, \cdots, 1) = L(1^d)$-labeling of a graph, where all vertices within distance $d \geq 1$ must have distinct colors have been studied in the literature. In [13], Nizisheki et al gave an $O(n^3)$ time algorithm to $L(1^d)$-label a graph of bounded treewidth k. As an outerplanar graph G is a graph of treewidth two, this algorithm can be used to give an optimal $L(1,1,1)$-labeling of G in $O(n^3)$ time. Let α be the chromatic number of the third power of a graph of bounded threewidth 2, then the multiplicative factor of this algorithm is $\alpha^{2^{31}}$; a number way too big to be of practical use.

In contrast, our approximation algorithm is linear of $O(n\Delta)$, and only within an additive constant of the optimum value.

For outerplanar graphs G, the $L(h, 1)$-labeling problem for $h \geq 1$ has also been studied. The $L(1, 1)$-labeling problem appeared in [3, 6] and the $L(2, 1)$-labeling in [3, 6, 9]. To the best of our knowledge, nothing is known for the $L(h, 1, 1)$-labeling of outerplanar graphs for $h \geq 2$.

The rest of the paper is organized as follows. The next section gives the preliminary materials on $L(h, 1, 1)$-labeling and outerplanar graphs. Section 3 describes the techniques and results of $L(1, 1, 1)$-labeling of outerplanar graphs. The same techniques are then used in section 4 to obtain results for $L(h, 1, 1)$-labeling for $h \geq 1$. The final section gives the conclusion and state some open problems.

2   Preliminaries

Let $G = (V, E)$ be a graph with vertex set V and edge set E. The number of vertices of the graph is denoted by n and the maximum degree by Δ. Throughout the paper we assume our graph is connected, loopless and simple.

2.1   L(h,1,1)-Labeling

Definition 1. *Let G be a graph and $h \geq 1$ be a non-negative integer. A $L(h, 1, 1)$-labeling of G is an assignment of colors (integers) to the vertices of G from the set of integers $\{0, \cdots, \lambda\}$ such that vertices of distance 1, have colors that differ by at least h and vertices of distance two or three have colors that differ by at least 1. The minimum value λ for which G has a $L(h, 1, 1)$-labeling is denoted by $\lambda_{h,1,1}$ and the minimum number of colors is denoted by $\chi_{h,1,1} = \lambda_{h,1,1} + 1$.*

In order to make easier the reading, we will deal with $\chi_{h,1,1}$ most of the time, even if most of the literature use $\lambda_{h,1,1}$.

2.2   Outerplanar Graphs

An *outerplanar* graph is a graph that has a planar embedding such that all the vertices lie on the exterior face.

We first state some known facts about outerplanar graphs, of which the first two are well-known.

Characterization by Minors. A graph G is outerplanar iff it does not contain the complete graph K_4 nor the complete bipartite graph $K_{2,3}$ as minors. (A *minor* of a graph is obtained by edge contractions, edge deletions, or deleting isolated vertices.)

Degree 1 or 2. An outerplanar graph G has a node of degree 1 or 2.

OBFT(G) [6]. An outerplanar graph G has an *ordered breadth first tree graph OBFT(G)*, constructed in the following manner. Choose a node r and induce a *total ordering* on the vertices clockwise on the exterior face of a planar embedding of G. Do a breadth first search starting with the root r and visit the vertices in order of the given ordering. We end up with an OBFT(G) with possibly some non-tree edges which have the following properties.

A non-tree edge can only exist between vertices x and y if :

1. they are adjacent vertices on the same level, i.e. $x = v_{l,i}$ and $y = v_{l,i+1}$ for some level $l \geq 1$ and $i \geq 1$, where $v_{l,i}$ denotes the vertex at level l and it is the ith vertex from the left,
2. they are vertices on adjacent levels, $x = v_{l,i}$ and $y = v_{l+1,j}$, and y must be the rightmost child of its parent $w = v_{l,k}$ and $k = i - 1$, i.e. vertex x must be the next vertex after w on the same level in the OBFT(G).

See Fig. 1 for an example of OBFT(G), where dotted lines denote non-tree edges.

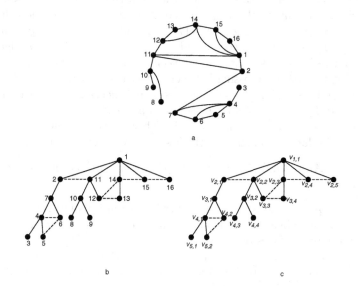

Fig. 1. Example of OBFT(G)

Given as input an outerplanar embedding of G, an OBFT(G) can be computed in $O(n)$ time.

We prove the following results concerning an OBFT(G) that will be useful to prove the upper bound of our algorithms.

Lemma 1. *Let G be an outerplanar graph with its associated OBFT(G), and two siblings x and y, $x < y$, in OBFT(G). Any node u in the subtree of OBFT(G) rooted at x is less than any node w in the subtree of OBFT(G) rooted at y.*

Proof. First observe that the parent of x and y, say r, can assume three possible relative positions with respect to x and y: $r < x < y$, $x < r < y$ and $x < y < r$ (see Fig. 2).

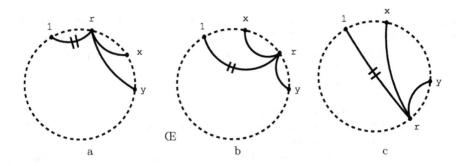

Fig. 2. Proof of Lemma 1. Lines with double bars are paths while simple lines represent edges.

In the first case (Fig. 2.a), u can lie either between r and x or between x and y, otherwise a crossing would be generated. Assume by contradiction that there exists a $w < u$. Now, w cannot lie between 1 and r (path $w \rightsquigarrow y$ would cross path $1 \rightsquigarrow r$); w cannot lie between r and x (path $w \rightsquigarrow y$ would cross edge (r, x)); so the only feasible interval for w is between x and y. Nevertheless, also in this interval, $w < u$ implies a crossing between paths $x \rightsquigarrow u$ and $y \rightsquigarrow w$. So $u < w$.

In the second case (Fig. 2.b) $1 < u < r$ as there is necessarily a path connecting root 1 to r, and $w > r$ for similar reasons. So $u < w$.

Finally, in the third case (Fig. 2.c) u is either between 1 and x or between x and y. With similar reasonings as in the first case, we prove again that $u < w$.   □

Theorem 1. *Any OBFT(G) of an outerplanar graph G is an outerplanar embedding of G.*

Proof. First observe that if the embedding is not outerplanar, then either there exists some node embedded inside an internal face, or there is some node on the boundary of internal faces only.

Given an OBFT(G), let us suppose first that there is a node v embedded inside an internal face f. In fact, if a whole subtree is embedded inside f then we can contract it to its root, say v. We will prove the claim by contradiction. The boundary of f is the cycle created in the OBFT(G) by at least one non-tree edge (u, w) (see fig. 3 a). Let us consider the lower common ancestor of u and w, $lca(u, w)$, and its two children on the boundary of f, let they be x and y. By OBFT(G) construction, it must be $x < y$. By Lemma 1, we have $u < v < w$ if (i) v is in the subtree rooted at x, ii) v is in the subtree rooted at y and (iii) v is a child of $lca(u, w)$. This configuration leads to an absurdity, as 1 must lie to the left of u and it is impossible to place path $1 \rightsquigarrow v$ not passing through u and not crossing edge (u, w). It follows that v does not exists.

Let us suppose now that a node v lies on the boundary of internal faces only and consider the simple cycle C constituted by the boundary of the union of all such faces. By construction, if v lies on level l of the OBFT(G), then on C there are necessarily a node w on a level strictly greater than l and a node u on a level strictly less than l such that there exist paths $w \rightsquigarrow v$ and $u \rightsquigarrow v$ not using nodes of C. As u and w both lie on C, then there are two distinct paths inside C connecting u and w both passing through a node at level l. This leads to an absurdity as we can construct the forbidden minor $K_{2,3}$: v represents the internal node, u and w are the degree 3 nodes and the two nodes on level l are the remaining degree 2 nodes.   □

Corollary 1. *In an OBFT(G) of an outerplanar graph G, for each node c, there exists at least one of c's children not having non-tree edges on both sides. (Refer to Fig. 3 b.)*

Proof. The claim directly follows from Theorem 1 as node c would be internal.   □

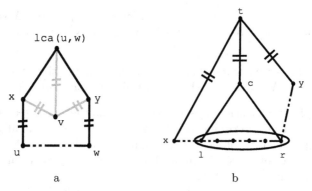

Fig. 3. a) Proof of Theorem 1. Lines with double bars are paths while simple lines represent edges. b) Proof of Corollary 1.

3   L(1,1,1)-Labeling

In this section we deal with the $L(1,1,1)$-labeling of outerplanar graphs. The technique used here will be generalized in the next section in order to handle the $L(h,1,1)$-labeling.

First we give the lower bound.

Theorem 2. *There exists an outerplanar graph of degree Δ that requires at least $3\Delta - 3$ colors to be $L(1,1,1)$-labeled.*

Proof. Consider the graph shown in Fig. 4 a; x, y and z are vertices of degree Δ. As all adjacent vertices of x, y and z are at mutual distance ≤ 3, it is easy to see that it requires at least $3\Delta - 3$ colors.   □

The greedy first-fit approach is a frequently used technique for labeling vertices of graphs and usually performs well in practise. This technique consists in considering nodes one by one in any order and assigning them the first color not used by any of their labeled neighbors satisfying the $L(1,1,1)$-labeling condition. If there is a tree-like structure, the followed order is typically the top-down left to right one. In our case, we can state the following theorem.

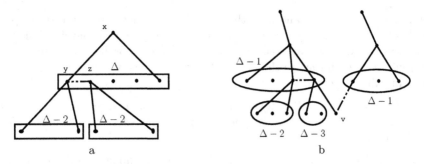

Fig. 4. a) Lower Bound. b) Greedy Algorithm.

Theorem 3. *There exists an outerplanar graph G of degree Δ such that the greedy first-fit approach requires at least $4\Delta - 2$ colors to $L(1,1,1)$-label G*

Proof. The worst case occurs when we have the configuration shown in Fig. 4 b where there are $4\Delta - 3$ vertices that are within distance three of a vertex v.   □

Since the gap between the lower bound on $\chi_{1,1,1}$ and the guaranteed performance of the greedy first-fit approach is rather large, we now present an algorithm that, given an outerplanar graph G of maximum degree $\Delta \geq 6$, finds an $L(1,1,1)$-labeling of vertices of G using at most $3\Delta + 9$ colors, and hence almost optimal as the lower bound is at least $3\Delta - 3$.

Let A contains colors $\{0, 1, \ldots \Delta + 2\}$, B colors $\{\Delta + 3, \Delta + 4, \ldots 2\Delta + 5\}$ and C colors $\{2\Delta + 6, 2\Delta + 7, \ldots, 3\Delta + 8\}$. The first step of the algorithm is to build an OBFT(G), rooted on a node of degree 1 or 2. Then the algorithm proceeds to assign a group of colors to the children of each node. Finally, it colors each node with a color from its color group.

Before describing how to assign groups of colors, in order to make easier the comprehension of the algorithm we introduce some definitions.

For a node v, let C_v denote the set of all the children of v in the OBFT(G) and $G(C_v)$ be the color group that is assigned to C_v. At each step we try to assign color groups in such a way as to avoid conflicting groups. By conflicting groups we mean that all the colors in a group may violate the $L(1,1,1)$-labeling condition.

For the algorithm refer to Fig. 5 a: Let v be a vertex assigned to a specific group of colors. All grandchildren of v are at distance ≤ 3 from C_v, hence we must forbid group $G(C_v)$ to all grandchildren of v and, in general, we are free to choose between the two remaining groups. Since v and possibly its left and right siblings (if they are adjacent to v), are at distance ≤ 3 from the grandchildren of v, we prefer to choose the color group different from that one assigned to v and its siblings when possible. Occasionally we will have no choice but to assign a specific color group because it is the only color group left that can be assigned without causing conflicts. This can occur for the grandchildren of a leftmost or rightmost child for a node. We call these color groups *fixed* (see Fig. 5 b and Fig. 6 b).

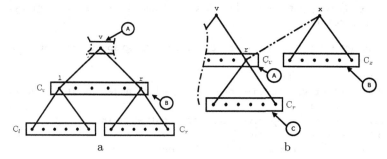

Fig. 5. a) Color group assignment. b) Fixed right color group.

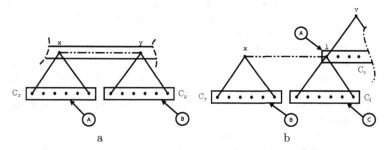

Fig. 6. a) Alternate color groups. b) Fixed left color group.

We now describe how to assign a color group. The color groups are assigned level by level top down from the root to the leaves and from the left to the right within each level of the tree, except in some special cases that will be explained later.

After we have assigned two separate color groups to the root and its children, we have two levels that are fully assigned and we have to assign group colors to the third level. At the general iteration we have already assigned color groups to level h and $h+1$, $h \geq 1$, we are now ready to assign color groups to level $h+2$. Refer to Fig 5 a: suppose v and its children C_v have been assigned groups, to assign color groups to v's grandchildren we first have to check C_r, where r is the rightmost child of v. The only case in which we do not follow the left to right order is depicted in Fig. 5 b: if there is edge (r, x) (i.e. the distance between any node in C_r and any node in C_x is ≤ 3) and $G(C_v) \neq G(C_x)$ then we have no choice but to assign the only color group available to C_r.

Afterwards, we have to check if node r is connected to its left sibling by a non-tree edge. If so, we have to assign groups from right to left, alternating with the only color groups left available (see Fig. 6 a), until there is a missing non-tree edge, which will occur due to Corollary 1. Next, we check the leftmost child l of node v. Again, if the color group to be assigned to C_l is fixed (Fig. 6 b), we have to assign the only available group and then check if node l is connected via a non-tree edge to its right sibling. If so, repeat the alternating group assignment as before (until a missing non-tree edge is encountered). After the two boundary

> Color Group Assignment Algorithm Let A, B and C be the three groups of distinct colors;
>
> Construct OBFS(G) tree T of an outerplanar graph G, rooted on a
>   node of degree 1 or 2;
> Assign $G(root)$=A and $G(C_{root})$=B;
> Suppose color groups have been assigned to nodes at levels h and
> $h+1$, $h \geq 1$.
> Visit nodes of T top down, left to right starting from the root;
> For each node v on level $h \geq 1$:
>   If $G(C_r)$ is *fixed* (see Fig. 5) then
>   Assign $G(C_r)$ to the only available color-group;
>   Proceed right to left, assign $G(C_x)$ (see Fig. 6) switch-
>   ing color-group until there is a missing non-tree edge
>   between x and y or until $G(C_l)$ is assigned;
>   If $G(C_l)$ is *fixed* then
>   Assign $G(C_l)$ to the only available color-group;
>   Proceed left to right, assign $G(C_y)$ (see Fig. 6) switch-
>   ing color-group until there is a missing non-tree edge
>   between x,y or until $G(C_r)$ is assigned;
>   Let z be the leftmost node in C_v such that $G(C_z)$ is not as-
>   signed.
>   While there is such a node z
>   Assign $G(C_z)$ to the available color-group not used by z;
>   Proceed from left to right alternating color group as in
>   Fig. 6.

Fig. 7. Color group assignment algorithm

groups have been assigned, we try to assign color groups from left to right using a color group that is different from G_v if possible, alternating color groups from left to right for any non-tree edge that is present.

A more formal description is given in Fig. 7.

Theorem 4. *There exists a linear time algorithm that $L(1,1,1)$-labels any outerplanar graph with $3\Delta + 9$ colors, where $\Delta \geq 6$.*

Proof. We have already described the first two steps of the algorithm (i.e. the construction of OBFT(G) and the color group assignment), so it remains to detail how to assign to each node a color from its color group.

Given a node group and its assigned color group, we can randomly choose a different color for each node, paying only attention to nodes that are at distance ≤ 3 from a node group having the same color group. So, we first assign colors to such nodes (there are no more than four: the leftmost, its right sibling, the rightmost and its left sibling) avoiding conflicts, and then we proceed with all other nodes.

It is straightforward to see that this algorithm correctly labels the graph in linear time. It remains to show that $\Delta+3$ colors in each group are always enough.

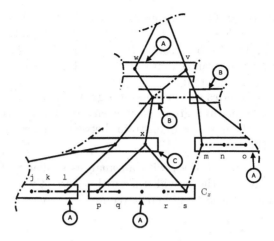

Fig. 8. Proof

By scanning in an exhaustive way all possible configurations in an OBFT(G), it is possible to see that the worst case is that in Fig. 8: at most $\Delta - 1$ nodes in C_x must be labeled using colors from a group (A in the figure) avoiding the colors assigned to m, l, w and v all at distance ≤ 3 from nodes in C_x. It follows that $\Delta + 3$ colors in each group are always enough.

Furthermore, observe that the color assigned to j cannot be used in p, the color assigned to k cannot be used neither in p nor in q, and similarly the color assigned to o cannot be used in s and the color assigned to n cannot be used neither in r nor in s. It follows that, after removing the 4 colors forbidden by m, l, w and v, the $\Delta - 1$ remaining colors must be at least 4. In the special case in which the color assigned to k is the same as the color assigned to n, one color more is necessary. Hence $\Delta \geq 6$.   □

4   L(h,1,1)-Labeling

In this section we show how to generalize to the $L(h,1,1)$-labeling the results obtained for the $L(1,1,1)$-labeling.

First, observe that Theorem 2 provides a lower bound of $3\Delta - 3$ for $\chi_{h,1,1}$, for any $h \geq 1$. Also Theorem 3 on the greedy first-fit approach applies to the general case $h \geq 1$.

In the following we get an $L(h,1,1)$-labeling by exploiting the Color-Group Assignment Algorithm and then by opportunely labeling nodes. Namely, we can prove the following theorem.

Theorem 5. *For any $h \geq 2$, there exists a linear time algorithm that $L(h,1,1)$-labels any outerplanar graph with $3\Delta + 2h + 7$ colors, $\Delta \geq h + 5$.*

Proof. (sketched) The reasonings are exactly the same as those presented in the proof of Theorem 4 with two main changes:

1. **Colors in a group.** Group A contains colors $\{0, 1, \ldots \Delta + 2\}$; group B contains colors $\{\Delta + h + 2, \Delta + h + 3, \ldots 2\Delta + h + 4\}$; and, group C colors $\{2\Delta + 2h + 4, 2\Delta + 2h + 5, \ldots, 3\Delta + 2h + 6\}$. The $h - 1$ colors in the gaps between color groups guarantee that the distance 1 constraint between adjacent group of nodes is respected;

2. **Assignment of colors to vertices of a group.** It is possible to prove that the number of colors in a group is sufficient to label vertices in a group so as to guarantee the distance 1 constraint between vertices connected to each other in a path.

It is straightforward to see that, with an analysis similar to the one used in the proof of Theorem 4, the provided algorithm correctly $L(h, 1, 1)$-labels the outerplanar graph in linear time and it requires at most $3\Delta + 2h + 7$ colors, where $\Delta \geq h + 5$. $\qquad\qquad\qquad\qquad\qquad\qquad\qquad\qquad\qquad\qquad\qquad\square$

5   Conclusion

In this paper we provide very close upper and lower bounds on $\chi_{h,1,1}$ for outerplanar graphs, showing also that the greedy first-fit technique does not work well in this case. In the literature, there is a known algorithm that optimally $L(1, 1, 1)$-labeling outerplanar graphs running in $O(n^3)$ time, but the multiplicative factor is extremely large to be of practical use. Our algorithm produces an approximate solution that only differs from the optimal solution by a constant additive factor and it is linear.

Some open problems arise from this work. First, there is a gap between the upper bound provided by the algorithm and the lower bound shown. It would be nice to close the gaps between the bounds.

Furthermore, for $L(h, 1^d) = L(h, 1, \cdots, 1)$, we have only studied the case when $d = 2$. It would be interesting also to study the $L(h, 1^d)$-labeling problem for outerplanar graphs for $d \geq 3$. The same technique of using color group assignments can be applied, but the number of cases to be considered increases quite a bit. The problem here is to find good estimates for $f(h, d)$ and $g(h, d)$ in the inequality $\chi_{h,1^d} \leq f(h, d)\Delta^{\lceil \frac{d}{2} \rceil} + g(h, d)$.

References

1. G. Agnarsson and M. Halldórsson. Coloring Powers of planar graphs. In Proc. *11th Ann. ACM-SIAM Symposium on Discrete Algorithms (SODA 2000)*: 654–662, 2000.
2. A.A. Bertossi, C.M. Pinotti and R.B. Tan. Channel assignment with separation for interference avoidance in wireless networks. *IEEE Transactions on Parallel and Distributed Systems* **14(3)**: 222–235, 2003. Preliminary version in *ACM Workshop DIAL M 2000*, 2000.
3. H.L. Bodlaender, T. Kloks, R.B. Tan and J. van Leeuwen. Approximations for λ-Colorings of Graphs. *The Computer Journal* **47**: 193-204, 2004. Preliminary version in Proc. *17th Annual Symp. on Theoretical Aspects of Computer Science (STACS 2000)*, Lectures Notes in Computer Science 1770: 395–406, 2000.

4. R.J. Bruce and M. Hoffmann. $L(p,q)$-labeling of outerplanar graphs. Tech. Rep. No. 2003/9, Department of Mathematics and Computer Science, University of Leicester, England.
5. T. Calamoneri. The $L(h,k)$-labeling problem: an annotated bibliography. Accepted to *The Computer Journal*, 2006. A continuously updated version is available online at http://www.dsi.uniroma1.it/~calamo/survey.html
6. T. Calamoneri and R. Petreschi. $L(h,1)$-Labeling Subclasses of Planar Graphs. *Journal on Parallel and Distributed Computing* **64(3)**: 414-426, 2004.
7. J.R. Griggs and R.K. Yeh. Labeling graphs with a Condition at Distance 2. *SIAM J. Disc. Math* **5**:586–595, 1992.
8. W. K. Hale. Frequency assignment: theory and applications. In Proc. *IEEE* **68**:1497–1514, 1980.
9. K. Jonas. *Graph Coloring Analogues With a Condition at Distance Two: $L(2,1)$-Labelings and List λ-Labelings*. Ph.D. thesis, University of South Carolina, Columbia, 1993.
10. S.T. McCormick. Optimal approximation of sparse Hessians and its equivalence to a graph coloring problem. *Math. Programming* **26**: 153–171, 1983.
11. R.K. Yeh. *A Survey on Labeling Graphs with a Condition at Distance Two*. Manuscript. 2004.
12. R.K. Yeh. *Labeling Graphs with a Condition at Distance Two*. Ph.D. Thesis, University of South Carolina, 1990.
13. X. Zhou, Y. Kanari and T. Nishizeki. Generalized vertex-coloring of partial k-trees. *IEICE Trans. Fundamentals of Electronics, Communication and Computer Sciences* **E83-A**: 671-678, 2000.

Combinatorial Algorithms for Compressed Sensing

Graham Cormode[1] and S. Muthukrishnan[2]

[1] Bell Labs, Lucent Technologies
cormode@lucent.com
[2] Rutgers University
muthu@cs.rutgers.edu

Abstract. In sparse approximation theory, the fundamental problem is to reconstruct a signal $\mathbf{A} \in \mathbb{R}^n$ from linear measurements $\langle \mathbf{A}, \psi_i \rangle$ with respect to a dictionary of ψ_i's. Recently, there is focus on the novel direction of *Compressed Sensing* [9] where the reconstruction can be done with very few—$O(k \log n)$—linear measurements over a modified dictionary if the signal is *compressible*, that is, its information is concentrated in k coefficients with the original dictionary. In particular, these results [9, 4, 23] prove that there exists a single $O(k \log n) \times n$ measurement matrix such that any such signal can be reconstructed from these measurements, with error at most $O(1)$ times the worst case error for the class of such signals. Compressed sensing has generated tremendous excitement both because of the sophisticated underlying Mathematics and because of its potential applications.

In this paper, we address outstanding open problems in Compressed Sensing. Our main result is an explicit construction of a non-adaptive measurement matrix and the corresponding reconstruction algorithm so that with a number of measurements polynomial in k, $\log n$, $1/\varepsilon$, we can reconstruct compressible signals. This is the first known polynomial time explicit construction of any such measurement matrix. In addition, our result improves the error guarantee from $O(1)$ to $1 + \varepsilon$ and improves the reconstruction time from $\text{poly}(n)$ to $\text{poly}(k \log n)$.

Our second result is a randomized construction of $O(k \, \text{polylog}(n))$ measurements that work for each signal with high probability and gives per-instance approximation guarantees rather than over the class of all signals. Previous work on Compressed Sensing does not provide such per-instance approximation guarantees; our result improves the best known number of measurements known from prior work in other areas including Learning Theory [20, 21], Streaming algorithms [11, 12, 6] and Complexity Theory [1] for this case.

Our approach is combinatorial. In particular, we use two parallel sets of group tests, one to filter and the other to certify and estimate; the resulting algorithms are quite simple to implement.

1   Introduction

We study a modern twist to a fundamental problem in sparse approximation theory, called *Compressed Sensing*, recently proposed in the Mathematics community.

Sparse Approximation Theory Background. The *dictionary* Ψ denotes an orthonormal basis for $\mathbb{R}^n$, i.e. Ψ is a set of n real-valued vectors ψ_i each of dimension n and $\psi_i \perp \psi_j$. The *standard basis* is the traditional coordinate system for n dimensions,

P. Flocchini and L. Gąsieniec (Eds.): SIROCCO 2006, LNCS 4056, pp. 280–294, 2006.

namely, for $i = 1, \ldots, n$, the vector $\psi_i = [\psi_{i,j}]$ where $\psi_{i,j} = 1$ iff $i = j$.[1] A *signal* vector $\mathbf{A}$ in $\mathbb{R}^n$ is transformed by this dictionary into a vector of *coefficients* $\theta(\mathbf{A})$ formed by inner products between $\mathbf{A}$ and vectors from Ψ. That is, $\theta_i(\mathbf{A}) = \langle \mathbf{A}, \psi_i \rangle$ and $\mathbf{A} = \sum_i \theta_i(\mathbf{A}) \psi_i$ by the orthonormality of Ψ.[2] From now on (for convenience of reference only), we reorder the vectors in the dictionary so $|\theta_1| \geq |\theta_2| \geq \ldots \geq |\theta_n|$.

In the area of sparse approximation theory [8], one seeks representations of $\mathbf{A}$ that are *sparse*, i.e., use few coefficients. Formally, $\mathbf{R} = \sum_{i \in K} \theta_i \psi_i$, for some set K of coefficients, $|K| = k \ll n$. Clearly, $\mathbf{R}(\mathbf{A})$ cannot exactly equal the signal $\mathbf{A}$ for all signals. The error is typically taken as $\|\mathbf{R} - \mathbf{A}\|_2^2 = \sum_i (\mathbf{R}_i - \mathbf{A}_i)^2$. By the classical Parseval's equality, this is equivalently $\|\theta(\mathbf{A}) - \theta(\mathbf{R})\|_2^2$. The optimal k representation of $\mathbf{A}$ under Ψ, $\mathbf{R}_{\text{opt}}^k$, therefore takes k coefficients with the largest $|\theta_i|$'s. The error then is $\|\mathbf{A} - \mathbf{R}_{\text{opt}}^k\|_2^2 = \sum_{i=k+1}^n \theta_i^2$. This is the error in representing the signal $\mathbf{A}$ in a compressed form using k coefficients from Ψ.

In any application (say audio signal processing), one has a "class" of input signals ($\mathbf{A}$'s) (e.g., sinusoidal waveforms comprising the audio signal), one chooses an appropriate dictionary Ψ (say discrete Fourier) so that most of the signals are "compressible" using that dictionary, and represents the signal using the adequate number ($k \ll n$) of coefficients ($\theta_1, \ldots, \theta_k$). There are different notions of a signal being compressible in a dictionary. In the past, e.g., in audio applications, researchers focused on the α-*exponentially decaying case* where the coefficients decay faster than any polynomial. That is, for some α, $|\theta_i| = O(2^{-\alpha i})$, for all i. More recently, there is focus on the *p-Compressible case*. Specifically the coefficients have a power-law decay: for some $p \in (0, 1)$, and for all i, $|\theta_i| = O(i^{-1/p})$. Consequently, $\|\mathbf{A} - \mathbf{R}_{\text{opt}}^k\|_2^2 \leq C_p k^{1-2/p}$ for some constant C_p. A simplification of these models is the *k-support case*, where the signal has at most k non-zero coefficients, so $\mathbf{R}_{\text{opt}}^k = \mathbf{A}$.

Study of sparse approximation problems involves the art of identifying suitable Ψ so the signals from an application are compressible, and studying their mathematical properties. This is a mature area of Mathematics with highly successful applications to signal processing, communication theory and compression [8].

Compressed Sensing. Recently, Donoho posed a fundamental question [9]: Since most of the information in the signal is contained in only a few coefficients and the rest of the signal is not needed for the applications, can one directly determine (acquire) only the relevant coefficients without reading (measuring) each of the coefficients? In a series of papers over the past year, the following result has emerged.

Theorem 1. *[9, 4, 23] There exists a non-adaptive set V of $O(k \log(n/k))$ vectors in $\mathbb{R}^n$ which can be constructed once and for all from the standard basis. Then, for fixed $p \in (0, 1)$ and any p-compressible signal $\mathbf{A}$ in the standard basis, given only* measurements $\langle \mathbf{A}, v_i \rangle$, $v_i \in V$, *a representation $\mathbf{R}$ can be determined in time polynomial in n such that* $\|\mathbf{A} - \mathbf{R}\|_2^2 = O(k^{1-2/p})$.

---

[1] Examples of other basis are *discrete Fourier* where $\psi_{i,j} = \frac{1}{\sqrt{n}} \exp(-2\pi\sqrt{-1}ij/n)$; and *Haar wavelet* where every ψ_i is a scaled and shifted copy of the same step like function. By applying an appropriate rotation to the basis and signal vectors, our problem can be thought of in the standard basis only.

[2] We refer to θ_i where $\mathbf{A}$ is implicitly clear.

There are several important points to note. First, since the worst case error for a p-compressible signal is $C_p k^{1-2/p}$, the representation above is *optimal, up to constant factors* for the class of all p-compressible signals, for a fixed p. Second, even if the signal consisted of precisely k nonzero coefficients $\theta_{i_1}, \ldots, \theta_{i_k}$, one needs k measurements $\langle \mathbf{A}, \psi_{i_j} \rangle$ for $j \in [1, k]$; hence, the set V of measurements is only a $\log(n/k)$ factor larger than the naive lower bound of measurements needed. Third, the proof shows existence of V by showing that a random set of V vectors will satisfy the theorem with nonzero probability. The proof immediately gives a *Monte Carlo* randomized algorithm by using such a random V.

This result has generated much interest, and a sequence of papers have improved different aspects of the result [9, 25, 4, 23]; found interesting applications including MR imaging wireless communication [23] and generated implementations [22]; found mathematical applications to coding and information theory [3]; and extended the results to noisy and distributed settings [2]. The interest arises for two main reasons. First, there is deep mathematics underlying the results, with interpretations in terms of high dimensional geometry [23], uncertainty principles [4], and linear algebra [9]. Second, there are serious applications—for example, in going from analog to digital representation of the signals, existing hardware chips can execute measurements $\langle \mathbf{A}, v_i \rangle$ extremely efficiently, so performing $O(k \log(n/k))$ measurements is significantly more efficient than measuring each component of the signal (hence "compressed sensing"). The results have inspired a number of workshops, meetings and talks [15, 18].

Outstanding Problems and Our Results. There are several outstanding questions in Compressed Sensing. The most fundamental issue is to explicitly construct the non-adaptive measurement set of vectors V (or equivalently, a *transformation matrix T* in which $T[i, j] = v_i[j]$) in the theorem. The existing results first show that if T satisfies certain conditions, the theorem holds; then they show that T chosen from an appropriate random distribution suffices. The necessary conditions are quite involved, such as computing the eigenvalues of every $O(k \log n)$ square submatrix of T [9], and testing that each such submatrix is an isometry, behaving like an orthonormal system [4]. No explicit construction is known to produce T's with these properties. Instead, algorithms for Compressed Sensing choose a random T, and assume that the conditions are met. Thus, these are *Monte Carlo* algorithms, with some probability of failure. This is a serious drawback for Compressed Sensing applications motivated by hardware implementations which will sense many, many signals over time. So it is highly desirable that there be an explicit construction of T suitable for Compressed Sensing. A natural approach is to take a random T and test whether it satisfies the necessary conditions. However, this is much too expensive, taking time at least $\Omega(n^{k \log n})$.

There are several other outstanding questions. For example, the time to obtain a representation from the measurements is significantly *superlinear* in n (it typically involves solving a Linear Program [9, 4, 23]). For large signals, this cost is overly burdensome. Since we make a small number of measurements, it is much better to find algorithms with running time polynomial in the number of measurements and hence, sublinear in n. Lastly, the guarantee given by the above theorem is not relative to the best possible for the given signal (i.e., *per-instance*), but to the *worst case* over the whole class of p-compressible signals. Clearly per-instance error guarantees (equivalently, true approximation algorithms) are preferable.

We address these questions and present the first known explicit algorithms for Compressed Sensing. Our approach is combinatorial, and yields a number of technical improvements such as sublinear time reconstruction, and tolerance to error. Our main results are twofold.

1. We present a deterministic algorithm that in time polynomial in k and n constructs a non-adaptive transformation matrix T of number of rows polynomial in $k \log n$, and present an associated reconstruction algorithm in the spirit of Theorem 1. More specifically, our algorithm outputs a representation $\mathbf{R}$ for a compressible signal $\mathbf{A}$ such that $\|\mathbf{R} - \mathbf{A}\|_2^2 < \|\mathbf{R}_{\text{opt}}^k - \mathbf{A}\|_2^2 + \varepsilon\|\mathbf{C}_{\text{opt}}^k\|_2$. Here, $\|\mathbf{C}_{\text{opt}}^k\|_2$ denotes the optimal error over the whole class of signals considered. This is the first explicit construction known for this problem in polynomial time.

In addition, this result leads to the following improvements: (a) the reconstruction time is subquadratic in the number of measurements (and hence sublinear in n), (b) the overall error is optimal up to $1 + \varepsilon$ of the worst case error $\|\mathbf{C}_{\text{opt}}^k\|_2$ for p-compressible signals, improving the $O(1)$ approximation factor in prior results, and (c) the approach applies to other cases of compressible signals with tighter bounds. For the exponentially decaying and k-sparse family, the size of T is only $O(k^2 \text{polylog}(n))$. The algorithms are simple and easy to implement, without linear programming and without running into precision-issues inherent in the choice of Gaussian random T in prior methods.

2. We address the issue of obtaining per-instance guarantees for each signal. We present a randomized algorithm that on any given $\mathbf{A}$, produces a T with $O(\frac{k}{\epsilon^2} \text{polylog}(n))$ rows such that in time linear in $O(k \text{polylog}(n))$, we can reconstruct a $\mathbf{R}$ with $\|\mathbf{A} - \mathbf{R}\|_2^2 \le (1 + \varepsilon)\|\mathbf{A} - \mathbf{R}_{\text{opt}}^k\|_2^2$, with probability at least $1 - \frac{1}{n^{O(1)}}$.

Notice crucially that this second result does not produce a T that works for all p-compressible signals, merely that on any given signal $\mathbf{A}$, we can produce a good $\mathbf{R}$ with high probability. In this regime, which is quite different from the regime in earlier papers on Compressed Sensing where a fixed T works for all p-compressible signals, many results in the Computer Science literature apply, in particular, from learning theory [20, 21], streaming algorithms [12, 11] and complexity theory [1]. Some of these results do not completely translate to our scenario: the learning theory approaches assume that the signal can be probed in the light of the results of prior measurements (this is similar to adaptive group testing). Other results can be thought of as producing a T with $O(k^{2+O(1)} \text{polylog}(n))$ rows which is improved by our result here. An exception is the result in [13] which works by sampling (that is, finding $\langle \mathbf{A}, v_i \rangle$ where $v_{i,j} = 1$ for some j and is 0 elsewhere) for the Fourier basis, but can be thought of as solving our problem using $O(k \text{polylog}(1/\varepsilon, \log n, \log \|\mathbf{A}\|))$ measurements. Our result improves [13] in the term $\text{polylog}(\|\mathbf{A}\|_2)$ which governs the number of iterations in [13]. Finally, we extend to the case when the measurements are noisy—an important practical concern articulated in [14]—and obtain novel results that give per-instance approximation results.

Technical Overview. The intuitive way to think about these problems is to consider combinatorial group testing problems. We have a set $U = [n]$ of items and a set D of *distinguished* items, $|D| \le k$. We identify the items in D by performing group tests on subsets $S_i \subseteq U$ whose output is 1 or 0, revealing whether that subset contains one or more distinguished items, that is $|S_i \cap D| \geq 1$. There exist collections of $O((k \log n)^2)$ nonadaptive tests which identify each of the distinguished items precisely.

There is a strong connection between this problem and Compressed Sensing. We can treat θ_i's as items and the largest (in magnitude) k as the members of D. Each test set S_i can be written as its characteristic vector χ_{S_i} of n dimensions. A difficulty arises in interpreting the outcome of $\langle \mathbf{A}, \chi_{S_i} \rangle$. The discussion so far has been entirely combinatorial, but the outcome of this linear-algebraic operation of inner product must be interpreted as a binary outcome to apply standard combinatorial group testing methods. In general, there is no direct connection between $\langle \mathbf{A}, \chi_{S_i} \rangle$ and presence or absence of the first k coefficients in S_i when the signal is from the p-compressible class. This is also the reason that prior work on this problem has delved into the linear-algebraic and geometric structure of the problem.

Our approach here is combinatorial. Our first results show that one can focus attention on some $k' > k$ coefficients, in order to meet our error guarantees. Then, we show that separating the k' coefficients using group testing methods serves as a filter and subsequently, using a different set of group tests serves to certify and estimate the largest k coefficients in magnitude. This use of two parallel sets of group tests is novel. For the second set of results, combinatorial group testing has been applied previously in Learning Theory [20, 21], Streaming Algorithms [11, 12, 6] and Complexity Theory [1]. Here, our contribution is to adapt the approach from our first set of results and provide a tighter analysis of the error in terms of $\|\mathbf{R}_{\mathrm{opt}}^k - \mathbf{A}\|_2$ rather than in terms of $\|\mathbf{A}\|_2$ as is more typical.

Note. Preliminary versions of this paper have appeared as technical reports [7], which are superseded by the results here. Several proofs have been omitted, for space reasons.

2   Preliminaries

Definition 1. *A collection* $\mathcal{S}$ *of* l *subsets of* $\{1 \ldots n\}$ *is called* k-selective *if for any* X *such that* $X \subset \{1 \ldots n\}$ *and* $|X| \leq k$, *there exists* $S_i \in \mathcal{S}$ *such that* $|S_i \cap X| = 1$, *i.e. there is a member of* X *which is separated from all other members of* X *in some* S_i.

Definition 2. *A collection* $\mathcal{S}$ *of* m *subsets of* $\{1 \ldots n\}$ *is called* k-strongly selective *if for any* X *with* $|X| \leq k$, *and for all* $x \in X$ *there exists* $S_i \in \mathcal{S}$ *such that* $S_i \cap X = \{x\}$, *i.e. every member of* X *occurs separated from all other members of* X *in some* S_i.

We note that the k-strongly selectivity is a stronger condition than k-selectivity, and so the former implies the latter. Explicit constructions of both collections of sets are known for arbitrary k and n. Strongly selective sets are used heavily in group testing [10], and can be constructed using superimposed codes [19] with $m = O((k \log n)^2)$. Indyk provided explicit constructions of k-selective collections of size $l = O(k \log^{O(1)} n)$, where the power depends on the degree bounds of constructions of disperser graphs [16]. Probabilistic constructions are also possible [5] of near-optimal size $O(k \log(n/k))$, which yield a more expensive Las Vegas-style algorithm for constructing such a set in $O(n^k \operatorname{poly}(k \log n))$: after randomly constructing a collection of sets, verify the required property holds for all $\binom{n}{k}$ choices of X.

We will also make use of the Hamming code matrix H_n, which is the $\lceil 1 + \log_2 n \rceil$ matrix whose ith column is 1 followed by the binary representation of i. We will combine matrices together to get larger matrices by (a) concatenating the rows of N to M and get matrix denoted $M \bigcup N$, or (b) a Tensor product-like operation we denote $\otimes$, defined as follows:

Definition 3. *Given matrices V and W of dimension $v \times n$ and $w \times n$ respectively, define the matrix $(V \otimes W)$ of dimension $vw \times n$ as $(V \otimes W)_{iv+l,j} = V_{i,j} W_{l,j}$.*

3 Non-adaptive Constructions

We must describe the construction of a set of m (row) vectors Ψ' that will allow us to recover sufficient information to identify a good set of coefficients. We treat Ψ' as an $m \times n$ matrix whose ith row is Ψ'_i. When given the vector of measurements $\Psi' \mathbf{A}$ we must find an approximate representation of $\mathbf{A}$. Ψ' is a function of Ψ, and more strongly (as is standard in compressed sensing) we only consider matrices Ψ' that can be written as a linear combination of vectors from the dictionary Ψ, i.e., $\Psi' = T\Psi$, for some $m \times n$ transform matrix T. Thus $\Psi' \mathbf{A} = T(\Psi \mathbf{A}) = T\theta$. Recall that the best representation under Ψ using k coefficients is given by picking k largest coefficients from θ. We use T to let us estimate k large coefficients from θ, and use these to represent $\mathbf{A}$; we show that the error in this representation can be tightly bounded.

Observe that we could trivially use the identity matrix I as our transform matrix T. From this we would have $T\theta = \theta$, and so could recover $\mathbf{A}$ exactly. However, our goal is to use a transform matrix that is much smaller than the n rows of I, preferably polynomial in k and $\log n$. In general for most classes of signals, the only way to achieve exact recovery of the optimal representation is to take a linear number of measurements:

Lemma 1. *Any deterministic construction which returns k coefficients and guarantees error exactly $\|\mathbf{R}^k_{\mathrm{opt}} - \mathbf{A}\|_2$ requires $\Theta(n)$ measurements.*

3.1 p-Compressible Signals

In the p-compressible case the coefficients (sorted by magnitude) obey $|\theta_i| = O(i^{-1/p})$ for appropriate scaling constants and some parameter p. Previous work has focused on the cases $0 < p < 1$ [4,9]. Integrating shows that $\sum_{i=k+1}^{n} \theta_i^2 = \|\mathbf{R}^k_{\mathrm{opt}} - \mathbf{A}\|_2^2 = O(k^{1-2/p})$. Our results, like those of [4,9], are stated with respect to the error due to the worst case over all signals in the class, which we denote $\|\mathbf{C}^k_{\mathrm{opt}}\|_2 = O(k^{1-2/p})$. For any signal that is p-compressible with fixed p and C_p it follows that $\|\mathbf{R}^k_{\mathrm{opt}} - \mathbf{A}\|_2 \leq \|\mathbf{C}^k_{\mathrm{opt}}\|_2$. We give two results on the p-compressible case, one that applies when $p < \frac{1}{2}$, the other that applies for all $0 < p < 1$ provided the p-compressible case is tight, i.e. $|\theta_i| = \theta(i^{-1/p})$. The measurements made are the same, but the analysis varies.

Our transform collects information based on two collections of strongly separating sets. The first ensures that sufficient separation occurs, allowing all large coefficients to be recovered. The second allows accurate estimates of the weight of each coefficient to be made.

Transform Definition. We define our transform matrix as follows. Let k' and k'' be functions of k, ε, $\log n$ to be defined later. Let $\mathcal{S}$ be a k'-strongly separating collection of sets (so that the number of sets in the collection is k''), and write T_1 as the matrix formed by the concatenation of χ_{S_i} for all S_i in $\mathcal{S}$. Similarly, let $\mathcal{R}$ be a k''-strongly separating collection of sets, and write T_2 as its characteristic matrix. We form our transform matrix T_p as $(T_1 \otimes H) \bigcup T_2$.

The intuition is that rather than ensuring separation for just the k largest coefficients, we will guarantee separation for the top-t coefficients (even though we do not know a priori which those top-t coefficients are), where t is chosen so that the remaining coefficients are so small that even if taken all together, the error introduced to the estimation of any coefficient is still within our allowable error bounds.

Reconstruction Algorithm. Our algorithm for recovering a representation from the results of the measurements $T_p \Psi A$ is as follows: for each set of $\lceil 1 + \log n \rceil$ measurements due to $S_i \otimes H$, we recover $x_0 \ldots x_{\lceil \log n \rceil} = (S_i \otimes H) \Psi A$, and decode identifier j_i as

$$j_i = \sum_{b=1}^{\log n} 2^{b-1} \frac{|x_b| - \min\{|x_b|, |x_0 - x_b|\}}{\max\{|x_b|, |x_0 - x_b|\} - \min\{|x_b|, |x_0 - x_b|\}}.$$

This generates a set of coefficients $J = \{j_1, j_2 \ldots j_{k''}\}$. We then use the measurements due to T_2 to estimate the weight of each coefficient named in J: for each $j \in J$, we set $\hat{\theta}_j = \chi_{R_i} \Psi A$ for $J \cap R_i = \{j\}$. The strong separation properties of $\mathcal{R}$ ensure that there will be at least one such R_i, and if there is more than one, then we can pick one arbitrarily. Our output is the set of k pairs $(j, \hat{\theta}_j)$ with the k largest values of $|\hat{\theta}_j|$.

Lemma 2. *Consider the case when the p-compressible case is tight within constant factors for all coefficients, i.e. $|\theta_i| = \Theta(i^{-1/p})$. Let $k' = c'(k\varepsilon^{-p})^{1/(1-p)^2}$ and $k'' = c''(k' \frac{\log n}{\log k'})^2$ for appropriately chosen c' and c''.*

Let K' denote the set of the k'^{1-p} largest coefficients.

1. *$\forall 1 \leq j \leq n : \theta_j^2 \geq \frac{\varepsilon^2}{25k} \|\mathbf{C}_{opt}^k\|_2^2 \Rightarrow j \in K'$*
2. *$\forall j \in K' : j \in J$.*
3. *$\forall j \in J : |\hat{\theta}_j - \theta_j| \leq \frac{\varepsilon}{5\sqrt{k}} \|\mathbf{C}_{opt}^k\|_2$.*

Proof. Observe that the square of the (absolute) sums of coefficients after removing the top t is $(\sum_{i=t+1}^{n} |\theta_i|)^2 = O(t^{2-2/p})$. Over the whole class of p-compressible signals, this is bounded by $O(t^{2-2/p}/k^{1-2/p}) \|\mathbf{C}_{opt}^k\|_2^2$. Substituting in $t = k'^{1-p} \geq C(k\varepsilon^{-p})^{1/(1-p)}$ for an appropriate constant C ensures $(\sum_{i=k'+1}^{n} |\theta_i|)^2 \leq \frac{\varepsilon^2}{25k} \|\mathbf{C}_{opt}^k\|_2^2$; Further, we have $|\theta_j| \geq \sum_{i=k'+1}^{n} |\theta_i|$, provided $j < k'^{1-p}$. This shows (1).

Now consider θ_j that satisfies the condition in the lemma. Although K' is unknown, we can be sure that, since $\mathcal{R}$ is k'-strongly separating, there is at least one set R_i such that $K' \cap R_i = \{j\}$, and more strongly, $K'' \cap R_i = \{j\}$, where K'' is the set of the k' largest coefficients. Consider the vector of measurements involving this set, $x = (\chi_{R_i} \oplus H) \Psi A$. When $H_{j,b} = 1$, $|x_b| \geq |\theta_j| - \sum_{l \neq j \in R_i} H_{l,b} |\theta_l|$ and $|x_0 - x_b| \leq \sum_{l \neq j \in R_i} (1 - H_{l,b}) |\theta_l|$. Since $\theta_j^2 > (\sum_{l=k'+1}^{n} |\theta_l|)^2$ we have $|\theta_j| > \sum_{l \neq j \in R_i} H_{l,b} |\theta_l| + (1 - H_{l,b}) |\theta_l|$. Hence $\min\{|x_b|, |x_0 - x_b|\} = |x_0 - x_b|$, and $\max\{|x_b|, |x_0 - x_b|\} = |x_b|$.

Thus $\frac{|x_b|-\min\{|x_b|,|x_0-x_b|\}}{\max\{|x_b|,|x_0-x_b|\}-\min\{|x_b|,|x_0-x_b|\}} = 1 = H_{j,b}$. Symmetrically, the results are reversed when $H_{j,b} = 0$, where $\frac{|x_b|-\min\{|x_b|,|x_0-x_b|\}}{\max\{|x_b|,|x_0-x_b|\}-\min\{|x_b|,|x_0-x_b|\}} = 0 = H_{j,b}$. Thus the decoded identifier $j_i = \sum_{b=1}^{\log n} 2^{b-1} H_{j,b} = j$ and so $j \in J$, showing (2).

For (3), observe that $|J| \leq k''$, since each $R_i \in \mathcal{R}$ generates at most one $j \in J$, and k'' is chosen as the number of sets forming the collection of k'-strongly separating sets. Hence, we can guarantee for each $j \in J$ there is at least one S_i such that $J \cap S_i = j$. We chose our k' to be sufficiently large that we can identify the $k'^{1/(1-p)} = O(k\varepsilon^{-p})^{1/1-p}$ largest coefficients. Since J contains the identities of the $(k\varepsilon^{-p})^{1/1-p}$ largest coefficients, we can choose the estimate of θ_j as any measurement of θ_j that avoids all other members of J. Thus, we can be sure that $|\hat{\theta}_j - \theta_j| = |\chi_{R_i}\Psi A - \theta_j| = |\sum_{l \in R_i, l \neq j} \theta_l| \leq \sum_{l=(k\varepsilon^{-p})^{1/1-p}+1}^{n} |\theta_l| \leq \frac{\varepsilon}{5\sqrt{k}}\|C_{\text{opt}}^k\|_2$.

Lemma 3. *Consider the p-compressible case with $p < \frac{1}{2}$. Let $k' = c'(k\varepsilon^{-p})^{1/(1-2p)}$ and $k'' = c''(k' \frac{\log n}{\log k'})^2$ for appropriately chosen c' and c''.*

Let K' denote the set of the k' largest coefficients.

1. $\forall 1 \leq j \leq n : \theta_j^2 \geq \frac{\varepsilon^2}{25k}\|C_{\text{opt}}^k\|_2^2 \Rightarrow j \in J$

2. $\forall j \in K' : \theta_j^2 > c_k k'^{2-2/p} \Rightarrow j \in J$, for appropriate scaling constant c_k.

3. $\forall j \in J : |\hat{\theta}_j - \theta_j| \leq \frac{\varepsilon}{5\sqrt{k}}\|C_{\text{opt}}^k\|_2$.

Proof. Consider $j \leq k'$. We know that $\mathcal{R}$ is k'-strongly separating, so there is some set R_i so that $K' \cap R_i = \{j\}$. From the vector of measurements involving this set, we know that the identity j will be recovered, as in the previous lemma, provided that j is the majority items in this set, i.e. if $|\theta_j| > \sum_{l \neq j \in R_i} |\theta_l|$. This can be at most $\sum_{l > k'} |\theta_l| \leq c_k k'^{1-1/p}$. Provided $|\theta_j| > c_k k'^{1-1/p}$, j will be found and so $j \in J$, showing (2). By our choice of k', $c_k k'^{2-2/p} \geq \frac{\varepsilon}{5\sqrt{k}}\|C_{\text{opt}}^k\|_2$, so (2) implies (1).

For (3), we consider the error in the estimation of θ_j. We have $|\hat{\theta}_j - \theta_j| \leq \sum_{l \notin J} |\theta_l|$, and from (2), we have that $l \notin J \Rightarrow \theta_j^2 \leq k'^{2-2/p} \vee l > k'$ (for $p \geq \frac{1}{2}$, this bound is not useful). Hence,

$$|\hat{\theta}_j - \theta_j| \leq \sum_{l < k', l \notin J} |\theta_l| + \sum_{l > k'} |\theta_l| \leq c_k(k'-1)k'^{1-1/p} + c_k k'^{1-1/p} \leq c_k k'^{2-1/p}.$$

By our choice of k' and c_k, we ensure that $c_k k'^{2-1/p} \leq \frac{\varepsilon}{5\sqrt{k}}\|C_{\text{opt}}^k\|_2$, as required.

Lemma 4 (Reconstruction accuracy). *Given $\hat{\theta}(A) = \{\hat{\theta}_i(A)\}$ such that $(\hat{\theta}_i - \theta_i)^2 \leq \frac{\varepsilon^2}{25k}\|C_{\text{opt}}^k\|_2^2$ if $\theta_i^2 \geq \frac{\varepsilon^2}{25k}\|C_{\text{opt}}^k\|_2^2$, picking the k largest coefficients from $\hat{\theta}(A)$ gives an error $\|R_{\text{opt}}^k - A\|_2^2 + \varepsilon\|C_{\text{opt}}^k\|_2^2$ k-term representation of A.*

Proof. As stated in the introduction, the error from picking the k largest coefficients exactly is $\|\theta(A) - \theta(R_{\text{opt}}^k)\|_2^2 = \sum_{i=k+1}^{n} \theta_i^2$ (where we index the θ_is in decreasing order of magnitude). We will write $\hat{\phi}_i$ for the ith largest approximate coefficient, and ϕ_i for its exact value. Let $\pi(i)$ denote the mapping such that $\hat{\phi}_i = \theta_{\pi(i)}$, and let $\sigma(i)$ denote a bijection satisfying $\sigma(i) = j \Rightarrow (i > k \wedge \pi(i) \leq k \wedge j \leq k \wedge \pi(j) > k)$.

Picking the k largest approximate coefficients has energy error

$$\|\mathbf{R} - \mathbf{A}\|_2^2 = \sum_{i=1}^{k}(\phi_i - \hat\phi_i)^2 + \sum_{i=k+1}^{n}\phi_i^2$$

$$= \sum_{i\le k}(\phi_i - \hat\phi_i)^2 + \sum_{i>k,\pi(i)>k}\phi_i^2 + \sum_{i>k,\pi(i)\le k}\phi_i^2$$

$$\le \sum_{i\le k}\frac{\varepsilon^2}{25k}\|\mathbf{C}_{\text{opt}}^k\|_2^2 + \sum_{i>k,\pi(i)>k}\phi_i^2 + \sum_{i>k,\pi(i)\le k}\phi_i^2$$

Consider i such that $i > k$ but $\pi(i) \le k$: this corresponds to a coefficient that belongs in the top-k but whose estimate leads us to not choose it. Then either $\phi_i^2 \le \frac{\varepsilon^2}{2k}\|\mathbf{C}_{\text{opt}}^k\|_2^2$, i.e. the top-$k$ coefficient is small compared to the optimal error, or else our estimate of $\phi_{\sigma(i)}^2$ was too high. In this case $\hat\phi_i^2 < \hat\phi_{\sigma(i)}^2$ but $\phi_{\sigma(i)}^2 \le \phi_i^2$. Assuming this, we can write

$$\phi_i^2 - \phi_{\sigma(i)}^2 = (\phi_i + \phi_{\sigma(i)})(\phi_i - \phi_{\sigma(i)})$$
$$= (|\phi_i| + |\phi_{\sigma(i)}|)(|\phi_i| - |\phi_{\sigma(i)}|)$$
$$= (2|\phi_{\sigma(i)}| + |\phi_i| - |\phi_{\sigma(i)}|)(|\phi_i + \hat\phi_i - \hat\phi_i| - |\phi_{\sigma(i)} + \hat\phi_{\sigma(i)} - \hat\phi_{\sigma(i)}|)$$
$$\le (2|\phi_{\sigma(i)}| + |\phi_i| - |\phi_{\sigma(i)}|)(|\phi_i - \hat\phi_i| + |\phi_{\sigma(i)} - \hat\phi_{\sigma(i)}| + |\hat\phi_i| - |\hat\phi_{\sigma(i)}|)$$
$$\le (2|\phi_{\sigma(i)}| + \frac{\varepsilon}{5\sqrt{k}}\|\mathbf{C}_{\text{opt}}^k\|_2)(\frac{2\varepsilon}{5\sqrt{k}}\|\mathbf{C}_{\text{opt}}^k\|_2)$$

In the case that $\phi_i^2 \le \frac{\varepsilon^2}{25k}\|\mathbf{C}_{\text{opt}}^k\|_2^2$ we can immediately write
$$\phi_i^2 - \phi_{\sigma(i)}^2 \le \phi_i^2 \le \frac{\varepsilon\|\mathbf{C}_{\text{opt}}^k\|_2}{5\sqrt{k}} \cdot \frac{\varepsilon\|\mathbf{C}_{\text{opt}}^k\|_2}{5\sqrt{k}} \le (2|\phi_{\sigma(i)}| + \frac{\varepsilon}{5\sqrt{k}}\|\mathbf{C}_{\text{opt}}^k\|_2)(\frac{2\varepsilon}{5\sqrt{k}}\|\mathbf{C}_{\text{opt}}^k\|_2)$$
Substituting this bound into the expression above, we use the facts that $\sum_{j=1}^{k}|a_j| \le \sqrt{k}(\sum_{j=1}^{k}a_j^2)^{1/2}$ and $\sum_{i>k,\pi(i)\le k}\phi_{\sigma_i}^2 = \sum_{j\le k,\pi(j)>k}\phi_j^2$, to bound $\|\mathbf{R} - \mathbf{A}\|_2^2$ by

$$\sum_{i\le k,\pi(i)\le k}\frac{\varepsilon}{25k}\|\mathbf{C}_{\text{opt}}^k\|_2^2 + \sum_{i>k,\pi(i)>k}\phi_i^2$$

$$+ \sum_{i>k,\pi(i)\le k}(\phi_{\sigma(i)}^2 + (2|\phi_{\sigma(i)}| + \frac{\varepsilon}{5\sqrt{k}}\|\mathbf{C}_{\text{opt}}^k\|_2)(\frac{2\varepsilon}{5\sqrt{k}}\|\mathbf{C}_{\text{opt}}^k\|_2))$$

$$\le \frac{\varepsilon}{25}\|\mathbf{C}_{\text{opt}}^k\|_2^2 + (2\sqrt{k} + \frac{\varepsilon\sqrt{k}}{5})\frac{2\varepsilon}{5\sqrt{k}}\|\mathbf{C}_{\text{opt}}^k\|_2^2 + \sum_{\pi(i)>k}\phi_i^2$$

$$\le \frac{23\varepsilon}{25}\|\mathbf{C}_{\text{opt}}^k\|_2^2 + \sum_{i>k}\theta_i^2 < \|\mathbf{R}_{\text{opt}}^k - \mathbf{A}\|_2^2 + \varepsilon\|\mathbf{C}_{\text{opt}}^k\|_2^2$$

Theorem 2. *We can construct a set of measurements for a signal $\mathbf{A}$ in time polynomial in k and n and return a $\mathbf{R}$ for $\mathbf{A}$ of at most k coefficients $\hat\theta$ under Ψ such that $\|\hat\theta - \theta\|_2^2 = \|\mathbf{R} - \mathbf{A}\|_2^2 < \|\mathbf{R}_{\text{opt}}^k - \mathbf{A}\|_2^2 + \varepsilon\|\mathbf{C}_{\text{opt}}^k\|_2^2$, and (i) if $p < \frac{1}{2}$, then the number of measurements is $O((k\varepsilon^p)^{4/(1-2p)}\log^4 n)$ and the time to produce the coefficients from the measurements is $O((k\varepsilon^p)^{6/(1-2p)}\log^6 n)$. (ii) if the p-compressible case is tight, then the number of measurements is $O((k\varepsilon^p)^{4/(1-p^2)}\log^4 n)$ and the time to find coefficients is $O((k\varepsilon^p)^{6/(1-p)^2}\log^6 n)$.*

Combining the above lemmas shows that the result of the algorithm has the desired accuracy. The reconstruction time can be broken down into the time to build J from the coefficients and the time to estimate the weight of each j in J. Building J takes time $O(k'' \log n)$, since it requires a linear pass over the results of the measurements. To choose the location to find estimates quickly, we can build a vector $y = T_2 \chi_J^T$ in time $O(|J|(k'' \log n)^2)$, by selecting and summing the necessary columns. Then for each $j \in J$, we find some i such that $y_i = T_{2j,i} = 1$ and return the measurement $(T_2 \Psi A)$ as $\hat{\theta}_j$. This takes $O((k'' \log n)^2)$ time per coefficient. Lastly, picking the k largest of the estimated coefficients can be done with a linear pass over them. The dominating cost is $O(|J|(k'' \log n)^2) = O((k'' \log n)^3)$

The number of measurements is polynomial in $k, \log n$ (recall that p is fixed independent of n and $\mathbf{A}$). We have not fully optimized the various polynomial factors, but still, our methods will not yield less than k^4 measurements, due to the use of the two collections of k-strongly separating sets. It is an open problem to further improve the number of measurements in explicit non-adaptive constructions. Note although we need to use p to define the measurements, we do not need the exact value of p. Rather, we need an upper bound on the true value of p (recall, the smaller the value of p, the faster the coefficients must reduce) — this is because our construction will simply take more coefficients than is necessary to get the required approximation accuracy.

3.2 Exponential Decay

As in the p-compressible case we state our results for the exponential decay case relative to the worst case error in the class for given α and C_α. In the case that $|\theta_i| \leq C_\alpha 2^{-\alpha i}$, we write $\|\mathbf{C}_{\text{opt}}^k\|_2^2 = \sum_{i=k+1}^n \theta_i^2$ as the worst case error over the class.

Measurements. The set of measurements we make is similar to the p-compressible case at the high level, but differs in the details. We set $k' = k + O(\frac{\log((k \log n)/\varepsilon)}{\alpha})$, and $k'' = O((k' \log n)^2)$ As before, we build S, a k'-strongly separating collection of sets, and write T_3 as the concatenation of χ_{S_i} for all $S_i \in S$ (k'' is chosen as the number of sets in the collection). However, we set Q to be a k''-separating collection of sets (not strongly separating), and write T_4 as its characteristic matrix. We form $T_\alpha = (T_3 \otimes H) \bigcup T_4$, and use $T_\alpha \Psi$ as the measurement matrix.

Reconstruction Algorithm. We recover a representation from the measurements from $T_3 \otimes H$ as before, to build a set J of identifiers. To make our estimates, we proceed iteratively to build $\hat{\theta}$, the vector of approximate coefficients. Initially $\hat{\theta} = 0$, and $M = \emptyset$. Let $j_1 \in (J \backslash M)$ satisfy $(J \backslash M) \cap Q_i = \{j_1\}$ (there will be at least one such Q_i and j_1). We set $\hat{\theta}_j = \chi_{Q_i}(\Psi A - \hat{\theta})$ and $M = M \cup \{j_1\}$. We now proceed to find a new $j_2 \in (J \backslash M)$ with $(J \backslash M) \cap Q_{i'} = \{j_2\}$ as the next coefficient to estimate, and proceed until $J = M$. We then return the k highest estimated coefficients as before.

Lemma 5. *Let K' denote the set of the k' largest coefficients.*

  1. $\forall 1 \leq j \leq n : \theta_j^2 > (\sum_{l=k'}^n |\theta_l|)^2 \Rightarrow j \in J$
  2. $\forall j \in K' : \theta_j^2 \geq \frac{\varepsilon^2}{25k} \|\mathbf{C}_{\text{opt}}^k\|_2 \Rightarrow j \in J$
  3. $\forall j \in J : |\hat{\theta}_j - \theta_j| \leq \frac{\varepsilon}{5\sqrt{k}} \|\mathbf{C}_{\text{opt}}^k\|_2.$

Proof. To show (1) and (2), we must bound the tail sums of coefficients of α-exponentially decaying signals. One can easily show that $\sum_{i=k+1}^{n} \theta_i^2 \leq c_\alpha 2^{-2\alpha k}$ and $(\sum_{i=k'+1}^{n} |\theta_i|)^2 \leq c'_\alpha 2^{-\alpha k'}$. Over the class of α-exponentially decaying signals, $(\sum_{i=k'}^{n} |\theta_i|)^2 \leq C_\alpha 2^{-\alpha(k'-k)} \|\mathbf{C}_{\text{opt}}^k\|_2^2$. Setting $k' = k + O(\frac{1}{\alpha} \log \frac{k}{\varepsilon})$ gives $(\sum_{i=k'}^{n} |\theta_i|)^2 \leq \frac{\varepsilon^2}{25k} \|\mathbf{C}_{\text{opt}}^k\|_2^2$. The remainder of the proof of (2) continues as in Lemma 2 (2), and (1) follows immediately as a consequence of the identification process.

To show (3), we scale ε by a factor of $O(kk'')$. Note that this does not affect the asymptotic sizes of k' or k''. This now ensures that the first coefficient j_1 is estimated with error $|\hat{\theta}_{j_1} - \theta_{j_1}| \leq k' \sum_{l=k'+1}^{n} |\theta_i| \leq \frac{\varepsilon}{k^{5/2}} \|\mathbf{C}_{\text{opt}}^k\|_2$. Now consider the estimation of the next coefficient j_2: it is possible that j_2 and j_1 occur in the same set $Q_{i'}$, in which case the error is bounded by $|\hat{\theta}_{j_2} - \theta_{j_2}| \leq |(\sum_{l \neq j_2, l \in Q_{i_2}} \theta_l) - \hat{\theta}_j| \leq \sum_{l \neq j_1, l \neq j_2, l \in Q_{i_2}} |\theta_l| + |\hat{\theta}_{j_1} - \theta_{j_1}| \leq \frac{2\varepsilon}{(k' \log n)^{5/2}} \|\mathbf{C}_{\text{opt}}^k\|_2$; else the error is bounded by $\frac{\varepsilon}{(k' \log n)^{5/2}} \|\mathbf{C}_{\text{opt}}^k\|_2$ as before. One can therefore show inductively that $|\hat{\theta}_{j_m} - \theta_{j_m}| \leq \frac{m\varepsilon}{5(k' \log n)^{5/2}} \|\mathbf{C}_{\text{opt}}^k\|_2$, and so, since $|J| \leq k'' = O((k' \log n)^2)$, we have $\forall j \in J.|\hat{\theta}_j - \theta_j| \leq \frac{\varepsilon}{5\sqrt{k}} \|\mathbf{C}_{\text{opt}}^k\|_2$, as required.

Theorem 3. *We can construct a set of $O(k^2 \text{ polylog}(n))$ measurements in time polynomial in k and n. For any α-exponentially decaying signal $\mathbf{A}$, from these measurements of $\mathbf{A}$, we can return a representation $\mathbf{R}$ for $\mathbf{A}$ of at most k coefficients $\hat{\theta}$ under Ψ such that $\|\hat{\theta} - \theta\|_2^2 = \|\mathbf{R} - \mathbf{A}\|_2^2 < \|\mathbf{R}_{\text{opt}}^k - \mathbf{A}\|_2^2 + \varepsilon \|\mathbf{C}_{\text{opt}}^k\|_2^2$. The time required to produce the coefficients from the measurements is $O(k^2 \text{ polylog}(n))$*

Proof. Using the results of Lemma 5 allows us to apply Lemma 4 and achieve the main theorem. For the time cost, we must first generate J, which takes time $O(k'' \log n)$, and then iteratively build the estimates. This can be done efficiently in time $O(k'' \text{ polylog}(n))$ per coordinate, a constant number of operations on each of the $O(k'' \text{ polylog}(n))$ measurements. For constant α and $\varepsilon = O(\text{poly}(1/n))$, we have $k' = O(k)$, $k'' = O((k \log n)^2)$ and the total number of measurements $= k'' \text{ polylog}(n) = O(k^2 \log^{O(1)} n)$.

k-**Support Case.** We note that the same approach can be used to give an explicit construction with $\tilde{O}(k^2)$ measurements for signals that have $\|\mathbf{R}_{\text{opt}}^k - \mathbf{A}\|_2 = 0$, i.e., there are at most k non-zero coefficients. This "k-support" case is a simplification of realistic signals, but has attracted interest in prior work (see [24] and references therein). The same approach outlined above, of using a combination of measurements based on k'-strongly separating sets and k''-separating sets, with an appropriate setting of $k' = \tilde{O}(k)$ and $k'' = \tilde{O}(k^2)$, is sufficient to recover the signal exactly.

4   Randomized Constructions

Here we focus on providing per-instance error estimates. For compressible signals (this section also works for arbitrary signals) one can give randomized constructions which guarantee to return a near-optimal representation for that signal, with high probability for each signal.

Transform Definition. Instead of using collections of sets with guaranteed separating properties, we make use of sets defined implicitly by hash functions to give a randomized separation property. We also use a random ± 1 valued vector to improve the accuracy of estimation of the coefficients. The necessary components are as follows:

Separation Matrix M. M is a $0/1$ $s \times n$ matrix with the property that for every column, exactly one entry is 1, and the rest are zero. We will define M based on a randomly chosen function $g : [n] \rightarrow [s]$, where $\Pr[g(i) = j] = 1/s$ for $i \in [n], j \in [s]$. Hence, $M_{i,j} = 1 \iff g(i) = j$, and zero otherwise. The effect is to separate out the contributions of the coefficients: we say i is separated from a set K if $\forall j \in K.g(i) \neq g(j)$. For our proofs, we require that the mapping g is only three-wise independent, and we set $s = O(\frac{k \log n}{\varepsilon^2})$. This will ensure sufficient probability that any i is separated from the largest coefficients.

Estimation Vector E. E is a ± 1 valued vector of dimension n so $\Pr[E_i = 1] = \Pr[E_i = -1] = \frac{1}{2}$. We will use the function $h : [n] \rightarrow \{-1, +1\}$ to refer to E, so that $E_i = h(i)$. For our proofs, we only require h to be four-wise independent.

Lastly, we compose T from M, Hamming matrix H and E by: $T = M \otimes H \otimes E$.

Reconstruction Procedure. We consider each set of inner-products generated by the row M_j. When composed with $(H \otimes E)$, this leads to $1 + \log_2 n$ inner products, $x_0 \dots x_{\log n} = (T\Psi A)_{j(1+\log n)} \cdots \theta'_{(j+1)(1+\log n)-1}$. From this, we attempt to recover a coefficient i by setting $i = \sum_{b=1}^{\log n} 2^{b-1} \frac{x_b^2 - \min\{x_b^2, (x_0 - x_b)^2\}}{\max\{x_b^2, (x_0 - x_b)^2\} - \min\{x_b^2, (x_0 - x_b)^2\}}$, and add i to our set of approximate coefficients, $\hat{\theta}$. We estimate $\hat{\theta}_i = h(i)x_0$, and output the k approximate coefficients obtaining the k largest values of $|\hat{\theta}_i|$.

Lemma 6 (Coefficient recovery). *(1) For every coefficient θ_i with $\theta_i^2 > \frac{\varepsilon^2}{25k}\|\mathbf{R}_{opt}^k - \mathbf{A}\|_2^2$, there is constant probability that the reconstruction procedure will return i (over the random choices of g and h).*
(2) We obtain an estimate of θ_i as $\hat{\theta}_i$ such that $(\theta_i - \hat{\theta}_i)^2 \leq \frac{\varepsilon^2}{25k}\|\mathbf{R}_{opt}^k - \mathbf{A}\|_2^2$ with constant probability.

Proof. The outline of the proof is as follows: for each coefficient θ_i with $\theta_i^2 > \frac{\varepsilon^2}{25k}\|\mathbf{R}_{opt}^k - \mathbf{A}\|_2^2$, we show that there is constant probability that it is correctly recovered. Let $x_b = (\Psi' A)_{g(i)(1+\log n)+b} = \sum_{g(j)=g(i)} H_{j,b}h(j)\theta_j$. One can show that
(i) $\mathsf{E}(x_b^2) \leq H_{i,b}\theta_i + O(\frac{\varepsilon^2}{k \log n})\|\mathbf{R}_{opt}^k - \mathbf{A}\|_2^2$ and
(ii) $\mathsf{Var}(x_b^2) \leq O(\frac{\varepsilon^2}{k \log n}\theta_i^2 H_{i,b}\|\mathbf{R}_{opt}^k - \mathbf{A}\|_2^2 + \frac{\varepsilon^4}{k^2 \log^2 n}\|\mathbf{R}_{opt}^k - \mathbf{A}\|_2^4)$.
Using the Chebyshev inequality on both x_b^2 and $(x_0 - x_b)^2$, and rearranging it can then be shown that $\Pr[\theta_b^2 - H_{i,b}(x_b^2) - (1 - H_{i,b})(x_0 - x_b)^2 \leq \frac{\theta_i^2}{2}] \leq \frac{2}{9 \log n}$ and $\Pr[(1 - H_{i,b})x_b^2 + H_{i,b}(x_0 - x_b)^2 \geq \frac{\theta_i^2}{2}] \leq \frac{2}{9 \log n}$. Combining these two results enables us to show that $\Pr[\frac{x_b^2 - \min\{x_b^2, (x_0 - x_b)^2\}}{\max\{x_b^2, (x_0 - x_b)^2\} - \min\{x_b^2, (x_0 - x_b)^2\}} \neq H_{i,b}] \leq \frac{4}{9 \log n}$. Thus, the probability that we recover i correctly is at least $\frac{5}{9}$.

For (2), we consider $\hat{\theta}_i = h(i)x_0 = h(i) \sum_{g(j)=g(i)} h(j)\theta_j$. One can easily verify that $\mathsf{E}(\hat{\theta}_i) = \theta_i$ and $\mathsf{Var}(\hat{\theta}_i) = \mathsf{E}(\sum_{g(j)=g(i), j \neq i} \theta_j^2)$. We argue that with constant probability none of the k largest coefficients collide with i under g, and so in expectation assuming this event $\mathsf{Var}(\hat{\theta}_i) = \frac{1}{s}\|\mathbf{R}^k_{\mathrm{opt}} - \mathbf{A}\|^2_2$. Applying the Chebyshev inequality to this, we show (2) with (better than) constant probability:

$$\Pr[|\hat{\theta}_i - \theta_i| > \sqrt{\tfrac{\varepsilon^2}{9k}}\|\mathbf{R}^k_{\mathrm{opt}} - \mathbf{A}\|_2] < \frac{\mathsf{Var}(\hat{\theta}_i)}{\frac{\varepsilon^2}{9k}\|\mathbf{R}^k_{\mathrm{opt}}-\mathbf{A}\|^2_2} \leq \frac{1}{9\log n}.$$

Lemma 7 (Failure probability). *By taking $O(\frac{ck\log^3 n}{\varepsilon^2})$ measurements we obtain a set of estimated coefficients $\hat{\theta}_i$ such that $(\theta_i - \hat{\theta}_i)^2 \leq \frac{\varepsilon^2}{25k}\|\mathbf{R}^k_{\mathrm{opt}} - \mathbf{A}\|^2_2$ with probability at least $1 - \frac{1}{n^c}$.*

Proof. In order to increase the probability of success from constant probability per coefficient to high probability over all coefficients, we will repeat the construction of T several times over using different randomly chosen functions g and h to generate the entries. We take $O(c \log n)$ repetitions: this guarantees that the probability of not returning any i with $\theta_i^2 > \frac{\varepsilon^2}{25k}\|\mathbf{R}^k_{\mathrm{opt}} - \mathbf{A}\|^2_2$ is n^{-c}, polynomially small. We also obtain $O(c \log n)$ estimates of θ_i from this procedure, one from each repetition of T. Each is within the desired bounds with constant probability at least $\frac{7}{8}$; taking the median of these estimates amplifies this to high probability using a Chernoff bounds. T has $m = s(\log n + 1) = O(\frac{k\log^2 n}{\varepsilon^2})$ rows, $O(c \log n)$ repetitions gives the stated bound. ∎

Theorem 4. *We can construct a dictionary $\Psi' = T\Psi$ of $O(\frac{ck\log^3 n}{\varepsilon^2})$ vectors, in time $O(cn^2 \log n)$. For any signal $\mathbf{A}$, given the measurements $\Psi'\mathbf{A}$, we can find a representation $\mathbf{R}$ of $\mathbf{A}$ under Ψ such that with probability at least $1 - \frac{1}{n^c}$ $\|\mathbf{R} - \mathbf{A}\|^2_2 \leq (1 + \varepsilon)\|\mathbf{R}^k_{\mathrm{opt}} - \mathbf{A}\|^2_2$. The reconstruction process takes time $O(\frac{c^2 k\log^3 n}{\varepsilon^2})$.*

The proof follows by combining the results of Lemma 6 with those of Lemma 4 to get the main result. We modify Lemma 4 to use $\|\mathbf{R}^k_{\mathrm{opt}} - \mathbf{A}\|_2$ in place of $\|\mathbf{C}^k_{\mathrm{opt}}\|_2$; the proof is essentially the same. It is easy to verify that the number of coefficients identified by the first part of the reconstruction process is $O(\frac{ck\log^2 n}{\varepsilon})$ (taking time linear in m). We find an accurate estimate of each recovered coefficient by taking the median of $O(c \log n)$ estimates of each one. If we spend linear time or more on reconstruction, we can work with fewer measurements:

Theorem 5. *We can construct a dictionary $\Psi' = T\Psi$ of $O(\frac{ck\log n}{\varepsilon^2})$ vectors, in time $O(cn^2 \log n)$. For any signal $\mathbf{A}$, given the measurements $\Psi'\mathbf{A}$, we can find a representation $\mathbf{R}$ of $\mathbf{A}$ under Ψ such that with probability at least $1 - \frac{1}{n^c}$ $\|\mathbf{R} - \mathbf{A}\|^2_2 \leq (1 + \varepsilon)\|\mathbf{R}^k_{\mathrm{opt}} - \mathbf{A}\|^2_2$. The reconstruction process takes time $O(cn \log n)$.*

The construction is similar to our main randomized result, but we do not use H and reduce s by a $\log n$ factor. Using only the separation and estimation matrices, we estimate *each* of the n coefficients, and take the k largest of them as before. By a similar argument to Lemma 6 (2), each coefficient is estimated with accuracy $\frac{\varepsilon^2}{25k}\|\mathbf{R}^k_{\mathrm{opt}} - \mathbf{A}\|^2_2$, and we can again apply Lemma 4.

Tolerance to Error. Several recent works have shown that compressed sensing-style techniques allow accurate reconstruction of the original signal even in the presence of error in the measurements (i.e. omission or distortion of certain $\theta_i's$). We adopt the same model of error as [3, 23][3] and show:

Lemma 8. *1. If a fraction $\rho = O(1)$ of the measurements are chosen at random to be corrupted in an arbitrary fashion, we can still recover a representation $\mathbf{R}$ with error $\|\mathbf{R} - \mathbf{A}\|_2^2 \leq (1 + \varepsilon)\|\mathbf{R}_{opt}^k - \mathbf{A}\|_2^2$ in time $O(cn \log n)$.*
2. If only a $\rho = O(\log^{-1} n)$ fraction of the measurements are corrupted we can recover a representation $\mathbf{R}$ with error $\|\mathbf{R} - \mathbf{A}\|_2^2 \leq (1 + \varepsilon)\|\mathbf{R}_{opt}^k - \mathbf{A}\|_2^2$ in time $O(\frac{kc^2 \log n}{\varepsilon^2})$.

Proof. 1. Consider the estimation of each coefficient in the process outlined in Theorem 5. Estimating θ_i takes the median of $O(\log n)$ estimates, each of which is accurate with constant probability. If the probability of an estimate being inaccurate or an error corrupting it is still constant, then the same Chernoff bounds argument guarantees accurate reconstruction. As long as ρ is less than a constant (say, $1/10$) then every coefficient is recovered with error $\varepsilon\|\mathbf{R}_{opt}^k - \mathbf{A}\|_2$, with high probability.

2. Consider the recovery of θ_i from T. We will be able to recover i provided the previous conditions hold, and additionally the some set of $\log n$ measurements of θ_i are not corrupted (we may still be able to recover i under corruption, but we pessimistically assume that this is not the case). Provided $\rho \leq 1/(3 \log n)$ then each set of $\log n$ measurements are uncorrupted with constant probability at least $2/3$ and with high probability i is recovered, and θ_i is estimated accurately (as in case (1)).

5 Concluding Remarks

We have presented a simple combinatorial approach of two sets of group tests with different separation properties that yields the first known polynomial time explicit construction of a non-adaptive transformation matrix and a reconstruction algorithm for the Compressed Sensing problem. The polynomial dependency is large, but we emphasize that no other construction with polynomial creation time is known, and the cost may be improved in future work. Our approach yields other results including sublinear reconstruction, improved approximation in error and others. Given the excitement about Compressed Sensing in the Applied Mathematics community, we expect many new results soon. The main open problem is to reduce the number of measurements used by explicit algorithms: our result here gives a cost polynomial in k, which is not close to the linear factor k in the existential results of [4, 9, 23]. For the case of k-sparse signals, (which have no more than k nonzero coefficients) Indyk recently developed a set of measurements, near linear in k in number (but has other superlogarithmic factors in n) [17]. Another outstanding question is to tease apart other properties of Compressed Sensing results—such as their ability to measure in one basis and reconstruct in another—and study their algorithmics.

Acknowledgments. We thank Ron Devore, Ingrid Daubechies, Anna Gilbert and Martin Strauss for explaining compressed sensing.

---

[3] These consider the exact recovery of a signal by taking $\Omega(n)$ measurements, and so do not compare to our result above of approximately recovering a signal using $o(n)$ measurements.

References

1. A. Akavia, S. Goldwasser, and S. Safra. Proving hard-core predicates by list decoding. In *FOCS*, pages 146–157, 2003.
2. E. Candès, J. Romberg, and T. Tao. Stable signals recovery from incomplete and inaccurate measurements. Unpublished Manuscript, 2005.
3. E. Candès, M. Rudelson, T. Tao, and R. Vershynin. Error correction via linear programming. In *FOCS*, 2005.
4. E. Candès and T. Tao. Near optimal signal recovery from random projections and universal encoding strategies. http://arxiv.org/abs/math.CA/0410542, 2004.
5. A. Clementi, A. Monti, and R. Silvestri. Selective families, superimposed codes, and broadcasting on unknown radio networks. In *SODA*, 2001.
6. G. Cormode and S. Muthukrishnan. What's hot and what's not: Tracking most frequent items dynamically. In *ACM PODS*, 2003.
7. G. Cormode and S. Muthukrishnan. Towards an algorithmic theory of compressed sensing. DIMACS Tech Report 2005-25, 2005.
8. R. Devore and G. G. Lorentz. *Constructive Approximation*, volume 303. Springer Grundlehren, 1993.
9. D. Donoho. Compressed sensing. Unpublished Manuscript, 2004.
10. D-Z Du and F.K. Hwang. *Combinatorial Group Testing and Its Applications*, volume 3 of *Series on Applied Mathematics*. World Scientific, 1993.
11. A. Gilbert, S. Guha, P. Indyk, Y. Kotidis, S. Muthukrishnan, and M. Strauss. Fast, small-space algorithms for approximate histogram maintenance. In *STOC*, 2002.
12. A. Gilbert, S. Guha, P. Indyk, S. Muthukrishnan, and M. Strauss. Near-optimal sparse Fourier representation via sampling. In *STOC*, 2002.
13. A. Gilbert, S. Muthukrishnan, and M. Strauss. Improved time bounds for near-optimal sparse Fourier representations. In *SPIE Conference on Wavelets*, 2005.
14. J. Haupt and R. Nowak. Signal reconstruction from noisy random projections. Unpublished Manuscript, 2005.
15. IEEE International Conference on Acoustics, Speech, and Signal Processing, 2005.
16. P. Indyk. Explicit constructions of selectors and related combinatorial structures, with applications. In *SODA*, 2002.
17. P. Indyk. Personal communication, 2005.
18. Integration of Sensing and Processing, Workshop at IMA, 2005.
19. W.H. Kautz and R.R. Singleton. Nonrandom binary superimposed codes. *IEEE Transactions on on Information Theory*, 10:363–377, 1964.
20. E. Kushilevitz and Y. Mansour. Learning decision trees using the fourier spectrum. *SIAM Journal on Computing*, 22(6):1331–1348, 1993.
21. Y. Mansour. Randomized interpoloation and approximation of sparse polynomials. *SIAM Journal of Computing*, 24(2), 1995.
22. Compressed sensing website. http://www.dsp.ece.rice.edu/CS/.
23. M. Rudelson and R. Vershynin. Geometric approach to error correcting codes and reconstruction of signals. Unpublished Manuscript, 2005.
24. J. Tropp and A. Gilbert. Signal recovery from partial information via orthogonal matching pursuit. Unpublished Manuscript, 2005.
25. Y. Tsaig and D. Donoho. Extensions of compressed sensing. Unpublished Manuscript, 2004.

On the Existence of Truthful Mechanisms for the Minimum-Cost Approximate Shortest-Paths Tree Problem*

Davide Bilò[1], Luciano Gualà[1], and Guido Proietti[1,2]

[1] Dipartimento di Informatica, Università di L'Aquila, Italy
[2] Istituto di Analisi dei Sistemi ed Informatica, CNR, Roma, Italy
{davide.bilo, guala, proietti}@di.univaq.it

Abstract. Let a communication network be modeled by a graph $G = (V, E)$ of n nodes and m edges, where with each edge is associated a pair of values, namely its *cost* and its *length*. Assume now that each edge is controlled by a selfish agent, which privately holds the cost of the edge. In this paper we analyze the problem of designing in this non-cooperative scenario a truthful mechanism for building a broadcasting tree aiming to balance costs and lengths. More precisely, given a root node $r \in V$ and a real value $\lambda \geq 1$, we want to find a *minimum* cost (as computed w.r.t. the edge costs) spanning tree of G rooted at r such that the *maximum* stretching factor on the distances from the root (as computed w.r.t. the edge lengths) is λ. We call such a tree the *Minimum-cost λ-Approximate Shortest-paths Tree* (λ-MAST).

First, we prove that, already for the unit length case, the λ-MAST problem is hard to approximate within better than a logarithmic factor, unless NP admits slightly superpolynomial time algorithms. After, assuming that the graph G is directed, we provide a $(1 + \varepsilon)(n - 1)$-approximate truthful mechanism for solving the problem, for any $\varepsilon > 0$. Finally, we analyze a variant of the problem in which the edge lengths coincide with the private costs, and we provide: (i) a constant lower bound (depending on λ) to the approximation ratio that can be achieved by any truthful mechanism; (ii) a $\left(1 + \frac{n-1}{\lambda}\right)$-approximate truthful mechanism.

Keywords: Algorithmic Mechanism Design, Bicriteria Network Design Problems, Broadcasting Tree, Truthful Mechanisms.

1 Introduction

Mechanisms are a classical concept of the theory of non-cooperative games [15]. In these games there are several independent agents that have to work together in order to optimize a global objective function. However, each agent has its

---

* Work partially supported by the Research Project GRID.IT, funded by the Italian Ministry of Education, University and Research, and by the European Union under IST FET Integrated Project 015964 AEOLUS and COST Action 293 GRAAL.

P. Flocchini and L. Gąsieniec (Eds.): SIROCCO 2006, LNCS 4056, pp. 295–309, 2006.

own valuation function and may lie in the hope of getting a higher profit. This leads to economically suboptimal resource allocation and is therefore undesirable. The main objective of *mechanism design* theory is to study how to incentive the agents in order to cooperate with the solving algorithm. A *mechanism* is a pair $\mathcal{M} = \langle g(\cdot), p(\cdot) \rangle$, where $g(\cdot)$ is an algorithm computing a solution, and $p(\cdot)$ specifies the payments provided to the agents. Informally, a mechanism is *truthful* if its payments guarantee that agents are not stimulated to lie.

Recently, there was a growing attention towards the question of designing a truthful mechanism by taking into account the computational complexity issues concerned with the underlying problem. This is exactly the topic of *algorithmic mechanism design* (AMD) for selfish agents. In their seminal paper concerned with AMD [14], Nisan and Ronen addressed the classic *shortest path* problem. This problem enjoys the property of being *utilitarian*. For utilitarian problems, there exists a well-known class of truthful mechanisms, i.e., the *Vickrey-Clarke-Groves (VCG) mechanisms* [20, 4, 6], and therefore the shortest path problem can be solved optimally. Another well-known class of truthful mechanisms is the class of *one-parameter mechanisms* [2]. Informally, a one-parameter mechanism applies to mechanism design problems where the information held by each agent can be expressed by a single parameter. Recently, in [11] the authors provided a general framework for designing truthful mechanisms for a subclass of one-parameter problems, called *binary demand games*, in which the agents' only available actions are to take part in the a game or not to.

By exploiting the results in [14, 2], in a sequel of papers efficient truthful mechanisms have been designed for solving several network design problems [1, 7, 8, 10, 14, 18]. In this paper we continue in this direction, by focusing on the following problem: Given a graph $G = (V, E)$ with n nodes and m edges, and with two functions $c(\cdot)$ and $l(\cdot)$ mapping edges to positive real numbers, called the *cost* and the *length* of an edge, respectively, we want to design a *broadcasting tree* network balancing total cost and distances from the source node. More precisely, given a root node $r \in V$ and a real value $\lambda \geq 1$, we want to find a *minimum* cost (w.r.t. to $c(\cdot)$) spanning tree of G rooted at r such that the *maximum* stretching factor on the distances (w.r.t. to $l(\cdot)$) from the root is λ. We call such a tree the *Minimum-cost λ-Approximate Shortest-paths Tree* (λ-MAST). We will address the problem under the assumption that each edge of G is controlled by a selfish agent, which privately holds the cost of the edge, while edge lengths are supposed to be public. This setting reflects a realistic scenario in which distances (i.e., transmission delays) between nodes are public, since induced by the everybody known physical layer of the network, while effective costs of transmission are unknown, since the various network components may be given in concession to private managers.

Leaving aside the non-cooperativeness aspect, our problem falls within the class of *bicriteria* network design problems [13]. In this class of problems, we are given a graph and two minimization objectives (under different edge-weight functions), with a budget specified on the first, and we have to find a subgraph – from a given set of feasible solutions – that minimizes the second objective subject to the budget on the first. In our specific case, the budget is defined by the parameter λ on the distances. To the best of our knowledge, despite of its apparent naturalness, the λ-MAST has been addressed in the past only when the two edge-weight functions coincide [12]. In the rest of the paper, we will refer to this special case as to the *single-weighted* case. With this restriction, in [12] the authors studied the more general (α, β)-*Light Approximate Shortest-path Tree* $((\alpha, \beta)$-LAST) problem, namely the problem of finding, given two values $\alpha, \beta \geq 1$ and a node $r \in V$, a spanning tree of G rooted at r such that: (i) for every vertex $v \in V$, the distance between r and v in T is at most α times the shortest distance from r to v in G, and (ii) the weight of T is at most β times the weight of a *minimum spanning tree* (MST) of G. For $\alpha > 1$ and $\beta \geq 1 + \frac{2}{\alpha - 1}$, this problem has been shown to be polynomial-time solvable. On the other hand, for $\alpha > 1$ and $1 \leq \beta < 1 + \frac{2}{\alpha - 1}$, the authors proved that deciding whether a given graph contains an (α, β)-LAST rooted at a given vertex is NP-complete. From these results, it immediately follows that: (i) the λ-MAST problem is NP-hard, and (ii) in its single-weighted formulation, it can be approximated within a factor of $(1 + \frac{2}{\lambda - 1})$, for any $\lambda > 1$.

In this paper, we first show that in our biweighted formulation, the λ-MAST problem is considerably hard. Indeed, we show that it has no polynomial time $(1 - o(1)) \ln \frac{n-2}{\lambda}$-approximate algorithm, unless $\mathsf{NP} \subseteq \mathsf{DTIME}\left((n/\lambda)^{\mathcal{O}(\log \log \frac{n}{\lambda})}\right)$. Afterwards, we turn our attention to the problem of designing a truthful mechanism for it, and we point out the existence of a polynomial time approximate one-parameter truthful mechanism, with a performance guarantee of $(1 + \varepsilon)(n - 1)$, for any $\varepsilon > 0$. Unfortunately, such mechanism applies to directed graphs only. Due to lack of space, we defer the detailed presentation of such mechanism to the full version of this paper. Finally, we also study the problem of designing a truthful mechanism for the single-weighted case, but also here we obtain rather negative results. Indeed, we prove a lower bound (depending on λ) to the approximation ratio that can be achieved by *any* (even exponential time) truthful mechanism. On a positive side, after showing that the algorithm in [12] cannot be used to design a truthful mechanism, we present a $\left(1 + \frac{n-1}{\lambda}\right)$-approximate truthful mechanism which can be computed in $\mathcal{O}(m + n \log n)$ time.

The paper is organized as follows: in Section 2 we give the definition of the problem and we recall some basic notions from the mechanism design theory. In Section 3 we present an inapproximability result for the problem, while in Section 4 we provide an approximate truthful mechanism for directed graphs. Finally, in Section 5 we analyze the existence of truthful mechanisms for the single-weighted version of the problem.

2 Preliminaries

2.1 Problem Definition and Notation

Let $G = (V, E)$ be a (either undirected or directed) graph, with n nodes and m edges, and with two different functions $c(\cdot)$ and $l(\cdot)$ mapping edges to positive real numbers. We will call $c(e)$ the *cost* of e, and $l(e)$ the *length* of e. A graph

$H = (V(H), E(H))$ is called a *subgraph* of G if $V(H) \subseteq V$ and $E(H) \subseteq E$. If $V(H) = V$ then H is called a *spanning subgraph* of G. A *simple path* P (or a *path* for short) from v_1 to v_k in G is a subgraph with $V(P) = \{v_1, \ldots, v_k \mid v_i \neq v_j$ for $i \neq j\}$ and $E(P) = \{e_i = (v_i, v_{i+1}) \mid 1 \leq i < k\}$, and it is denoted by $(v_1, \ldots, v_k)$. Given a source node r and a destination node s, a path in G from r to s is a *shortest path*, say $P_G(r, s)$, if the sum of its edge lengths (called *distance*, and denoted by $d_G(r, s)$) is minimum. We define the *total cost* of a spanning subgraph H of G as $c(H) = \sum_{e \in E(H)} c(e)$.

In this paper, we will focus on the following problem: Given a source node r, and a real value $\lambda \geq 1$, the λ-MAST problem asks for computing a cheapest (i.e., of minimum total cost) spanning tree T of G which satisfies the following distance constraint: for each node $v \in V$, the distance from r to v in T is at most λ times the distance in G between the same nodes, i.e., $d_T(r, v) \leq \lambda d_G(r, v)$.

2.2 Algorithmic Mechanism Design

Algorithmic mechanism design deals with algorithmic problems in a non-cooperative setting, in which part of the input is owned by selfish agents. As such agents may lie about their parts of input, they are capable of manipulating the algorithm. The main task of the mechanism design theory is the study of how to pay the agents in order to convince them to cooperate with the algorithm. We will deal with the case in which each agent controls a single link of a communication network. We provide a simplified formalization below, and we refer the interested reader to [14, 2].

For an edge e of G owned by a selfish agent a_e, we denote by t_e the private information held by a_e. We call t_e the (private) *type* of the agent a_e. Each agent has to declare a (public) *bid* b_e to the mechanism. We will denote by t the vector of private types, and by b the vector of bids.

For a given optimization problem defined on G, let $\mathcal{F}$ denote the corresponding set of feasible solutions. For each feasible solution $x \in \mathcal{F}$, some measure function $\mu(x, t)$ is defined, which depends on the true types. A *mechanism* is a pair $\mathcal{M} = \langle g(b), p(b) \rangle$, where $g(b)$ is an algorithm that, given agents' bids, computes a solution, and $p(b)$ is a scheme which describes the payments provided to the agents. A mechanism has a runtime of $\mathcal{O}(f(n))$ if $g(\cdot)$ and $p(\cdot)$ are computable in $\mathcal{O}(f(n))$ time. For each solution x, a_e incurs a cost $\nu_e(t_e, x)$ (sometimes called *valuation* of a_e w.r.t. x). The *utility* of an agent is defined as the difference between the payment provided by the mechanism and its cost w.r.t. the computed solution. Each agent tries to maximize its utility, while an *exact* mechanism aims to compute a solution which optimizes $\mu(x, t)$, but of course it does not know t directly. Similarly, if we denote by $\varepsilon(n)$ a positive real function of the input size n, an $\varepsilon(n)$-*approximate* mechanism returns a solution $g(b)$ whose measure comes within a factor $\varepsilon(n)$ from the optimum. In a *truthful* mechanism this tension between the agents and the system is resolved, since each agent maximizes its utility when it declares its type, regardless of what the other agents do. Moreover, a mechanism design problem is called *utilitarian* if its measure function satisfies $\mu(x, t) = \sum_{e \in E} \nu_e(t_e, x)$. For utilitarian problems, there exists a well-known class of truthful mechanisms, i.e., the *Vickrey-Clarke-Groves (VCG) mechanisms* [20, 4, 6].

Basically, VCG-mechanisms handle arbitrary valuation functions, but only utilitarian problems. In [2], Archer and Tardos have shown how to design truthful mechanisms for non-utilitarian problems under the assumption that the problem is *one-parameter*. A problem is said one-parameter if: (i) the type of each agent a_e can be expressed as single parameter $t_e \in \mathbb{R}$, and (ii) each agent's valuation has the form $\nu_e(t_e, x) = t_e w_e(b)$, where $w_e(b)$ is called *work curve* for agent a_e, i.e., the amount of work for a_e depending on the output specified by the mechanism algorithm, which in its turn is a function of the bid vector b. In [2], it is shown that for one-parameter problems, a sufficient condition for truthfulness is given by a particular monotonicity property of the mechanism algorithm. Let b be the vector of the bids of the agents, and let b_{-e} denote the vector of all bids besides $c(e)$; the pair $(b_{-e}, c(e))$ will denote the vector b. If we fix b_{-e}, a monotone algorithm defines a threshold value θ_e such that if a_e bids no more than θ_e, then e will be selected, while if a_e bids above θ_e, e will not be selected. Then, the following holds:

Theorem 1 ([2]). *A one-parameter mechanism $\mathcal{M} = \langle g(\cdot), p(\cdot) \rangle$ is truthful if and only if $g(\cdot)$ is monotone and the payment for each agent is defined as its threshold value if it owns a selected edge, and 0 otherwise.*

We will consider the case in which the type of each agent a_e represents the true cost incurred for forwarding a message through the link e, denoted by $\hat{c}(e) \in \mathbb{R}^+$. Under these assumptions, the non-cooperative λ-MAST problem can be handled through a one-parameter mechanism. Indeed, for each agent a_e, we can rewrite the valuation of a_e as $\nu_e(\hat{c}(e), g(b)) = \hat{c}(e) w_e(b)$, where $w_e(b)$ is equal to 1 if e belongs to the solution computed by the mechanism, 0 otherwise. Notice that this assumption on the work curve also implies that the λ-MAST problem is a binary optimization demand game [11].

3   Hardness of the λ-MAST Problem

In this section we prove an inapproximability result for the λ-MAST problem. This result is obtained by a reduction (preserving the approximation) from the SET COVER PROBLEM (SCP). An instance $I = \langle O, \mathcal{S} \rangle$ for the SCP consists of a set $O = \{o_1, \ldots, o_h\}$ of h objects, and a set $\mathcal{S} = \{S_1, \ldots, S_\ell\}$ of ℓ subsets of O. The objective is to find a minimum-size collection of subsets in $\mathcal{S}$ whose union is O. In [5] it is shown that SCP cannot be approximated within $(1 - o(1)) \ln h$, unless $\mathsf{NP} \subseteq \mathsf{DTIME}(h^{\mathcal{O}(\log \log h)})$. The same result holds even for the case $\ell \leq h$ [5]. The following holds:

Theorem 2. *Let $\lambda > 1$ be a real. Then the λ-MAST problem has no polynomial-time approximate algorithm with a performance guarantee better than $(1 - o(1)) \ln \frac{n-2}{\lambda}$, where $n \geq \lambda + 2$, unless $\mathsf{NP} \subseteq \mathsf{DTIME}\left((n/\lambda)^{\mathcal{O}(\log \log \frac{n}{\lambda})}\right)$, even for the unit length case.*

300   D. Bilò, L. Gualà, and G. Proietti

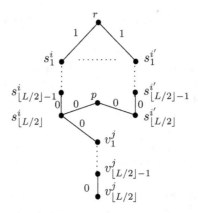

Fig. 1. The reduction of Theorem 2 when L is odd. Edges with cost greater than ℓ are omitted. Note that here o_j belongs to S_i, but not to $S_{i'}$.

Proof. Let $L = \lfloor \lambda \rfloor$ and let $I = \langle O, \mathcal{S} \rangle$ be an instance for the SCP with $\ell \leq h$, and h such that $L \leq h^k$ for some integer k. From I we build an instance $\mathcal{I} = \langle G, c, l, r, \lambda \rangle$ for the λ-MAST problem in the following way (see Figure 1). Graph G is complete and its node set is defined as follows:

- a node r, which is the source;
- $\lfloor L/2 \rfloor$ nodes $s_1^i, \ldots, s_{\lfloor L/2 \rfloor}^i$, for each $S_i \in \mathcal{S}$;
- $\lfloor L/2 \rfloor$ nodes $v_1^j, \ldots, v_{\lfloor L/2 \rfloor}^j$, for each $o_j \in O$;
- a node p only if L is odd.

The costs of the edges in G are defined as follows:

- $c(r, s_1^i) = 1$ for each $S_i \in \mathcal{S}$;
- $c(s_k^i, s_{k+1}^i) = 0$, $k = 1, \ldots, (\lfloor L/2 \rfloor - 1)$, for each $S_i \in \mathcal{S}$;
- $c(v_k^j, v_{k+1}^j) = 0$, $k = 1, \ldots, (\lfloor L/2 \rfloor - 1)$, for each $o_j \in O$;
- $c(s_{\lfloor L/2 \rfloor}^i, v_1^j) = 0$, for each S_i and for each o_j such that $o_j \in S_i$;
- if L is even then $c(s_{\lfloor L/2 \rfloor}^i, s_{\lfloor L/2 \rfloor}^j) = 0$, for each $S_i, S_j \in \mathcal{S}$, otherwise $(s_{\lfloor L/2 \rfloor}^i, p) = 0$, for each $S_i \in \mathcal{S}$;
- all other edges have a cost greater than ℓ.

Every edge has length 1 and the stretch factor is λ. Notice that, in any λ-MAST, the distance between any node v and r must be less or equal than L. We claim that G has a λ-MAST of cost k if and only if the original instance of the SCP has a solution of size k.

It is easy to see that a solution $\mathcal{C}$ for the SCP instance provides a solution for $\mathcal{I}$ with the same total cost. Indeed, the feasible solution is given by the spanning tree T defined as follows:

- begin with $H = (V(G), \emptyset)$;
- for each set $S_i \in \mathcal{C}$, add to H the path $(r, s_1^i, s_2^i, \ldots, s_{\lfloor L/2 \rfloor}^i)$, and all the paths $(s_{\lfloor L/2 \rfloor}^i, v_1^j, v_2^j, \ldots, v_{\lfloor L/2 \rfloor}^j)$, for each $o_j \in S_i$;

- for each set $S_j \notin \mathcal{C}$, add to H either the path $\left(s^i_{\lfloor L/2 \rfloor}, s^j_{\lfloor L/2 \rfloor},\right.$
  $\left. s^j_{\lfloor L/2 \rfloor - 1}, \ldots, s^j_1 \right)$ if L is even, or the path $\left(s^i_{\lfloor L/2 \rfloor}, p, s^j_{\lfloor L/2 \rfloor}, s^j_{\lfloor L/2 \rfloor - 1}, \ldots, s^j_1 \right)$
  if L is odd, where $S_i \in \mathcal{C}$.
- Let T be a *shortest-paths tree* (SPT) of H.

Now, let T be a solution for $\mathcal{I}$ of total cost $k \leq \ell$ (notice that such a solution always exists). It is not hard to see that for each $v^j_{\lfloor L/2 \rfloor}$, T must have a path $\left(r, s^i_1, s^i_2, \ldots, s^i_{\lfloor L/2 \rfloor}, v^j_1, v^j_2, \ldots, v^j_{\lfloor L/2 \rfloor} \right)$, for some $S_i \in \mathcal{S}$ such that $o_j \in S_i$, otherwise T has a total cost greater than ℓ. Hence, $\mathcal{C} = \{ S_i \in \mathcal{S} \mid (r, s^i_1) \in E(T) \}$ is a solution for I of size no more than the total cost of T. Since $\ell \leq h$, the number of nodes n is at most $2\lfloor L/2 \rfloor h + 2$. The claim now follows from the inapproximability result for the SCP proved in [5]. □

We conclude the section by noticing that Theorem 2 holds for directed graphs as well.

4 An Approximate Mechanism in Directed Graphs

In this section we claim the existence of an approximate one-parameter mechanism for the λ-MAST problem in directed graphs. Due to lack of space, we only provide the monotone algorithm representing the core of the mechanism, while we defer the formal proof of the approximation guarantee and the time complexity of the mechanism to the full version of the paper.

Our algorithm uses the monotone algorithm in [3] (denoted by $\mathcal{A}$) for the *constrained shortest path problem*. In such a problem one looks for a cheapest path from a source node r to a destination node s among paths of length L or less. This problem is (weakly) NP-hard, and it admits a pseudo-polynomial time algorithm, which is transformed by scaling techniques to a fully polynomial-time approximation scheme for the problem [9, 17]. Moreover, in [3] it is shown how to make the algorithm in [17] monotone.

The algorithm computes a set $\Pi = \{ \pi_v : v \in V \setminus \{r\} \}$ of "light" paths, meaning that π_v is a $(1+\xi)$-approximation of a cheapest path of length at most $\lambda\, d_G(r, v)$, for any $\xi > 0$. Then, it considers the subgraph H of G consisting of the union of all these paths. As H may not be an arborescence, then the algorithm needs to remove edges from H so that a feasible arborescence of G is returned. We point out that the removal steps (Lines 5–7) must be defined in order to guarantee both the monotonicity property and the feasibility of the solution. Indeed, for instance, computing a SPT of H yields to a non-monotone algorithm, since the composition of monotone algorithms is not necessarily monotone.

In the full version of the paper, we will show that Algorithm 1 can be used to prove the following:

Theorem 3. *Given any $\varepsilon > 0$, there exists a $(1 + \varepsilon)(n - 1)$-approximate truthful mechanism for the λ-MAST problem on directed graphs, running in* $\mathcal{O}\left(\frac{mn^4}{\sqrt{1+\varepsilon}-1} \cdot \log \frac{n}{\log(1+\varepsilon)} \cdot \log \frac{n}{\sqrt{1+\varepsilon}-1} \cdot \log \frac{n}{2-\sqrt{1+\varepsilon}} \right)$ *time.* □

Algorithm 1

Input: $G = (V, E, c, l)$, $r \in V$, $\lambda > 1$, $\xi > 0$.
Output: An arborescence T of G rooted at r.
1: **for** each $v \in V$ **do**
2:    find a $(1 + \xi)$-apx π_v of a cheapest path from r to v with $l(\pi_v) \leq \lambda d_G(r, v)$;
3: **end for**
4: Let T be the digraph made up of the union of all edges in π_v, $\forall v \in V$
5: **for** each $v \in V$ **do**
6:    remove all edges in T entering in v but that belonging to the shortest path in Π from r to v
7: **end for**
8: **return** T

---

5 The Non-cooperative Single-Weighted λ-MAST Problem

In this section we analyze the problem of designing a truthful mechanism for the single-weighted λ-MAST problem.

5.1 A Constant Lower Bound

We start by proving the following negative result:

Theorem 4. *Let $\lambda > 1$, and let $p > x > 0$. Then, it does not exist any truthful σ-approximate mechanism for the single-weighted λ-MAST problem, for any $\sigma < L(\lambda, p, x) = 1 + \min \left\{ \frac{x}{\lambda(x+p)+p-x}, \frac{p-x}{\lambda(p+x)} \right\}$.*

Proof. Let $0 < \varepsilon < x$. Suppose by contradiction that $\mathcal{M}$ is a σ-approximate mechanism for the single-weighted λ-MAST problem, with $\sigma < L(\lambda, p, x)$, and consider the instance of Figure 2, where all the edges on the path joining u and v have a small enough cost.

It is easy to see that the MST M of G consists of all the edges of G except e_2 and e_3. Note that M is not a feasible solution because the node v is unfeasible in M, i.e., $d_M(r, v) = \lambda(x + p) + (\lambda - 1)\varepsilon > \lambda(x + p) = \lambda d_G(r, v)$.

In fact, any feasible solution must contain at least one of the edges e_2 or e_3. Since $x > \varepsilon$, the optimal solution is given by the tree $T_{\neg e_1}$, obtained from M by adding the edge e_2 and by removing e_1. The following lemma shows that the mechanism must compute $T_{\neg e_1}$ as a solution.

Lemma 1. *The mechanism $\mathcal{M}$ returns $T_{\neg e_1}$ as solution.*

Proof. Let T be a feasible solution containing e_1, and suppose by contradiction that $\mathcal{M}$ computes T as solution. It is easy to see that $c(T) \geq \lambda(x + p) + p$. Then, the approximation ratio achieved by the mechanism is

$$\rho = \frac{c(T)}{c(T_{\neg e_1})} \geq \frac{\lambda(x + p) + p}{\lambda(x + p) + (\lambda - 1)\varepsilon + p - x},$$

which goes to $1 + \frac{x}{\lambda(x+p)+p-x}$ for ε that goes to 0. This means that by choosing ε small enough, we can make ρ arbitrary close to the value $1 + \frac{x}{\lambda(x+p)+p-x}$. Since $\sigma < L(\lambda) \leq 1 + \frac{x}{\lambda(x+p)+p-x}$, we obtain a contradiction (since $\mathcal{M}$ was supposed to be σ-approximate), and the lemma follows. □

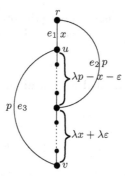

Fig. 2. The scheme of the instance of Theorem 4

Consider now the same instance in which the cost of e_1 becomes $x' = x + \varepsilon$. Since $g(\cdot)$ is monotone, the solution computed by the mechanism remains $T_{\neg e_1}$, while the optimal solution is changed. Indeed, M is now a feasible solution. Thus, the approximation ratio achieved by the algorithm is

$$\rho = \frac{c(T_{\neg e_1})}{c(M)} = \frac{\lambda(x + p) + (\lambda - 1)\varepsilon + p - x}{\lambda(x + p + \varepsilon)},$$

which goes to $1 + \frac{p-x}{\lambda(p+x)}$ for ε that goes to 0. Once again, we can choose ε small enough to make $\rho > \sigma$, which contradicts the σ-approximation assumption, and the claim follows.   □

By choosing x and p in order to maximize the function $L(\lambda, x, p)$, we obtain the following:

Corollary 1. *Let $\lambda \geq 1$. Then, it does not exist any truthful σ-approximate mechanism for the single-weighted λ-MAST problem, for any $\sigma < L(\lambda)$, where*

$$L(\lambda) = 1 + \frac{5\lambda - \sqrt{\lambda(9\lambda + 8)}}{\lambda(3\lambda - 4 + \sqrt{\lambda(9\lambda + 8)})}.$$ □

Since it can be shown that $L(\lambda)$ remains constantly below the $1 + \frac{2}{\lambda-1}$ threshold of approximability established in [12] for the single-weighted λ-MAST problem, we cannot conclude that the mechanism design version of the problem is harder than its corresponding optimization version. However, we stress the fact that our inapproximability result holds unconditionally, i.e., even for exponential time mechanisms. Besides, we want to point out an interesting peculiarity of the non-cooperative single-weighted λ-MAST problem: the set of feasible solutions depends on the agents' types. For this reason, even if in this problem the objective function is the sum of the agents' costs, as in utilitarian problems, it cannot be handled by VCG-mechanisms. Indeed, the algorithm of a VCG-mechanism is exactly the optimal algorithm of the corresponding optimization problem, but from Theorem 4 this algorithm cannot be monotone. It is also interesting to notice that this reasoning does not apply to the biweighted formulation of the λ-MAST problem, which in fact might be solved optimally by a (computationally unfeasible) VCG-mechanism.

5.2   An Approximate Truthful Mechanism

We start by pointing out the following:

Fact 1. *The algorithm given in [12] does not enjoy the monotonicity property.*

Proof. The high-level description of the algorithm in [12] is the following. The algorithm computes a MST M of G, then it performs a *depth-first search* of M by starting from the root r and in such a way, it fixes an ordering of the nodes. Then, at each step it maintains a current tree T. Initially, T is equal to M. The algorithm visits the nodes by following the (depth-first) fixed order. When a node v which is unfeasible in T is visited, the algorithm adds the shortest path $P_G(r,v)$ to T, and yields a new current tree by computing a SPT of the graph consisting of all the edges of M and all the edges of every added shortest path. Then the algorithm goes on to visit nodes by following the fixed ordering.

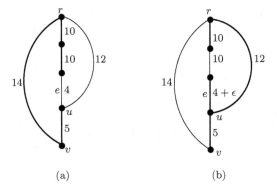

Fig. 3. A counterexample to the monotonicity of the algorithm in [12]. Bold edges are edges in the solution. In (a) it is shown the solution computed by the algorithm when e has cost 4, while in (b) e raises its cost of ε.

Consider now the instance in Figure 3. Let $\lambda = 2$, and let ε be a tiny positive value. Clearly, the (unique) MST is the line joining r and v. When the cost of the edge e is 4, the node u is feasible, and the algorithm does not select e (see Figure 3.a), while when e raises its cost to $4 + \varepsilon$, the node u becomes unfeasible and the algorithm adds the shortest path $P_G(r,u)$ (instead of $P_G(r,v)$), and e enters into the solution (see Figure 3.b). This contradicts the monotonicity. □

Thus, for an approximate mechanism, we need a different approach, as described in the following.

A monotone algorithm. The algorithm starts by computing a MST M_b and a SPT S_b of the graph, and then it looks at the set of *unfeasible nodes* in M_b, i.e., all the nodes $v \in V$ for which $d_{M_b}(r,v) > \lambda \, d_G(r,v)$. For each unfeasible node, say v, it adds to M_b the *winning edge* e_v^* *for* v of the path $P_{S_b}(r,v)$, i.e., the (unique) edge of $P_{S_b}(r,v)$ incident to v, in order to make v feasible. As M_b may be cyclic, then in Lines 6–8 edges are removed until there is no cycle left.

Algorithm 2

Input: $G = (V, E)$, $r \in V$, $\lambda \geq 1$, $b = (c(e_1), \ldots, c(e_m))$
Output: A spanning tree T
 1: Let S_b be a SPT of G, b, r
 2: Let M_b be a MST of G, b
 3: Let U be the set of unfeasible nodes in M_b
 4: $F = \{e_v^* \mid v \in U\}$
 5: Let $\mathcal{H}$ be the digraph obtained as follows
    – edges of M_b are directed from r towards the leaves
    – every winning edges e_v^* is directed towards v
 6: **while** $\exists \, v \in V$ with in-degree 2 in $\mathcal{H}$ **do**
 7:    delete from $\mathcal{H}$ all edges entering in v but the winning one e_v^*
 8: **end while**
 9: Let T be the undirected version of $\mathcal{H}$
10: **return** T

From now on, we will assume that edge costs are distinct.[1] It is easy to see that the computed solution satisfies the following property:

Lemma 2. *Algorithm 2 returns a $(1 + \frac{n-1}{\lambda})$-approximate solution for the single-weighted λ-MAST problem.*

Proof. First of all, we prove that Algorithm 2 returns a feasible solution. Hence, we must prove that T is a spanning tree satisfying the distance constraints. Since T has $n-1$ edges, then it suffices to prove that for each node v, $d_T(r,v) \leq \lambda \, d_{S_b}(r,v)$, which implies T to be connected. The proof is by induction on the number k of the edges on the path $P_{S_b}(r,v)$. The basic case is $k = 0$, which is trivial. Now assume that for every node $u \in V$ such that $P_{S_b}(r,u)$ is made up of $h < k$ edges, the claim is true. Let v be an unfeasible node in M_b such that $P_{S_b}(r,v)$ consists of k edges. Consider the winning edge $e_v^* = (u,v)$ for node v. From the fact that $P_{S_b}(r,u)$ has $k-1$ edges, by the inductive hypothesis u is feasible in T and thus

$$d_T(r,v) = d_T(r,u) + c(e_v^*) \leq \lambda \, d_{S_b}(r,u) + c(e_v^*) \leq \lambda \, (d_{S_b}(r,u) + c(e_v^*)) = \lambda \, d_{S_b}(r,v),$$

where the last inequality holds because $\lambda \geq 1$.

Now we show that Algorithm 2 returns a $(1 + \frac{n-1}{\lambda})$-approximate solution. Let $\mathcal{T}$ be an optimal solution, and let $v_1, v_2, \ldots, v_k$ be the unfeasible nodes in M_b. Then, the algorithm adds the winning edge $e_{v_i}^*$ whose cost is at most

---

[1] This is not a restrictive hypothesis, as we can suppose that each edge e has a unique index i_e, and we can assume that for every pair of distinct edges e, e', $c(e) \prec c(e')$ iff $c(e) < c(e')$, or $c(e) = c(e')$ and $i_e < i_{e'}$.

$d_{S_b}(r, v_i)$, for each $i = 1, \ldots, k$. Moreover, as v_i is unfeasible in M_b, we have that $\lambda d_{S_b}(r, v_i) < d_{M_b}(r, v_i)$. Summing over i, we can bound the cost of the added edges as follows:

$$\lambda \sum_{i=1}^{k} d_{S_b}(r, v_i) < \sum_{i=1}^{k} d_{M_b}(r, v_i) \leq k\, c(M_b) \leq (n-1)\, c(M_b),$$

from which we obtain that the cost of the added edges is less than $\frac{n-1}{\lambda} c(M_b) \leq \frac{n-1}{\lambda} c(T)$. □

Lemma 3. *Algorithm 2 is monotone.*

Proof. Let e be a non-selected edge in b. We have to show that e is still a non-selected edge in $b' = (c_{-e}, c'(e))$, for any $c'(e) \geq c(e)$. The proof breaks into the following cases:

Case 1: $e \in E(S_b) \setminus E(M_b)$. Let $e = (u, v)$, such that u is closer to r in S_b than v. Then, by construction, v must be feasible in M_b, otherwise e was the winning edge for v, and hence, by the simple observation that $F \subseteq E(T)$, it follows that e was in T. When the cost of e raises, the distance in $S_{b'}$ from r to v does not decrease, while the distance between the same nodes in $M_{b'}$ remains the same as $e \notin M_b$. Then, v will still be feasible, and the algorithm will still not select e.

Case 2: $e \in E(M_b) \setminus E(S_b)$. Notice that the distance between r and v in S_b and $S_{b'}$ is the same. We can assume that $e \in E(M_{b'})$, otherwise e will be clearly a non-selected edge. Let $e = (u, v)$, such that u is closer to r in M_b than v. It is easy to see that v must be unfeasible in M_b, otherwise e was in T. As v is unfeasible in $M_{b'}$, then the algorithm will still not select it.

Case 3: $e \in E(S_b) \cap E(M_b)$. We can assume that e appears in $\mathcal{H}$ in both directions, otherwise e was in the selected solution. Let $e = (u, v)$, such that u is closer to r in M_b than v, while v is closer to r in S_b than u. It is easy to see that v must be unfeasible in M_b, while u must be feasible in M_b. We have two subcases:

- $e \in E(M_{b'})$. Then, v remains unfeasible in $M_{b'}$, since $d_{M_{b'}}(r, v) \geq d_{M_b}(r, v)$, while the distance in $S_{b'}$ between r and v remains the same, since $P_{S_{b'}}(r, v)$ does not depend on e. Moreover, it is clear that u is still feasible in $M_{b'}$, hence e will still be not selected.
- $e \notin E(M_{b'})$. Once again, it is easy to see that u remains feasible in M'_b. Then, the algorithm will still not select e. □

Computing the payments. We now show how to efficiently compute the payments for the agents. Let T be the solution returned by the algorithm. Then, we have to compute the threshold value θ_e, for each selected edge $e \in E(T)$.

Assume b_{-e} is fixed. By $\theta_e(M, b_{-e})$ and by $\theta_e(S, b_{-e})$, we denote the threshold values for the edge e w.r.t. M_b and S_b, respectively. Note that such values are always defined, since any algorithm computing the MST or the SPT is monotone.

Henceforth, for the purpose of lightening the notation, whenever b_{-e} is clear from the context, we will simply write $\theta_e(M)$ and $\theta_e(S)$ instead of $\theta_e(M, b_{-e})$ and $\theta_e(S, b_{-e})$.

Let $e = (u, v) \in E(T)$ be e selected edge. We have three cases:

1. $e \in E(S_b) \setminus E(M_b)$. W.l.o.g. assume that u is closer to r than v in S_b. Since e is a selected edge, v must be unfeasible, i.e., $d_{M_b}(r, v) > \lambda d_{S_b}(r, v)$. Then e exits from the solution T if either e exits from the SPT, or e still remains in the SPT but v becomes feasible. This latter condition holds whenever e declares a value x_e which satisfies $d_{M_b}(r, v) \leq \lambda d_{S_b}(r, u) + \lambda x_e$. By solving this inequality for x_e and by taking the minimum, we obtain

$$\theta_e = \min \left\{ \theta_e(S), \frac{d_{M_b}(r, v)}{\lambda} - d_{S_b}(r, u) \right\}. \tag{1}$$

2. $e \in E(M_b) \setminus E(S_b)$. W.l.o.g. assume that u is closer to r than v in M_b. It is easy to see that v must be feasible, i.e., $d_{M_b}(r, v) \leq \lambda d_{S_b}(r, v)$. Then e exits from T if either e exits from the MST, or v becomes unfeasible, which happens whenever e declares a value x_e such that $d_{M_b}(r, u) + x_e > \lambda d_{S_b}(r, v)$. By taking the minimum value for x_e satisfying the latter inequality, we have that

$$\theta_e = \min \{\theta_e(M), \lambda d_{S_b}(r, v) - d_{M_b}(r, u)\}. \tag{2}$$

3. $e \in E(S_b) \cup E(M_b)$. We have two subcases:

   (a) u is closer to r than v in M_b but not in S_b. It is easy to see that v feasible implies u feasible, and u unfeasible implies v unfeasible.

   Let u and v be feasible. Since $d_{M_b}(r, u)$ does not depend on e, u remains feasible when a_e raises its bid. This implies that e exits from T if either e exits from the MST, or v becomes unfeasible. Hence, θ_e is defined as in (2).

   Now assume u, v are unfeasible. Since $d_{S_b}(r, v)$ does not depend on e, node v remains unfeasible when a_e raises its bid and e remains in the MST. This implies that e exits from T if either e exits from the SPT, or u becomes feasible. Hence, similarly to (1), θ_e is defined as follows:

   $$\theta_e = \min \left\{ \theta_e(S), \frac{d_{M_b}(r, u)}{\lambda} - d_{S_b}(r, v) \right\}.$$

   (b) u is closer to r than v both in M_b and S_b. Note that, since in $\mathcal{H}$ the unique edge entering in v is e, it is easy to see that

   $$\min\{\theta_e(M), \theta_e(S)\} \leq \theta_e \leq \max\{\theta_e(M), \theta_e(S)\}.$$

   We have three subcases:
   i. $\theta_e(M) = \theta_e(S)$. Clearly, $\theta_e = \theta_e(M) = \theta_e(S)$.
   ii. $\theta_e(M) < \theta_e(S)$. Let M_{-e} be the (unique) MST when a_e declares a value $x_e > \theta_e(M)$. The only case of interest is when $x_e \leq \theta_e(S)$. It is easy to see that v is feasible in M_{-e} iff $x_e \geq \frac{d_{M_{-e}}(r, v)}{\lambda} - d_{S_b}(r, u)$.

Then e exits from T if either v becomes feasible or e exits from S_b. We obtain:

$$\theta_e = \min\left\{\theta_e(S), \max\left\{\theta_e(M), \frac{d_{M_{-e}}(r,v)}{\lambda} - d_{S_b}(r,u)\right\}\right\}.$$

iii. $\theta_e(M) > \theta_e(S)$. Let S' be the SPT when a_e declares a value $x_e > \theta_e(S)$. Notice that $d_{S'}(r,v) = d_{S_b}(r,u) + \theta_e(S)$. The only case of interest is when $x_e \leq \theta_e(M)$. It is easy to see that v is unfeasible in M iff $x_e > \lambda\left(d_{S_b}(r,u) + \theta_e(S)\right) - d_M(r,u)$. Then e exits from T if either v becomes unfeasible or e exits from M_b. We obtain:

$$\theta_e = \min\left\{\theta_e(M), \max\left\{\theta_e(S), \lambda\left(d_{S_b}(r,u) + \theta_e(S)\right) - d_M(r,u)\right\}\right\}.$$

The mechanism. By using the results in [8, 16, 19], we can finally prove the following:

Theorem 5. *There exists a $\left(1 + \frac{n-1}{\lambda}\right)$-approximate truthful mechanism for the single-weighted λ-MAST problem running in $\mathcal{O}(m + n\log n)$ time.*

Proof. Truthfulness and approximation ratio follow from Lemma 3 and Lemma 2, respectively.

Concerning the time complexity, the algorithm first computes a SPT of G in $\mathcal{O}(m + n\log n)$ time, next it computes a MST of G in $\mathcal{O}(m\,\alpha(m,n)) = \mathcal{O}(m + n\log n)$ time [16], where $\alpha(m,n)$ is the classical inverse of Ackermann's function [19]. The other steps of the algorithm takes $\mathcal{O}(n + m)$ time. Concerning the payments, the mechanism first computes $\theta_e(M), \theta_e(S)$ for each selected edge e. This can be done in $\mathcal{O}(m\,\alpha(m,n))$ time [8]. Next, for each selected edge $e = (u,v) \in E(M_b)$, it computes $d_{M_{-e}}(r,v)$, where v is farther to r than u and M_{-e} is the (unique) MST computed in $G - e$. Once again, this can be accomplished in $\mathcal{O}(m\,\alpha(m,n))$ time by using the algorithm for the sensitive analysis [19]. Indeed, such algorithm computes the *swap edge* f_e for each edge $e \in E(M_b)$, where $f_e = (u',v')$ is the unique edge belonging to M_{-e} but not to M_b. It is not hard to prove that

$$d_{M_{-e}}(r,v) = d_{M_b}(r,u') + c(f_e) + \left(d_{M_b}(r,v') - d_{M_b}(r,v)\right),$$

where u' is closer to r than v' in M_{-e}. From this, the claim follows.   □

Acknowledgements. The authors would like to thank the anonymous referees for their very helpful comments.

References

1. C. Ambühl, A. Clementi, P. Penna, G. Rossi, and R. Silvestri, Energy consumption in radio networks: selfish agents and rewarding mechanisms, *Proc. 10th Int. Colloquium on Structural Information Complexity (SIROCCO'03)*, Proceedings in Informatics 17, Carleton Scientific, 1–16, 2003.

2. A. Archer and É. Tardos, Truthful mechanisms for one-parameter agents, *Proc. 42nd IEEE Symp. on Foundations of Computer Science (FOCS'01)*, 482–491, 2001.
3. P. Briest, P. Krysta, and B. Vöcking, Approximation techniques for utilitarian mechanism design, *Proc. 37th Ann. ACM Symp. on Theory of Computing (STOC'05)*, 39–48, 2005.
4. E. Clarke, Multipart pricing of public goods, *Public Choice*, 8:17–33, 1971.
5. U. Feige, A threshold of $\ln n$ for approximating set cover, *J. of the ACM*, 45(4): 634-652, 1998.
6. T. Groves, Incentives in teams, *Econometrica*, 41(4):617–631, 1973.
7. L. Gualà and G. Proietti, A truthful $(2\text{-}2/k)$-approximation mechanism for the Steiner tree problem with k terminals, *Proc. 11th Int. Computing and Combinatorics Conference (COCOON'05)*, Vol. 3595 of Lecture Notes in Computer Science, Springer-Verlag, 390–400, 2005.
8. L. Gualà and G. Proietti, Efficient truthful mechanisms for the single-source shortest paths tree problem, *Proc. 11th Int. Euro-Par Conf. (Euro-Par'05)*, Vol. 3648 of Lecture Notes in Computer Science, Springer-Verlag, 941–951, 2005.
9. R. Hassin, Approximation schemes for restricted shortest path problems, *Math. Oper. Res.*, 17(1):36–42, 1992.
10. J. Hershberger and S. Suri, Vickrey prices and shortest paths: what is an edge worth?, *Proc. 42nd IEEE Symp. on Foundations of Computer Science (FOCS'01)*, 252–259, 2001.
11. M.-Y. Kao, X.-Y. Li, and W. Wang, Towards truthful mechanisms for binary demand games: a general framework, *Proc. 6th ACM Conference on Electronic Commerce (EC'05)*, 213–222, 2005.
12. S. Khuller, B. Raghavachari, and N.E. Young, Balancing minimum spanning trees and shortest-path trees, *Algorithmica*, 14(4):305–321, 1995.
13. M.V. Marathe, R. Ravi, R. Sundaram, S.S. Ravi, D.J. Rosenkrantz, and H.B. Hunt III, Bicriteria network design problems, *J. Algorithms* 28(1):142–171, 1998.
14. N. Nisan and A. Ronen, Algorithmic mechanism design, *Games and Economic Behaviour*, 35:166–196, 2001.
15. M.J. Osborne and A. Rubinstein, *A Course in Game Theory*, MIT Press, 1994.
16. S. Pettie and V. Ramachandran, An optimal minimum spanning tree algorithm, *J. of the ACM*, 49(1):16–34, 2002.
17. C.A. Phillips, The network inhibition problem, *Proc. 25th Ann. ACM Symp. on Theory of Computing (STOC'93)*, 776–785, 1993.
18. G. Proietti and P. Widmayer, A truthful mechanism for the non-utilitarian minimum radius spanning tree problem, *Proc. 17th ACM Symp. on Parallelism in Algorithms and Architectures (SPAA'05)*, 195–202, 2005.
19. R.E. Tarjan, Sensitivity analysis of minimum spanning trees and shortest path problems, *Inform. Proc. Lett.*, 14:30–33, 1982.
20. W. Vickrey, Counterspeculation, auctions and competitive sealed tenders, *J. of Finance*, 16:8–37, 1961.

Dynamic Asymmetric Communication

Travis Gagie

Department of Computer Science
University of Toronto
Toronto, Canada
travis@cs.toronto.edu

Abstract. We present four new asymmetric communication protocols, with which a server with high bandwidth can help clients with low bandwidth send it messages. Three of our protocols are the first to use only a single round of communication for each message. Unlike previous authors, we do not assume the server knows the messages' distribution.

1 Introduction

Internet users usually download more than they upload, and many technologies have asymmetric bandwidth — greater from servers to clients than from clients to servers. Adler and Maggs [3] considered whether a server can use its greater bandwidth to help clients send it messages. They proved it can, assuming it knows the messages' distribution. We argue this assumption is often both unwarranted and, fortunately, unnecessary.

Suppose a number of clients want to send messages to a server. At any point, the server knows all the messages it has received so far; each client only knows its own messages and does not overhear communication between other clients and the server. Thus, the server may be able to construct a good code but the clients individually cannot. Adler and Maggs assumed the server, after receiving a sample of messages, can accurately estimate the distribution of *all* the messages. This assumption let them simplify the problem: Can the server help a *single* client send it a message drawn from a distribution known to the server? Given a representative sample of messages and a protocol for this simpler problem, the server can just repeat the protocol for each remaining message. In fact, it can even do this in parallel.

Adler and Maggs gave protocols for the simpler problem in which the server uses its knowledge to reduce the expected number of bits the client sends to roughly the entropy of the distribution. Their work has been improved and extended by several authors [17, 10, 13, 5], whose results are summarized in Table 1,[1] and used in the Infranet anti-censorship system [7, 8]. However, while

---

[1] Table 1 does not include a recent paper by Adler [1], in which he considered a harder version of the original problem with many clients: Can the server take advantage of correlations between messages? He showed it can, but used the even stronger assumption that the server knows the probability distribution over entire sequences of messages.

P. Flocchini and L. Gąsieniec (Eds.): SIROCCO 2006, LNCS 4056, pp. 310–318, 2006.

Table 1. Suppose a server tries to help a client send it one of N messages, chosen according to a distribution with entropy H that is known to the server but not the client. Adler and Maggs [3], Watkinson, Adler and Fich [17], Ghazizadeh, Ghodsi and Saberi [10] and Bose, Krizanc, Langerman and Morin [5] gave protocols for this problem whose expected-case upper bounds appear above; the last three protocols take a parameter $k \geq 1$. This table is based on one given by Bose *et al.* but, to be consistent with the data compression literature (particularly [4]), we use a different notation.

References	Bits sent by Server	Bits sent by Client	Rounds
[3, 13]	$3\lceil \log N \rceil$	$1.09H + 1$	$1.09H + 1$
[3]	$O(\log N)$	$O(H + 1)$	$O(1)$
[17]	$(H + 2)\lceil \log N \rceil$	$H + 2$	$H + 2$
[17]	$O(2^k H \log N)$	$H + 2$	$(H + 1)/k + 2$
[10]	$kH\lceil \log N \rceil + 1$	$H \log_{k-1} k + 1$	$H/\log k + 1$
[5]	$(k + 2)\lceil \log N \rceil$	$\frac{H \log(k+2)}{\log(k+2)-1} + \log(k + 2)$	$\frac{H}{\log(k+2)-1} + 1$

implementing Infranet, Wang [16] found the distribution of the messages (webpage requests) changed over time — the sample was unreliable.

We return to the original problem but without the assumption of a representative sample. We present and analyze four asymmetric communication protocols based on techniques from data compression for handling changing and unknown distributions. Our results are summarized in Table 2. To make it easier to compare our results and previous results, we show the average cost of each message; notice H is same in both tables, because the distribution of messages in a representative sample is the same as in the whole sequence. In Section 2 we give a dynamic version of Watkinson, Adler and Fich's Bit-Efficient Split protocol [17]. In Section 3, we present and analyze our TreeQuery and ListQuery protocols, which are the first to use only a single round of communication for each message. This is desirable because, as Adler, Demaine, Harvey and Pătraşcu [2] wrote:

> Any time savings obtained from reducing the number of bits sent by the client could easily be lost by the extra latency cost induced by multiple rounds in the protocol, particularly in long-distance networks, such as satellites, where communication has very high latency.

Finally, in Section 4 we show how Bentley, Sleator, Tarjan and Wei's Move-to-Front compression algorithm [4] can be turned into an elegant single-round asymmetric communication protocol, QueueQuery. Our protocols can be implemented so that each party's computation is proportional to the number of bits it sends and receives.

Consider an everyday example of asymmetric communication — placing a call on a cellular telephone. Because reading information from the phone's display is much faster than typing that information on its keypad, we can say the phone has greater bandwidth than the user. One of the ways the phone helps the user place calls faster is by storing a list of recently called numbers. This feature is common and frequently used, even though the phone may only store the 10 most recently called numbers, which demonstrates the advantage of dynamic

Table 2. Suppose a server tries to help clients send it m messages whose distribution — known to neither the server nor the clients — has entropy H; of N possible distinct messages, n occur. We present four protocols for this problem and prove upper bounds on the average cost of sending each message, which appear above; the last two protocols take a parameter $k > 1$.

Protocol	Bits sent by Server	Bits sent by Client	Rounds
DBES	$(H + O(1)) \log N$	$H + \frac{n \log N}{m} + O(1)$	$H + O(1)$
TreeQuery	$2N - 1$	$H + \frac{n \log N}{m} + O(1)$	1
ListQuery	$\lfloor N^{1/k} \rfloor \lceil \log N \rceil$	$kH + \frac{n \log N}{m} + O(1)$	1
QueueQuery	$\lfloor N^{1/k} \rfloor \lceil \log N \rceil$	$kH + \frac{n \log N}{m} + O(1)$	1

asymmetric communication protocols; it would be difficult or impossible to accurately estimate the called numbers' distribution from a preliminary sample because, for most users, that distribution changes over time. Let m be the number of calls made from the phone, n be the number of distinct phone numbers called, H be the entropy of the called numbers' distribution, and N be the number of phone numbers in the world. If the phone only stores the 10 most recently called numbers, then it displays about $10 \log_{10} N$ digits per call. To call the ith number in this list, the user types about $\log_{10} i$ digits and, to call a number not stored, he or she types about $\log_{10} N$ digits. Notice this takes a single round. We speculate this protocol is, in practice, so efficient that many users type far fewer than $H / \log_{10} 2 + \frac{n \log_{10} N}{m}$ digits on average per call. Unfortunately, in theory, it is almost useless — if the user calls 11 numbers in turn over and over, then $H / \log_{10} 2 = \log_{10} 11 \approx 1.04$ but he or she has to type every number in full.

The drawback of our single-round protocols is the large bound on the number of bits the server sends; this seems unavoidable if we want to prove good upper bounds. For the simpler problem with a single client and message and the distribution known to the server, Adler and Maggs showed that single-round protocols in which the client sends $O(H + 1)$ bits cannot have an $N^{o(1)}$ upper bound on the number of bits the server sends. Adler, Demaine, Harvey and Pǎtraşcu showed that protocols that use $o\left(\frac{\log \log N}{\log \log \log N}\right)$ rounds with high probability and in which the client sends $O(H + 1)$ bits cannot have a $2^{(\log N)^{1-\epsilon}}$ upper bound on the number of bits the server sends, for any $\epsilon > 0$.[2]

2   Dynamic Bit-Efficient Split

Let $S = s_1, \ldots, s_m$ be a sequence of messages some clients want to send a server. Let N be the number of possible distinct messages and let n be the number that occur in S. Let $H = \sum_{a \in S} \frac{\#_a(S)}{m} \log \frac{m}{\#_a(S)}$, where $a \in S$ means message a occurs

---

[2] This does not contradict the second row of Table 1; Adler and Maggs' second protocol uses $O(1)$ rounds in the expected case but not with high probability.

in S, $\#_a(S)$ is a's frequency in S and log means $\log_2$; i.e., H is the entropy of the messages' distribution (sometimes called the 0th-order empirical entropy of S).

We now present Watkinson, Adler and Fich's *Bit-Efficient Split* protocol [17]. For the moment, assume the server has a representative sample and uses it to build a leaf-oriented binary search tree T on the distinct messages in S; the jth leaf of T stores the jth lexicographically largest message $a \in S$, at depth at most $\left\lceil \log \frac{m}{\#_a(S)} \right\rceil + 1$; each internal node has exactly two children (i.e., T is *strictly* binary) and stores the lexicographically largest message in its left subtree. Gilbert and Moore's algorithm [11], for example, will build such a tree in linear time. For each message s_i, the server starts at the root of T and descends to the leaf v of T storing s_i, as follows. At each proper ancestor u of v, the server sends the active client the message a stored at u; if $a \geq s_i$, then the client responds with 0 and the server descends to u's left child; if $a > s_i$, then the client responds with 1 and the server descends to u's right child. By *active* we mean the client currently sending its message to the server. Without loss of generality, assume messages are $\lceil \log N \rceil$ bits long; otherwise, we use their indices. For $a \in S$, the server descends $\#_a(S)$ times to the leaf storing a. Thus, there are at most

$$\sum_{a \in S} \#_a(S) \left(\left\lceil \log \frac{m}{\#_a(S)} \right\rceil + 1 \right) < (H + 2)m$$

rounds. During each round, the server sends $\lceil \log N \rceil$ bits and the active client sends 1 bit.

In contrast, our *Dynamic Bit-Efficient Split* (DBES) protocol does not require the server to know the distribution beforehand. Next, we give a simple implementation of DBES; for each message, the server sends $(H + O(1)) \log N$ bits, on average, and performs $O(N)$ computations. It is possible to reduce the number of computations the server makes to $O((H+1) \log N)$ using a technique developed for dynamic alphabetic coding [9].[3]

For each message s_i, the server builds a leaf-oriented binary search tree T_i on all N possible distinct messages; the jth leaf of T_i stores the jth lexicographically largest possible message a, at depth at most $\left\lceil \log \frac{i}{\#_a(s_1,\ldots,s_{i-1})+1/N} \right\rceil + 1$. (Notice $\sum_a \#_a(s_1,\ldots,s_{i-1}) = i - 1$, so $\sum_a (\#_a(s_1,\ldots,s_{i-1}) + 1/N) = i$.).

The server starts at the root of T_i and descends to the leaf v storing s_i. If v is high in the tree, the server descends as in Bit-Efficient Split; however, if the server reaches an internal node at depth $\lceil \log i \rceil + 1$, then it knows the active client's message must be one it has not seen before. In the latter case, to cut short the protocol and save rounds, the server signals the client by sending the same message twice (notice it never does this otherwise); the client responds with s_i. Our analysis relies on the following technical lemma.

---

[3] Gilbert and Moore's algorithm builds T as a binary trie, with the path to the leaf storing a labeled by a prefix of the binary representation of $\sum_{a' < a} \frac{\#_{a'}(S)}{m} + \frac{\#_a(S)}{2m}$; we can use an augmented splay-tree [15] as a dynamic partial-sum data structure to implicitly represent T, and update it as frequencies change.

Lemma 1. $\displaystyle\sum_{i=1}^{m} \log \frac{i}{\max\left(\#_{s_i}(s_1,\ldots,s_{i-1}),1\right)} < (H+2)m.$

Proof. Shannon [14] showed that, once we have recorded the frequency of each distinct message in S, we can encode S in less than $(H+1)m$ bits. However, the frequencies tell us nothing about how the messages are ordered in S; since there are $m!/\prod_{a \in S} \#_a(S)!$ possible orderings,

$$(H+1)m > \log \frac{m!}{\prod_{a \in S} \#_a(S)!} = \sum_{i=1}^{m} \log i - \sum_{a \in S} \log(\#_a(S)!) .$$

Notice

$$\sum_{a \in S} \log(\#_a(S)!)$$

$$= \sum_{a \in S} \sum_{j=1}^{\#_a(S)-1} \log j + \sum_{a \in S} \log \#_a(S)$$

$$= \sum_{a \in S} \sum_{s_i = a} \log \max\left(\#_{s_i}(s_1,\ldots,s_{i-1}),1\right) + \sum_{a \in S} \log \#_a(S)$$

$$= \sum_{i=1}^{m} \log \max\left(\#_{s_i}(s_1,\ldots,s_{i-1}),1\right) + \sum_{a \in S} \log \#_a(S) ,$$

so

$$(H+1)m > \sum_{i=1}^{m} \log \frac{i}{\max\left(\#_{s_i}(s_1,\ldots,s_{i-1}),1\right)} - \sum_{a \in S} \log \#_a(S) .$$

Since $\sum_{a \in S} \log \#_a(S) \leq n \log \frac{m}{n} < m$, the claim follows. $\square$

Using Lemma 1, it is easy to bound the number of bits the server sends, the number the clients send and the number of rounds in DBES.

Theorem 1. *Suppose some clients send S to a server using DBES. On average, each message takes $H + O(1)$ rounds, during which the server sends $(H + O(1)) \log N$ bits and the active client sends $H + \frac{n \log N}{m} + O(1)$ bits.*

Proof. There is one round for each level the server descends in a tree. For each message s_i, the server descends at most

$$\min\left(\left\lceil \log \frac{i}{\#_{s_i}(s_1,\ldots,s_{i-1})} \right\rceil + 1, \lceil \log i \rceil + 1\right)$$

$$= \left\lceil \log \frac{i}{\max(\#_{s_i}(s_1,\ldots,s_{i-1}),1)} \right\rceil + 1$$

times so, by Lemma 1, there are a total of $(H+O(1))m$ rounds for all m messages. The server sends $\lceil \log N \rceil$ bits during each round. If s_i has occurred before, then the client sends 1 bit during each round; otherwise, it sends 1 bit during each round except the last, when it may send $\lceil \log N \rceil$ bits. $\square$

As an aside, we note DBES can easily be modified so the trees are only based on the distribution of recent messages — a data compression technique for increasing robustness. The server maintains a queue of messages; for each message s_i, it builds a tree based on the distribution of messages in the queue; after receiving s_i, it dequeues the oldest message and enqueues s_i. Knuth [12] discusses "sliding windows" such as this.

3   TreeQuery and ListQuery

Our next protocol, *TreeQuery*, is a simple modification of DBES. Instead of querying the active client repeatedly to find a path in a tree, the server encodes and sends the whole tree; the client finds the path and sends back all of what would have been its responses. To encode the tree, the server performs a preorder traversal, recording each internal node as a 1 and each leaf as a 0. Since Gilbert and Moore's algorithm is linear, both the server and the client perform $O(N)$ computations.

Theorem 2. *Suppose some clients send S to a server using TreeQuery. For each message, the server sends $2N - 1$ bits and, on average, the active client sends $H + \frac{n \log N}{m} + O(1)$ bits.*

Proof. For each message, the server encodes and sends a strict binary tree on N leaves, which takes $2N - 1$ bits; the number of bits the clients send is bounded as in Theorem 1.   □

For *ListQuery*, the server keeps a list of the possible messages, in non-increasing order by their frequency in the prefix of S it has received so far. If the server stores the list in a standard balanced binary-search tree implementation of a priority-queue, then updating it takes $O(\log N)$ time after each message.

For each message s_i, the server sends the active client the first $\lfloor N^{1/k} \rfloor$ messages in the list, where $k > 1$ is a parameter. The client makes a single pass through this sublist and, if s_i is rth in the sublist, responds with 1 followed by the codeword for r in Elias' delta code [6]; if s_i is not in the sublist, the client responds with 0 followed by s_i. The codeword for $r \geq 1$ in the delta code consists of a self-delimiting, $(2\lfloor \log(\log r + 1) \rfloor + 1)$-bit encoding of $\lfloor \log r \rfloor + 1$ followed by the $(\lfloor \log r \rfloor + 1)$-bit binary representation of r with the leading 1 removed.

Theorem 3. *Suppose some clients send S to a server using ListQuery with parameter $k > 1$. For each message, the server sends at most $\lfloor N^{1/k} \rfloor \lceil \log N \rceil$ bits and, on average, the active client sends $kH + \frac{n \log N}{m} + O(1)$ bits.*

Proof. Suppose the client's message s_i appears rth in the sublist it receives from the server; then $r \leq N^{1/k}$ and $\#_{s_i}(s_1, \ldots, s_{i-1}) \leq (i-1)/r$. Since $k > 1$, the length of the codeword for r in the delta code — i.e., $\lfloor \log r \rfloor + 2\lfloor \log(\log r + 1) \rfloor + 1$ — is bounded by

$$k \log r + O(1) \leq \min\left(k \log \frac{i-1}{\#_{s_i}(s_1, \ldots, s_{i-1})}, \log N \right) + O(1) .$$

Now suppose s_i does not appear in the sublist; then $\#_{s_i}(s_1, \ldots, s_{i-1}) \leq (i - 1)/N^{1/k}$, so $\log N \leq k \log \frac{i-1}{\#_{s_i}(s_1,\ldots,s_{i-1})}$. Again, the number of bits the client sends is bounded by

$$\min\left(k\log\frac{i-1}{\#_{s_i}(s_1,\ldots,s_{i-1})}, \ \log N\right) + O(1) \ .$$

Therefore, by Lemma 1 and straightforward calculation, the average number of bits a client sends is $kH + \frac{n\log N}{m} + O(1)$.   □

4   QueueQuery

ListQuery is reminiscent of Bentley, Sleator, Tarjan and Wei's Move-to-Front (MTF) compression algorithm [4]. To encode S, MTF keeps a list of the possible messages in increasing order by the time since their last occurrence; e.g., the most recent message is first in the list. For each message s_i, if s_i is the rth message in the list, then MTF records the codeword for r in the delta code and moves s_i to the front of the list. Bentley *et al.* proved MTF encodes S using $(H + o(H))m + n\log N$ bits. In fact, if the messages' distribution changes, MTF may use significantly fewer than H bits. Notice MTF's list is essentially a reversed queue; the tail is first, the head is last and when messages occur, they move to the tail.

Inspired by MTF and our cell phone example in the introduction, we modify ListQuery to obtain *QueueQuery*. The server keeps a queue of the $\lfloor N^{1/k} \rfloor$ most recent distinct messages, where $k > 1$ is a parameter. This is the only thing the server stores, so QueueQuery is more space-efficient than our previous protocols.

For each message s_i, the server sends the active client this queue. The client makes a single pass through the queue and, if s_i is rth from the tail, responds with 1 followed by the codeword for r in the delta code; if s_i is not in the queue, it responds with 0 followed by s_i. The server then puts s_i at the tail and, if that lengthens the queue, dequeues the message at the head.

Notice that, if s_i is in the queue, then apart from the leading 1 indicating s_i's presence the client sends as many bits as MTF would use to encode s_i. If s_i is not in the queue, then apart from the leading 0 indicating s_i's absence the client sends at most about k times as many bits as MTF would use; i.e., $\lceil \log N \rceil$ instead of at least $\lfloor \log N^{1/k} \rfloor + 2\lfloor \log(\log N^{1/k} + 1) \rfloor + 1$ bits. Thus, although our analysis below of QueueQuery is self-contained, it is naturally very similar to that of MTF.

Theorem 4. *Suppose some clients send S to a server using QueueQuery with parameter $k > 1$. For each message, the server sends $\lfloor N^{1/k} \rfloor \lceil \log N \rceil$ bits and, on average, the active client sends $kH + \frac{n\log N}{m} + O(1)$ bits.*

Proof. Suppose the client's message s_i is the first occurrence of that distinct message a in S; then it responds with $\log N + O(1)$ bits. Now suppose s_i is not the first occurrence of a and let s_h be the preceding occurrence. If the number of distinct messages in $s_h, \ldots, s_{i-1}$ is greater than $N^{1/k}$, then s_i is no longer in the queue and the client responds with $\log N + O(1)$ bits; since $i - h$ must be greater than $N^{1/k}$, this is bounded by $k \log(i - h) + O(1)$. Otherwise, s_i is still in the queue and the client responds with $\log(i - h) + 2 \log \log(i - h) + O(1)$ bits which, since $k > 1$, is also bounded by $k \log(i - h) + O(1)$.

Let $s_{i_1}, \ldots, s_{i_{\#_a(S)}}$ be the occurrences of a in S. The clients with these messages send a total of

$$\log N + \sum_{j=2}^{\#_a(S)} k \log(i_j - i_{j-1}) + O(\#_a(S))$$

$$\leq \log N + \#_a(S) k \log \frac{m}{\#_a(S)} + O(\#_a(S))$$

bits to communicate them to the server. Summing over the distinct messages in S, the clients send $kHm + n \log N + O(m)$ bits in total; the average number of bits a client sends is $kH + \frac{n \log N}{m} + O(1)$. □

Acknowledgments

Many thanks to Faith Ellen Fich, Giovanni Manzini and Charlie Rackoff, who supervised this research, and to the anonymous referees, for helpful comments.

References

1. M. Adler. Collecting correlated information from a sensor network. In *Proceedings of the 16th Symposium on Discrete Algorithms*, pages 479–488, 2005.
2. M. Adler, E.D. Demaine, N.J.A. Harvey, and M. Pătraşcu. Lower bounds for asymmetric communication complexity and distributed source coding. In *Proceedings of the 17th Symposium on Discrete Algorithms*, pages 251–260, 2006.
3. M. Adler and B.M. Maggs. Protocols for asymmetric communication channels. *Journal of Computer and System Sciences*, 64(4):573–596, 2001.
4. J.L. Bentley, D.D. Sleator, R.E. Tarjan, and V.K. Wei. A locally adaptive data compression scheme. *Communications of the ACM*, 29(4):320–330, 1986.
5. P. Bose, D. Krizanc, S. Langerman, and P. Morin. Asymmetric communication protocols via hotlink assignments. *Theory of Computing Systems*, 36(6):655–661, 2003.
6. P. Elias. Universal codeword sets and representations of the integers. *IEEE Transactions on Information Theory*, 21(2):194–203, 1975.
7. N. Feamster, M. Balazinska, G. Harfst, H. Balakrishnan, and D. Karger. Infranet: Circumventing web censorship and surveillance. In *Proceedings of the 11th USENIX Security Symposium*, pages 247–262, 2002.
8. N. Feamster, M. Balazinska, W. Wang, H. Balakrishnan, and D. Karger. Thwarting web censorship with untrusted messenger discovery. In *Proceedings of the 3rd International Workshop on Privacy Enhancing Technologies*, pages 125–140, 2003.
9. T. Gagie. Dynamic Shannon coding. In *Proceedings of the 12th European Symposium on Algorithms*, pages 359–370, 2004.

10. S. Ghazizadeh, M. Ghodsi, and A. Saberi. A new protocol for asymmetric communication channels: Reaching the lower bounds. *Scientia Iranica*, 8(4):297–302, 2001.
11. E.N. Gilbert and E. Moore. Variable-length binary encodings. *Bell System Technical Journal*, 38:933–968, 1959.
12. D.E. Knuth. Dynamic Huffman coding. *Journal of Algorithms*, 6(2):163–180, 1985.
13. E.S. Laber and L.G. Holanda. Improved bounds for asymmetric communication protocols. *Information Processing Letters*, 83(4):205–209, 2002.
14. C.E. Shannon. A mathematical theory of communication. *Bell System Technical Journal*, 27:379–423, 623–656, 1948.
15. D.D. Sleator and R.E. Tarjan. Self-adjusting binary search trees. *Journal of the ACM*, 32:652–686, 1985.
16. W. Wang. Implementation and security analysis of the Infranet anti-censorship system. Master's thesis, Massachusetts Institute of Technology, 2003.
17. J. Watkinson, M. Adler, and F. Fich. New protocols for asymmetric communication channels. In *Proceedings of the 8th International Colloquium on Structural Information and Communication Complexity*, pages 337–350, 2001.

Approximate Top-k Queries in Sensor Networks[*]

(Extended Abstract)

Boaz Patt-Shamir and Allon Shafrir

Dept. of Electrical Engineering
Tel Aviv University
Tel Aviv 69978, Israel
boaz@eng.tau.ac.il, shafrir@eng.tau.ac.il

Abstract. We consider a distributed system where each node has a local count for each item (similar to elections where nodes are ballot boxes
and items are candidates). A top-k query in such a system asks which
are the k items whose sum of counts, across all nodes in the system, is
the largest. In this paper we present a Monte-Carlo algorithm that outputs, with high probability, a set of k candidates which approximates the
top-k items. The algorithm is motivated by sensor networks in that it focuses on reducing the individual communication complexity. In contrast
to previous algorithms, the communication complexity depends only on
the global scores and not on the partition of scores among nodes. If the
number of nodes is large, our algorithm dramatically reduces the communication complexity when compared with deterministic algorithms.
We show that the complexity of our algorithm is close to a lower bound
on the cell-probe complexity of any non-interactive top-k approximation
algorithm. We show that for some natural global distributions (such as
the Geometric or Zipf distributions), our algorithm needs only polylogarithmic number of communication bits per node.

1 Introduction

Possibly one of the clearest examples of the difference between "global" and "local" can be seen in elections: each ballot box has a local score for each candidate,
but the result we care about is the global scores, i.e., how many votes does each
candidate have overall. A top-k query in this case is "Which are the k globally
most popular candidates?". Other examples for the top-k task abound: in peer-
to-peer file-sharing networks (such as Gnutella), users may wish to find which
are today's most popular downloads; in sensor networks, a sensor may count the
number of occurrences of different species of birds, and a user might be interested in the most frequent species observed over the whole instrumented area; in
a server farm with several gateways, denial-of-service (DoS) attacks are a major
concern. The first question to be answered in this case is which are the most
frequent sources of requests; and many others.

---

[*] This research was supported in part by Israel Ministry of Science and Technology
contract 3-941.

In general, a top-k query returns the k items having largest global score in a distributed system, where each item has a set of local scores. The global score of an item is just the sum of its local scores, over all locations. The main difficulty is that the scores may be divided arbitrarily among the different locations. In elections, for example, it may be the case that the most popular global candidate has the lowest (positive) count in each ballot box.

When computing top-k queries in a distributed system, a key question is how to minimize the *communication complexity* required to provide an answer. This issue is particularly important in sensor networks, where the communication subsystem is by far the largest energy consumer at the nodes. An algorithm which allows us to trade communication for local computation may have a decisive effect on the longevity of node batteries and hence on the usability of the system (see, e.g. [14]). This observation has established the measure of *individual* communication complexity as a key performance criterion in sensor networks [14, 16, 24, 7, 20]. In this work, we adopt this measure to evaluate the complexity of top-k computation. Observe that deterministic algorithms are sensitive to the way scores are partitioned among the different nodes and for some partitions they may communicate all scores to a single node.

Our Results. In this paper we propose a simple and effective way to overcome the problem of adversarial partition of the scores among the nodes. Our algorithm is Monte Carlo (it may err with some arbitrarily small probability), and its results are only approximate: using very little communication, the algorithm can tell, roughly, which items are in the top-k set. We focus on the worst-case individual communication complexity, i.e., our goal is to minimize the maximal number of bits communicated (sent or received) by any single node.

Our basic tool is random sampling. Done in the right way, sampling strips away the difficulties due to geographical distribution of scores which are the main difficulties for deterministic and Las Vegas algorithms. The basic idea is compounded with techniques adapting it to the specific input at hand. The performance of the algorithm depends on how popular are the top-k items, and on how "flat" is the distribution of scores. Specifically, suppose that the global scores adhere to the Zipf distribution with parameter $a > 1$ (namely the relative popularity of the ith popular item is proportional to i^{-a}). Then our Algorithm R guarantees that the communication complexity is bounded by $O(\frac{k}{\varepsilon^2})$ times a polylogarithmic factor, where ε is the required approximation accuracy. This case is quite important, as it is widely believed that the statistics of many phenomena are well approximated by the Zipf distribution (see, e.g., [3, 12]).

We note that the communication complexity of our algorithms scales very well compared with previous algorithms [11, 6]. Our simulations demonstrate that the performance of our algorithm is significantly superior to the best previously known algorithms.

We give some evidence showing that our algorithm is close to optimal. In particular, we demonstrate the optimality of its *cell-probe* complexity [23, 13] among a limited class of single-round Monte Carlo algorithms.

Previous Work. In [11], Fagin, Lotem and Naor introduced the Threshold Algorithm (TA) in the context of databases. They define a notion of 'instance-optimality,' and prove that TA incurs at most n accesses times the optimum, where n is the number of nodes in the system (they also show that an $\Omega(n)$ factor blowup is unavoidable for any deterministic or Las-Vegas algorithm). In [6], Cao and Wang propose the TPUT algorithm to reduce latency and save communication for the case where the *local* inputs are generated by Zipf-like distributions. As expected from deterministic algorithms, the performance of TPUT and TA depend crucially on the partition of scores to nodes. Other related work include variations of TA and TPUT optimized for certain network models [17, 25, 5].

From the sensor networks perspective, top-k queries are viewed as a special case of aggregate queries (see, e.g., [16, 24]). Typically, it is assumed that data is routed on a spanning tree, and each node does some aggregation en-route. Simple aggregates (such as counting the number of items, summing numbers etc.) can be done with $O(\log n)$ bits per node. Considine et al. [7], and Gibbons et al. [18], present methodologies for robust approximation of aggregates in sensor networks. In [18] they also present a sketch of a top-k-approximation algorithm that appears promising, but the algorithm is not fully specified, and no formal statement or analysis is given.

Techniques for efficient monitoring of aggregates in sensor networks are studied by [21, 4, 8, 2]. The main question in these works is how to efficiently update the results under some assumptions on the way the input changes.

Paper Organization. The remainder of the paper is organized as follows. Section 2 describes our model, problem definition and a few known results about efficient counting. Section 3 presents our algorithms with formal analysis results. Simulation results are presented in Sect. 4. In Sect. 5, we present our lower bound, and we conclude in Sect. 6. Due to lack of space, proofs are omitted from this extended abstract.

2   Model and Preliminaries

System Model. The system is modeled as a communication graph $G(V, E)$ with $n \triangleq |V|$, where each node models a classical RAM machine with access to its local input and to an infinite tape of random bits. A distinguished node $v^* \in V$ is the *root node* and is assumed to have a special write-once output register.

The system executes distributed algorithms according to the standard asynchronous message passing model (see, e.g., [1, 15]). Very briefly, in this model an event (such as arrival of a message to node u) triggers a state-transition (e.g., u computes a response message and inserts it to the link buffer). An execution in our model is considered terminated when the root-node has written the result to the output register.

Let M denote some finite string of bits. We assume that the system contains a message passing infrastructure supporting the following facilities:

- Each node $u \in V$ may send a message M to any other node $v \in V$. This causes each node along a path from u to v to send and receive $\Theta(|M|)$ bits.

- Each node $u \in V$ can broadcast a message M to all other nodes. This causes every node in V to send and receive $\Theta(|M|)$ bits.

While the particular way in which these actions are implemented is immaterial for our purposes, we note that these assumptions can be justified by the existence of a spanning-tree of constant degree for message passing (see, e.g., [14, 16, 24, 7, 20]).

Input Model and Problem Statement. Let $\mathcal{I}$ denote the set of possible *items*. We assume $\mathcal{I}$ is finite. An instance of the problem, denoted by X, is a vector of multisets of $\mathcal{I}$: one multiset, denoted X_v, for each node $v \in V$. We sometimes slightly abuse notation and use X to also denote the multiset $\bigcup_{v \in V} X_v$.

It is convenient to imagine each multiset as a set of *cells*, where each cell contains a single item, so that an item with multiplicity w has w replicas, one in a cell. The *weight* of item i in node v, denoted $w_v(i)$, is the multiplicity of i in X_v. The *weight of a node* v is the total number of cells in v, formally $W_v \triangleq |X_v|$. The *weight of an item* $i \in \mathcal{I}$, is the sum of its multiplicities over all nodes, i.e., $w(i) \triangleq \sum_{v \in V} w_v(i)$. The input size is defined to be the total number of cells, $W(X) \triangleq \sum_{v \in V} |X_v|$. The *empirical probability*, or *frequency*, of an item $i \in \mathcal{I}$ in X, is defined by $p_X(i) \triangleq \frac{w(i)}{W}$. When the context is clear we omit the subscript.

Using this notation, we define the top-k set of X as follows.

Definition 1. *Let k be a natural number, and suppose that $|\mathcal{I}| \geq k$. The top-k set of X, denoted $\mathrm{top}(k, X)$, is a subset of $\mathcal{I}$ of size k containing the items with the maximal weights.*

Following [11], we extend Definition 1 to the concept of approximate top-k sets.

Definition 2. *Let $\varepsilon \geq 0$. An ε-approximation of the top-k set of X is a set top_ε of k items, such that for all items $i \in \mathrm{top}_\varepsilon$ and $j \notin \mathrm{top}_\varepsilon$, we have $p_X(j) \leq (1 + \varepsilon)p_X(i)$.*

We will mainly be interested in small values of ε so without loss of generality, we assume henceforth that $\varepsilon \leq 1$.

It turns out that the following quantity has a central role in the complexity of computing top-k (and approximate top-k) queries. For a given input, the *critical frequency* of the instance, denoted $p^*(X, k)$, is the empirical probability of the least popular item in $\mathrm{top}(k, X)$, i.e., given input X and a natural number k, we define $p^*(X, k) \triangleq \min\{p_X(i) \mid i \in \mathrm{top}(k, X)\}$.

Throughout the paper, we assume instances having n nodes, total weight W, and critical frequency p^*. We denote the set of all such instances by $\mathcal{X}(W, n, p^*)$.

Complexity Measures. We evaluate the performance of certain algorithms using a worst-case measure per node. Specifically, the communication complexity of an algorithm is the maximum, over all inputs and over all nodes, of the total number of bits transmitted and received by a node throughout the execution of the algorithm. Formally, $c_A(X, v)$ denotes the total number of bits transmitted and received by node v, throughout the execution of algorithm A, for the input

X; $c_A(X)$ denotes the maximal node-communication of algorithm A on input X, i.e., $c_A(X) \triangleq \max\{c_A(X,v) \mid v \in V\}$. Finally, given a collection $\mathcal{X}$ of possible inputs, $C_A(\mathcal{X})$ denotes the worst-case communication complexity of algorithm A over all inputs in $\mathcal{X}$, i.e., $C_A(\mathcal{X}) \triangleq \max\{c_A(X) \mid X \in \mathcal{X}\}$.

Note that our communication complexity measure is individual in the sense that we measure the maximal number of bits communicated by any single node. The motivation for such a measure is that in sensor networks, each node has an individual energy source, and the longevity of the system often depends on the longevity of the weakest sensors (see, e.g., [14]). Furthermore, assuming a spanning tree of bounded degree, we can disregard many aspects of wireless communication and focus on the net communication used by the algorithm.

Loglog Counting. Let us present a known result which we use. First we define the following concept.

Definition 3. *Let Z be a positive number we wish to estimate, let $\varepsilon \geq 0$ and $\sigma \geq 0$ be real numbers. A random variable $\hat{Z}$ is a (ε, σ^2)-estimate of Z if $\frac{1}{Z}|E[\hat{Z}] - Z| \leq \varepsilon$, and $\frac{1}{Z^2}Var[\hat{Z}] \leq \sigma^2$.*

Durand and Flajolet [10] prove a result which, specialized to our system model, can be stated as follows.

Fact 4 ([10]). *There exists an algorithm A_{loglog} which outputs an (ε, σ^2)-estimate of W with $\varepsilon = 10^{-6}$ and $\sigma = 1$, using $O(\log \log W)$ bits of communication.*

To get bounds that hold with high probability, we iterate Algorithm A_{loglog} and use Bernstein's Inequality (see, e.g., [9]).

Algorithm BoundCount (Input: ε, δ, i)

1. $M \leftarrow (6/\varepsilon^2)\ln 1/\delta$.
2. Broadcast a filtering message indicating that only input cells holding item i should be considered in Step 3.
3. for $\ell = 1$ to M, run A_{loglog} obtaining an independent estimate $\hat{w}_\ell$ of $w(i)$.
4. Output $\hat{w} \triangleq \frac{1}{M}\sum_{\ell=1}^{M} \hat{w}_\ell$.

Corollary 5 (high-probability estimates). *For $10^{-5} \leq \varepsilon \leq 1$ and $\delta > 0$, the output $\hat{w}$ of Algorithm BoundCount satisfies $\Pr\left\{\frac{1}{w(i)}|\hat{w} - w(i)| < \varepsilon\right\} \geq 1 - \delta$. The individual communication complexity of the algorithm is of order $O\left(\log|\mathcal{I}| + \frac{1}{\varepsilon^2}\log\frac{1}{\delta}\log\log w(i)\right)$. Also, if the algorithm ran M iterations in Step 3, then for any $\zeta > 10^{-5}$, $\Pr\left\{\frac{1}{w(i)}|\hat{w} - w(i)| < \zeta\right\} \geq 1 - \exp\left(-\Omega\left(M\zeta^2\right)\right)$.*

3   Algorithms

In this section we present our main result, namely a randomized algorithm for computing top-k. In our algorithm, the basic idea is to view each cell (representing a unit of weight, or score) as a "vote," and to *sample each vote independently*.

Thus, the expected number of sampled votes for candidate i is proportional to the total number of votes candidate i has in the input *regardless of their partition into nodes*. The sampling results provide a good indication which items are globally popular, so that counting can be applied only to these items.

Next, we need to determine the sample size. Let p^* denote the frequency of the least popular of the top-k items. Clearly, if we sample once, the sample size should be proportional to $O(1/p^*)$ (or else the sample will fail to find all top k items). A more refined analysis shows what should be the sample size as a function of p^*, the approximation parameter ε, and the confidence parameter δ. To deal with unknown p^*, we augment the basic sampling algorithm with a technique to find the right sample size. Intuitively, we have a simple test which can prove whether the sample size is sufficiently large; if it isn't, we double the sample size.

Finally, we address the issue of very small p^* values: while $\Omega(1/p^*)$ sample size cannot be avoided for worst-case inputs, a much better bound can be obtained if the popularity of items decreases relatively rapidly. Consider, for example, the case where the global scores are close to the geometric distribution, e.g., when the frequency of the ℓth popular element is about $2^{-\ell}$. Then we have $p^* = 2^{-k}$, and therefore the sample size should be $\Omega(2^k)$. This cost can be reduced exponentially by utilizing the following simple idea: Whenever a sample is taken, the top item in the sample is by far the most popular (it is expected to have half of the weight in the sample). Therefore it is safe to add the top item to the output list, remove it from further consideration, and take another sample *of the same size*. In the geometric case, this approach has communication cost linear in k. This intuition leads us to our final algorithm, called Algorithm R. In essence, the idea is to iteratively discover the very top items, "shave them off," and to continue recursively with the remainder of the population. The algorithm combines this idea with an additional way to verify that the top-k items have been discovered. Algorithm R is far better than naive sampling for some common input distributions, such as Zipf distribution.

We start, in Sect. 3.1, by describing the basic algorithm we later use as a building block. Section 3.2 presents Algorithm S which uses adaptive sample size. Section 3.3 presents Algorithm R, which is our main result.

3.1   Algorithm B: Basic Sampling

Consider basic sampling: if the top-k items occupy a constant fraction of the total weight, then a log-size sample is sufficient to detect them for any input size (the logarithm is of the inverse of the error probability).

It is convenient to first analyze the following sampling routine.

Algorithm A (Input: P_{SAMPLE})

1. The root sends P_{SAMPLE} to all other nodes.
2. Each node sends each cell to the root with probability P_{SAMPLE}.

Lemma 6. *There exists a function* $S^*(p^*, \varepsilon, \delta) = \Theta\left(\frac{1}{p^* \varepsilon^2} \cdot \ln \frac{1}{p^* \delta}\right)$, *such that for any input* $X \in \mathcal{X}(W, n, p^*)$, *the top-k elements of a random sample of size at least* $S^*(p^*, \varepsilon, \delta)$ *is a top-k ε-approximation of* X *with probability at least* $1 - \delta$.

The proof (like all others) is omitted from this extended abstract. Intuitively, the argument is as follows. Define a 'swap' to be a pair of items i, j such that i is more popular than j in the input but less popular than j in the sample. We identify which item-pairs may not be swapped in an ε-approximation, and bound the probability that such a swap occurs using the Chernoff-like bound for *self-weakening* random variables presented in [19]. The probability bound is used to deduce the required sample size. The exact definition of S^* used by our algorithms is $S^*(p^*, \varepsilon, \delta) \triangleq \frac{g(\varepsilon)}{p^*} \cdot \left(\ln \frac{1+\varepsilon}{p^* \delta} + 4\right)$, where $g(\varepsilon) \triangleq \frac{(1+\varepsilon) \ln(1+\varepsilon)}{\varepsilon \ln\left(\frac{\varepsilon}{\ln(1+\varepsilon)}\right) + \ln(1+\varepsilon) - \varepsilon}$.

Algorithm B, presented below, first determines the sampling probability P_{SAMPLE} using the function S^* from Lemma 6. Each cell (i.e., unit of weight) is then sent to the root with probability P_{SAMPLE}, and the root outputs the top k items in the sample as an approximation of the top k items in the complete input.

Algorithm B (Input: $W, p^*, k, \varepsilon, \delta$)

1. The root computes $P_{\text{SAMPLE}} \leftarrow 2 \cdot S^*(p^*, \varepsilon, \frac{\delta}{2})/W$.
2. Execute Algorithm A with parameter P_{SAMPLE} to get a sample $\mathcal{S}$.
3. Output top$(k, \mathcal{S})$.

When running Algorithm B as described above, each sampled vote is sent to the root separately incurring communication $\log |\mathcal{I}|$. An obvious optimization is to aggregate votes for the same candidate along the way, for example by sending the *count* of votes for each candidate. While such optimization is very worthwhile to implement, it would not help much when the partition of votes to nodes is adversarial. We therefore ignore such optimizations in our upper bounds.

Theorem 7. *Let* $X \in \mathcal{X}(W, n, p^*)$ *be an instance. Provided that p^*, ε, δ and W are known, Algorithm B outputs a top-k ε-approximation with probability at least $1 - \delta$ and communication* $O\left(\frac{1}{p^* \varepsilon^2} \cdot \ln \frac{1}{p^* \delta} \cdot \log |\mathcal{I}|\right)$.

The proof is rather standard; we omit it due to lack of space. Note the $1/p^*$ factor: It is unavoidable because if the sample is to contain the top k elements, it should contain the least popular of them, and hence the sample size must be $\Omega(1/p^*)$. A stronger bound is proved in Sect. 5.

3.2   Algorithm S: Adaptive Sample Size

Algorithm B requires knowing the values of W and p^*. While estimating W is straightforward and cheap (by deterministic or randomized counting, at the cost of $O(\log W)$ or $O(\log \log W)$ communication, respectively), obtaining a lower bound on p^* seems less trivial. We solve this problem as follows.

First, we note that by counting the weight of *any* k items, we obtain a lower bound on p^*: the least popular among any k items is certainly no more popular than the least popular among the top-k items. Second, we note that exact counting is not necessary: we can use high-probability estimates as described in Corollary 5 to get a lower bound on p^* that holds with high probability. However, if we simply use an arbitrary set of k items to bound p^*, that value can be smaller than the true value of p^* by an arbitrary factor, resulting in communication cost that is higher than the bound in Theorem 7 by an arbitrary factor.

Our solution, in Algorithm S below, combines the ideas described above with an iterative approach that avoids unbounded 'overshoots.' The algorithm uses a variable $\hat{p}$ as an estimate of p^*. The algorithm repeatedly halves $\hat{p}$ while improving its lower bound on p^*, stopping when $\hat{p}$ is smaller than the lower bound.

Algorithm S (Input: k, ε, δ)

1. $\hat{W} \leftarrow$ BoundCount($\varepsilon = 1, \delta/5$, 'ALL')
2. $\hat{p} \leftarrow 1$ ($\hat{p}$ is the current estimate of p^*)
3. Repeat
   (a) $\hat{p} \leftarrow \frac{\hat{p}}{2}$
   (b) Execute Algorithm B with parameters ($\hat{W}/2, \hat{p}, \varepsilon, \delta/5$), to get a candidate top-k-set T.
   (c) For each item $i \in T$,
       $\hat{w}(i) \leftarrow$ BoundCount($\varepsilon = 1, \delta/5, i$)
   Until $\hat{p} < \frac{1}{8} \min \left\{ \hat{w}(i)/\hat{W} \mid i \in T \right\}$.
4. Output the set T computed at Step 3b of the last iteration.

Theorem 8. *For any input* $X \in \mathcal{X}(W, n, p^*)$, *with probability at least* $1 - \delta$, *Algorithm S outputs a top-k ε-approximation with communication complexity*
$$C_S = O\left(\frac{1}{p^* \varepsilon^2} \cdot \log \frac{1}{p^* \delta} \cdot \log |\mathcal{I}| \ + \ k \log \frac{1}{p^*} \log \frac{1}{\delta} \log \log W \right).$$

Note that when $|\mathcal{I}| \geq \log W$ and $k < O(1/(-p^* \log p^*))$, the communication complexity of Algorithm S is within a constant factor of the complexity of Algorithm B.

The general idea in the proof is that the high-probability estimates ensure that we run approximately $\log \frac{1}{p^*}$ iterations. As a result, the last iteration is similar to an execution of Algorithm B with the correct parameters and its communication complexity is as specified in Theorem 7.

3.3   Algorithm R: Sample and Remove

As already mentioned above, it appears that the $1/p^*$ factor is unavoidable when we want all the top-k items to be included in the sample (because p^* is the empirical probability of the kth popular item). However, when the popularity of the popular items is far from uniform, one can do much better, as mentioned for the geometric distribution example in the beginning of this section: the $1/p^*$ factor in the communication complexity can be replaced by a factor of k. Algorithm R

Algorithm R (Input: k, ε, δ)

1. $Q \leftarrow \emptyset$ (Q holds all items whose weights were already estimates)
2. $\hat{p} \leftarrow 1$, $\ell \leftarrow 0$. ($\hat{p}$ is the current estimate of p^*, ℓ is the iteration index)
3. Repeat
   (a) $\hat{W} \leftarrow \text{BoundCount}(\varepsilon = 1, \delta/7, \text{'ALL'})$.
   (b) $\ell \leftarrow \ell + 1$; $\hat{W}[\ell] \leftarrow \hat{W}$.
   (c) $\hat{p} \leftarrow \hat{p} \cdot \frac{1}{2} \cdot \hat{W}[\ell - 1]/\hat{W}[\ell]$.
   (d) $\text{P}_{\text{SAMPLE}} \leftarrow 2S^*(\hat{p}, \varepsilon/4, \delta/7) / \hat{W}$.
   (e) Execute Algorithm A with parameter P_{SAMPLE} to get sample $\mathcal{S}$.
   (f) $\eta \leftarrow (\delta/7) \cdot (1/ - \log p_{lo}(\text{top}(k, Q), \hat{W}))$;
       $T^{\mathcal{S}} \leftarrow \text{top}(|\mathcal{S}| \log |\mathcal{I}| / (\log |\mathcal{I}| + \log \log \hat{W}), \mathcal{S})$.
   (g) For each $i \in T^{\mathcal{S}}$, $\hat{w}(i) \leftarrow \text{BoundCount}(\varepsilon = 1, \eta, i)$.
   (h) $Q \leftarrow Q \cup T^{\mathcal{S}}$; remove elements of $T^{\mathcal{S}}$ from the input.
   Until $\text{safe}(\hat{W}, Q, \mathcal{S}, \hat{p}, \varepsilon, \delta)$.
4. If $|T^{\mathcal{S}}| < k$, then for each $i \in \text{top}(k, \mathcal{S})$, do $\hat{w}(i) \leftarrow \text{BoundCount}(\varepsilon = 1, \eta, i)$ and add i to Q.
5. $Q \leftarrow \left\{ i \in Q \mid \hat{w}(i)/\hat{W} > p_{lo}(\text{top}(k, Q), \hat{W}) \right\}$
6. For each $i \in Q$, $\hat{w}(i) \leftarrow \text{BoundCount}(\varepsilon/4, \frac{\delta}{5|Q|}, i)$.
7. Output the top k items in Q according to the new estimates.

Function p_{lo} (Input: Q, W)

1. Return $\frac{1}{4W} \min \{\hat{w}(i) \mid i \in Q\}$.

Predicate safe (Input: $W, Q, \mathcal{S}, \hat{p}, \varepsilon, \delta$)

1. If $\hat{p} \cdot (W[\ell]/W[1]) < p_{lo}(\text{top}(k, Q), W)$ return TRUE.
2. $r_{hi} \leftarrow p_{lo}(\mathcal{S} \cap Q, W)$
3. $q \leftarrow (1 + \varepsilon/2) \cdot p_{lo}(\text{top}(k, Q), W)$; $\alpha \leftarrow q/r_{hi}$.
4. If $\alpha > 1$ AND $\exp\left(-|\mathcal{S}|q\left(\frac{\alpha-1}{\alpha}\right)^2\right) < \frac{q\delta}{5}$ then return TRUE else return FALSE.

Fig. 1. Algorithm R

achieves this improvement for both geometric and Zipf distributions. In a nutshell, the idea is still to cut the estimate of p^* by half in each iteration; however, while Algorithm S achieves this by doubling the sample size, Algorithm R tries also to reduce the size of the relevant population.

In more detail, the algorithm works as follows (see Fig. 1). In each iteration, the algorithm samples with probability P_{SAMPLE}, obtained from the current estimate $\hat{p}$ of p^*. Then, in Step 3f, the algorithm counts (approximately) some of the top items in the sample: the number of items is such that the cost of counting equals the cost of sampling. The counted items (recorded in Q) are not considered part of the input anymore. In Step 3c, $\hat{p}$ is adjusted so that its value *in the full input* is halved (but since the input is smaller now, $\hat{p}$ is multiplied by a factor larger than $1/2$). The loop is executed until one of the stopping rules specified in Predicate safe is met. These rules use a lower bound on p^* computed by function p_{lo}: it is a high-probability lower bound on the kth smallest value among all values counted so far. Using p_{lo}, the stopping rules are defined as follows.

First (Step 1 of Predicate $\mathtt{safe}$), we can stop if the value of $\hat{p}$ w.r.t. the full input is smaller than the lower bound on p^*. This test is done in Step 1. Second, we bound the probability that an "important" item did not arrive at the top of the current sample and therefore was not counted. More specifically, we bound the probability that an item with frequency q, where q is sufficiently larger than p^*, was not counted, while an item whose frequency is r_{hi} was counted. If the probability of this event is low, we know that the top items counted by the algorithm contain an ε-approximation to the top-k set (Step 4). This test is useless in some cases (say, the uniform distribution), but it is effective in some distributions (such as Zipf).

Theorem 9 shows some general bounds on its complexity. As expected, it shows that for some inputs, Algorithm R has very little advantage over Algorithm S. Theorem 10, however, presents an alternative analysis which, when applied to some natural input distributions, attains much better bounds than those of Theorem 9.

Theorem 9. *For any input $X \in \mathcal{X}(W, n, p^*)$, with probability at least $1 - \delta$, Algorithm R outputs a top-k ε-approximation while the communication complexity satisfies $C_R = O\left(\frac{1}{p^* \varepsilon^2} \log \frac{1}{\delta p^*} \left(\log |\mathcal{I}| + \log \log W\right)\right)$.*

Again, when $|\mathcal{I}| < \log W$, the communication complexity of Algorithm R is within a constant factor of the complexity of Algorithm B.

Next we refine the analysis to depend more closely on the global distribution (rather than only on p^*). Let π_ℓ denote the probability that a random item has global frequency at most $2^{-\ell}$, namely π_ℓ is the fraction of the votes for candidates whose popularity is at most $2^{-\ell}$.

Theorem 10 (Main result). *Let $X \in \mathcal{X}(W, n, p^*)$. Then with probability at least $1 - \delta$, Algorithm R outputs a top-k ε-approximation of X while using $O\left(\frac{1}{\varepsilon^2} \log \frac{1}{p^* \delta} \left(\log |\mathcal{I}| + \log \log W\right) \cdot \sum_{\ell=1}^{\log \frac{1}{p^*}} 2^\ell \pi_\ell\right)$ communication bits per node.*

Using Theorem 10, we can prove the following corollaries for specific distributions. The proof of Corollary 11 is straightforward; the proof of Corollary 12 is based on approximating the Zipf distribution by the Pareto distribution.

Corollary 11. *Let X be an instance where item frequencies are geometrically distributed, i.e., $p_x(i) = (1 - \lambda)\lambda^{i-1}$ for some constant $0 < \lambda < 1$. Then Algorithm R, when run on input X with parameters $\delta, \varepsilon > 0$ has communication complexity $c_R(X) = O\left(\frac{k}{\varepsilon^2} \log \frac{1}{p^* \delta} \left(\log |\mathcal{I}| + \log \log W\right)\right)$.*

For the important case of Zipf distribution, we have the following corollary.

Corollary 12. *Let X be an instance where item frequencies have Zipf distribution with parameter $a > 1$, i.e., $p_x(i) \triangleq \frac{i^{-a}}{h}$, where $a > 1$ and h is the normalization* *factor. Then Algorithm R, when run on input X with parameters $\delta, \varepsilon > 0$ has communication complexity $c_R(X) = O\left(\frac{a+1}{a-1} \cdot \frac{k}{\varepsilon^2} \cdot \log \frac{1}{p^* \delta} \cdot (\log |\mathcal{I}| + \log \log W)\right)$.*

The results in this section analyze the behavior of R for asymptotically large inputs. As described in the next section, we have tested our algorithms on realistic-size inputs to study the effect of the hidden constants. We have also compared R with other known algorithms from the literature. We remark that Algorithm R turns out to have significantly superior performance.

4   Simulations Results

To evaluate the performance of our algorithm in more realistic scenarios, we ran simulations examining the actual number of bits communicated throughout the execution. We compared our algorithm R with the best known algorithms, namely TA [11] and TPUT [6]. We compared the performance of the algorithms when the input is partitioned randomly and when it is partitioned adversarially. Finally, we evaluated the performance of R for various values of p^* (popularity of the kth most popular item) and various values of ϵ (the approximation parameter). Informally, we find that Algorithm R offers a dramatic improvement over TA and TPUT unless the system is very small and the distribution is particularly favorable to TA and TPUT. Moreover, for typical inputs, R behaves better than predicted by our analytical bounds.

Simulated Instances. We used randomly generated inputs of various sizes, and our main focus was on inputs generated by Zipf distribution with exponent between 0.8 and 3; all shown charts used Zipf distribution with exponent 1.5 (different exponent values showed qualitatively similar behavior). Typical other parameters were $k = 5$, $\varepsilon = 0.5$ and $\delta = 0.05$. The total number of votes (W) was either 10^6 or 10^7 and the varying parameter is the number of nodes (n). The number of candidate-items was equal to the number of nodes $(|\mathcal{I}| = n)$. All simulations measured the worst-case individual communication (in bits).

Comparison with Existing Algorithms. In Fig. 2 we compare algorithms R, TA and TPUT. We used the approximation version of the TA algorithm. We note

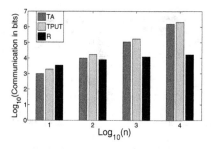

Fig. 2. *Communication costs of algorithm R compared to TA and TPUT for random partition of scores in various system scales*

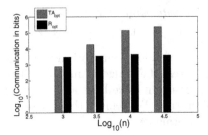
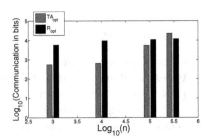

Fig. 3. *Communication costs of optimized algorithms for various system scales. Left: with an adversarial partition of scores among nodes. Right: with a random partition of scores among nodes.*

that TPUT has no approximation version, but in all simulations we ran, TPUT appeared less efficient than TA, even for high accuracy such as $\varepsilon = 0.01$.

Results for adversarial and random partition of scores were similar. As shown in Fig. 2, the performance of TA and TPUT for a random partition is rather poor for large systems and Algorithm R is superior even for a system of 100 nodes. R is overwhelmingly superior (by more than two orders of magnitude) for $n > 10000$.

Next, we modified the deterministic algorithms by using loglog counting and allowing them to err. We also assumed the existence of a bounded degree spanning tree, and used unlimited memory at each node (which allows simple aggregation while traversing the tree). These modifications resulted in drastic improvement in TA performance: see Fig. 3. The optimized version of TPUT is not shown because it performed far worse than the optimized version of TA. Note that when the input is adversarially partitioned, R beats TA for $n > 2000$. But even when the input is partitioned randomly (which is close to the best case of TA), R performs better for n large enough, where large enough means $n > 200000$ in this case.

The Effect of Accuracy and Critical Frequency on Performance. We examined the performance of Algorithms R and algorithm S while varying ε and p^*

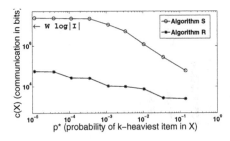
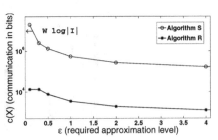

Fig. 4. *Communication costs of algorithms S and R. Left: as a function of p^*. Right: as a function of ε. The communication cost of reading the entire input is $W \log |\mathcal{I}|$, (W is the input size and $|\mathcal{I}|$ is the number of items).*

(for n fixed). The results are given in Fig. 4. Obviously, Algorithm R was always superior to Algorithm S. The results for p^* (Fig. 4, left) show that Algorithm R is less sensitive to p^* when the input is generated by Zipf distribution. We note that the flat left segment in the graph of S is due to the fact that for small values of p^*, the sample of Algorithm S consists of the entire input.

The effect of the approximation factor ε (Fig. 4, right) on the communication of both algorithms is proportional to $\left(\frac{1+\varepsilon}{\varepsilon}\right)^2$, as suggested by our upper bounds. Algorithm R is more efficient due to its improved termination criterion.

5   A Lower Bound on the Total Cell-Probe Complexity

In this section we give a lower bound on the *total cell probe complexity* [23, 13] of randomized algorithms for the top-k problem. For a restricted settings, we can deduce a lower bound on the individual communication complexity of such randomized algorithms.

We define the total cell probe complexity of an algorithm as the total number of input cells accessed by a all nodes for the worst-case input. Formally, $\mathrm{cp}_A(X,v)$ denotes the number of input cells accessed by node v, throughout the execution of A for input X; $\mathrm{cp}_A^T(X)$ denotes the total number of cell-probes for an instance X, i.e., $\mathrm{cp}_A^T(X) \triangleq \sum_{v \in V} \mathrm{cp}_A(X,v)$. The total cell-probe complexity of A is defined by $\mathrm{CP}_A^T(\mathcal{X}) \triangleq \max \left\{ \mathrm{cp}_A^T(X) \mid X \in \mathcal{X} \right\}$. Note that the set of all probed cells can be viewed as a sample whose size is the total number of cell-probes.

Theorem 13. *For some natural k, let $\varepsilon > 0$, $0 < p^* \leq \frac{1}{2k}$, $0 < \delta < 0.014$, and $W = \Omega\left(\frac{1}{\delta(p^*)^2} \cdot \frac{(1+\varepsilon)^2}{\varepsilon^4}\right)$ and let A be any Monte Carlo algorithm. If A outputs a top-k ε-approximation with probability at least $1 - \delta$ for any input in $\mathcal{X}(W, n, p^*)$, then its total cell-probe complexity satisfies $E\left[\mathrm{CP}_A^T\right] = \Omega\left(\frac{1}{p^* \varepsilon^2} \cdot \log \frac{1}{\delta}\right)$.*

The proof (omitted from this extended abstract) applies Yao's Minimax Principle to obtain a lower bound on Monte Carlo algorithms. To this end, we prove a lower bound on the expected sample size used by any deterministic algorithm when the input is drawn from a certain distribution we construct, so that there is a single correct output to the top-k ε-approximation query. Furthermore, under our distribution, the local view of each node is just a random subset of the items. We show that any algorithm which does not output the top-k set of its sample is doomed to err with probability at least $1/2$. Finally, we bound the required sample size for other algorithms.

Star Topology and "Smart Dust" Systems. Theorem 13 demonstrates the optimality of Algorithm B with respect to the *total* cell-probe complexity but not the *individual* cell-probe complexity. Consider a setting where the system has star topology, i.e., all nodes are connected to a root node, and suppose that each node holds a single cell. This model is a reasonable abstraction of the setting in sensor networks using *passive communication*. In these systems, the only communication is between a powered base-station and a sensor, forming a star topology (as opposed to the common spanning tree). Passive communication is suitable for very small devices, such as "smart dust" systems (see e.g., [22]). Now, for these systems, Theorem 13 says that any algorithm that satisfies the conditions of the theorem has individual communication complexity at least $\Omega\left(\frac{1}{p^* \varepsilon^2} \cdot \log \frac{1}{\delta}\right)$. We stress that while our model assumes certain routing capabilities, our algorithms only use a broadcast-convergecast scheme rooted by the root node. Such communication is suitable for star topology.

6   Conclusions and Future Work

In this paper we have proposed algorithms solving the top-k problem by adaptive sampling. The communication complexity of our algorithms does not depend on the way the input is partitioned in the network: only the global statistics affect the complexity. Our final algorithm performs particularly well when the global statistics are far from flat. We have tested our algorithm by simulation and found empirical support for our analytical claims. Although our study is mainly theoretical, simulation results indicate that our algorithm is rather practical and can be very useful in real-life scenarios. Future work may extend our algorithm to specific models, where spatial and temporal dependence among different sensors holds (e.g., nearby nodes have similar readings).

From the theoretical viewpoint, we think that it is very interesting to extend our lower bound to the case of interactive algorithms. We conjecture that our Algorithm R is nearly optimal in this more general model.

References

[1] H. Attiya and J. Welch. *Distributed Algorithms.* McGraw-Hill Publishing Company, UK, 1998.

[2] B. Babcock and C. Olston. Distributed top-k monitoring. In *Proc. 2003 ACM SIGMOD.*

[3] P. Bak. *How Nature Works: The science of self-organized criticality.* Springer-Verlag, New York, 1996.

[4] W.-T. Balke, W. Nejdl, W. Siberski, and U. Thaden. Progressive distributed top k retrieval in peer-to-peer networks. In *Proc. 21st Int. Conf. on Data Engineering,* 2005.

[5] N. Bruno, L. Gravano, and A. Marian. Evaluating top-k queries over web-accessible databases. In *Proc. 18th Int. Conf. on Data Engineering,* 2002.

[6] P. Cao and Z. Wang. Efficient top-k query calculation in distributed networks. In *Proc. 23rd Ann. ACM Symp. on Principles of Distributed Computing,* 2004.

[7] J. Considine, F. Li, G. Kollios, and J. Byers. Approximate aggregation techniques for sensor databases. Apr. 2004.

[8] G. Cormode, M. N. Garofalakis, S. Muthukrishnan, and R. Rastogi. Holistic aggregates in a networked world: Distributed tracking of approximate quantiles. In *Proc. 2005 ACM SIGMOD,* 2005.

[9] P. Dagum, R. M. Karp, M. Luby, and S. Ross. An optimal algorithm for Monte Carlo estimation. *SIAM J. Comput.,* 29(5), 2000.

[10] M. Durand and P. Flajolet. Loglog counting of large cardinalities (extended abstract). In *Algorithms: ESA 11th Ann. European Symp.*, 2003.

[11] R. Fagin, A. Lotem, and M. Naor. Optimal aggregation algorithms for middleware. In *Proc. 20th ACM Symp. on Principles of Database Systems*, 2001.

[12] M. Faloutsos, P. Faloutsos, and C. Faloutsos. On power-law relationships of the internet topology. In *Proc. SIGCOMM '99*, New York, NY, USA. ACM Press.

[13] M. Fredman and M. Saks. The cell probe complexity of dynamic data structures. In *Proceedings of the 21st Annual ACM Symposium on Theory of Computing*, May 1989.

[14] M. Greenwald and S. Khanna. Power-conserving computation of order-statistics over sensor networks. In *Proc. 23rd ACM Symp. on Principles of Database Systems*, 2004.

[15] N. Lynch. *Distributed Algorithms*. Morgan Kaufmann, San Mateo, CA, 1995.

[16] S. Madden, M. J. Franklin, J. M. Hellerstein, and W. Hong. The design of an acquisitional query processor for sensor networks. In *Proc. 2003 ACM SIGMOD*.

[17] S. Michel, P. Triantafillou, and G. Weikum. Klee: A framework for distributed top-k query algorithms. In *Proc. 31st Int. Conf. on Very Large Data Bases*, 2005.

[18] S. Nath, P. B. Gibbons, S. Seshan, and Z. R. Anderson. Synopsis diffusion for robust aggregation in sensor networks. In *SenSys '04: Proc. 2nd international conference on Embedded networked sensor systems*, 2004.

[19] A. Panconesi and A. Srinivasan. Fast randomized algorithms for distributed edge coloring (extended abstract). In *Proc. 11th Ann. ACM Symp. on Principles of Distributed Computing*, 1992.

[20] B. Patt-Shamir. A note on efficient aggregate queries in sensor networks. In *Proc. 23rd Ann. ACM Symp. on Principles of Distributed Computing*, 2004.

[21] A. Silberstein, R. Braynard, C. Ellis, K. Munagala, and J. Yang. A sampling-based approach to optimizing top-k queries in sensor networks. In *Proc. 22nd Int. Conf. on Data Engineering*, 2006.

[22] B. Warneke. Miniaturizing sensor networks with mems. In M. Ilyas and I. Mahgoub, editors, *Handbook of Sensor Networks: Compact Wireless and Wired Sensing Systems*. CRC Press, 2004.

[23] A. C.-C. Yao. Should tables be sorted? *J. ACM*, 28(3), 1981.

[24] Y. Yao and J. Gehrke. The Cougar approach to in-network query processing in sensor networks. *ACM SIGMOD Record*, 31(3):9–18, Sept. 2002.

[25] D. Zeinalipour-Yazti, Z. Vagena, D. Gunopulos, V. Kalogeraki, V. Tsotras, M. Vlachos, N. Koudas, and D. Srivastava. The threshold join algorithm for top-k queries in distributed sensor networks. In *Proc. 2nd Int. Workshop on Data Management for Sensor Networks*, 2005.

Self-stabilizing Space Optimal Synchronization Algorithms on Trees

Doina Bein, Ajoy K. Datta, and Lawrence L. Larmore

University of Nevada, Las Vegas, USA
{siona, datta, larmore}@cs.unlv.edu

Abstract. We present a space and (asymptotically) time optimal self-stabilizing algorithm for simultaneously activating non-adjacent processes in a rooted tree (Algorithm $SSDST$). We then give two applications of the proposed algorithm: a time and space optimal solution to the local mutual exclusion problem (Algorithm $LMET$) and a space and (asymptotically) time optimal distributed algorithm to place the values in min-heap order (Algorithm $HEAP$). All algorithms are self-stabilizing and uniform, and they work under any unfair distributed daemon. In proving the time complexity of the heap construction, we use the notion of *pseudo-time*. Pseudo-time is similar to *logical time* introduced by Lamport [12].

Keywords: heap, local mutual exclusion, self-stabilization.

1 Introduction

Fault-tolerance is the ability of a system to withstand transient faults. A fault-tolerant system is guaranteed to continue to perform its function when a number of transient errors has occurred. In 1973 [8], Dijkstra defined a distributed system to be *self-stabilizing* when, "regardless of its initial state, it is guaranteed to arrive at a legitimate state in a finite number of steps."

Self-stabilizing algorithms aim to achieve performance comparable to that of non-stabilizing distributed algorithms when transient faults or arbitrary initialization cause the system to enter a state where a non-stabilizing algorithm cannot continue to perform its task properly. In this paper, we propose a general synchronization scheme for a rooted tree, and use this scheme to solve two fundamental problems: heap construction and local mutual exclusion.

Related Work. The self-stabilizing heap problem has been studied in [1, 4, 5, 10, 13]. The first self-stabilizing binary-search tree construction algorithm was proposed in [4]. In [1], the self-stabilizing algorithm for a min-heap construction improves the algorithm of [5] in three ways: no global reset is required, the time complexity is reduced from $O(nh)$ to $O(h)$ (h is the height of the tree with n nodes), and the space complexity per node is reduced from $O(degL)$ to $O(deg + L)$ (deg is the degree of the process and L is the maximal size of the initial values in the tree). Synchronization among the nodes is achieved by using the global rooted synchronizer defined in [2], plus two additional bits. In

P. Flocchini and L. Gąsieniec (Eds.): SIROCCO 2006, LNCS 4056, pp. 334–348, 2006.
© Springer-Verlag Berlin Heidelberg 2006

[13], the self-stabilizing max-heap protocol that uses a neighborhood synchronizer protocol [11] reduces the memory requirement further to $2L + 3$ bits; its time complexity is $O(h)$. A heap construction that supports insert and delete operations in arbitrary states of a variant of the standard binary heap [7] with capacity K is proposed in [10]. It takes $O(m \log K)$ heap operations to stabilize (m is the initial number of items in the heap). The space complexity per node i is $O(h_i)$, where h_i is the height of the subtree T_i rooted at i.

Bein *et al.* [4] proposed the first snap-stabilizing binary search tree (BST) and the first snap-stabilizing heap construction algorithm. (A snap-stabilizing algorithm is a self-stabilizing algorithm with stabilization time of 0 rounds). The algorithms use a PIF scheme [6] to synchronize the nodes in the tree. The space complexity of the snap-stabilizing heap construction algorithm is $3L + 3$.

Contributions. We propose a space and (asymptotically) time optimal self-stabilizing algorithm for simultaneously activating non-adjacent processes in a rooted tree (Algorithm $\mathcal{SSDST}$). It uses $1 + \lceil log(deg) \rceil$ bits in each node (deg is the node degree); during the first $2h + 2t - 1$ rounds, every node is enabled at least t times, *i.e.*, on the average, once every second round. For a synchronous system, after at most $2h$ steps, every node is enabled every second step. If the synchronous network starts in a normal starting configuration, then a node is active every other step from the beginning.

We then give two applications on rooted trees of the proposed algorithm: a time and space optimal solution to the local mutual exclusion problem (Algorithm $\mathcal{LMET}$), and a space and (asymptotic) time optimal solution to the heap problem (Algorithm $\mathcal{HEAP}$). Algorithm $\mathcal{LMET}$ uses only $2 + \lceil log(deg) \rceil$ bits per node and stabilizes in 0 rounds (it is *snap*-stabilizing). During the first $2h + 2t - 1$ rounds, a node enters its CS at least t times. Algorithm $\mathcal{HEAP}$ arranges n values, not necessarily distinct, in non-decreasing order from top to bottom (min-heap), in at most $4(7h/2 - 4)$ rounds ($h =$ height). Each process holds only one value at any moment, and uses a total of $1 + \lceil log(deg) \rceil$ bits per node, not counting the bits needed to store the value being sorted ($deg =$ node degree) which is optimal, thus an improvement over [13, 4].

In proving the time complexity of heap-building, we use the notion of *pseudo-time*. Each node in the network has a "local clock" which has the property that when any action must be executed between the node and its children, the local clocks of all the nodes involved in the action have the same value.

Outline of the Paper. In Section 2, we briefly introduce self-stabilization and the topological models used by the proposed algorithms. Section 3 contains a description of Algorithm $\mathcal{SSDST}$, followed by a sketch of its proof of correctness. Algorithm $\mathcal{LMET}$ is presented in Section 4. In Section 5 we first present a min-heap algorithm for an abstract model of communication (Algorithm $\mathcal{A\_HEAP}$), and then show how the min-heap will be built using the usual shared-memory model of communication (Algorithm $\mathcal{HEAP}$). A sketch of correctness proof of Algorithm $\mathcal{A\_HEAP}$ is given in 5.3. The reduction of Algorithm $\mathcal{A\_HEAP}$ to Algorithm $\mathcal{HEAP}$ is given in 5.4. We finish with concluding remarks in Section 6.

2   Computational Models

We consider an asynchronous, rooted tree of n processors, with height h. The root node is denoted by R. We assume that an underlying self-stabilizing spanning tree construction protocol maintains the parent pointer p_v and the set of children D_v of a node v. For the root node R, $p_R = \bot$. For a leaf node v, $D_v = \bot$.

If the topology of the network that is given as the input to the spanning tree construction algorithm changes, the spanning tree may change. This will change the input to our protocols (local mutual exclusion and heap). In that sense, the proposed protocols can deal with dynamic trees. The model of communication among the neighboring nodes is shared memory — a process can read and write its own memory, but can only read the memory of its neighbors.

The program of every processor consists of a finite set of guarded actions of the form: $<$ label $>$::$<$ guard $> \rightarrow <$ action $>$, where each guard is a function of the variables of the processor and its direct neighbors. The *state* of a process is defined by the values of its variables. The *system state (configuration)* is the Cartesian product of all the nodes' states. If an action has its guard, a Boolean expression, evaluated to *true*, then it is called *enabled*. A node with at least one enabled guard is called *enabled*. A daemon will non-deterministically select a non-empty subset of enabled nodes to execute one of its enabled actions. Guard evaluation and execution of the its action are done in one atomic step.

We assume an asynchronous system. In order to compute the time complexity, we use the definition of *round* [9]. A round is a minimal sequence of computation steps during which each processor that was enabled in the first configuration of the sequence executes at least once during this sequence.

We consider the strongest distributed daemon, the *unfair* daemon. The *unfair daemon* does not have a fairness mechanism: a continuously enabled process will not necessarily be selected for execution unless it is the only enabled process.

Let $\mathcal{C}$, the set of all possible states, and a predicate $\mathcal{P}$ over $\mathcal{C}$. We denote by $\mathcal{L_P} \subseteq \mathcal{C}$ the set of all *legitimate states with respect to* $\mathcal{P}$. Let $\mathcal{C}_1, \mathcal{C}_2 \subseteq \mathcal{C}$. $\mathcal{C}_2$ is a *closed attractor* for $\mathcal{C}_1$ if (i) every execution starting in $\mathcal{C}_1$ eventually reaches a configuration in $\mathcal{C}_2$, and (ii) every execution starting in $\mathcal{C}_2$ remains in $\mathcal{C}_2$.

Definition 1 (Self-stabilization). *If $\mathcal{P}$ is a predicate, a protocol S is called self-stabilizing to $\mathcal{P}$ if $\mathcal{L_P}$ is a closed attractor for $\mathcal{C}$.*

3   Self-stabilizing Distributed Simultaneous Execution of Non-adjacent Nodes in a Rooted Tree $\mathcal{SSDST}$

Each node v holds a variable $S \in \{A, B\}$ and a pointer $i \in 0..|D_v| - 1$ to some child of v. Thus, the total memory requirement of node v is $1 + \lceil \log(deg) \rceil$ bits (deg is the node degree). (For a binary tree, Algorithm $\mathcal{HEAP}$ uses at most three bits per node.)

For simplicity we write $S = S.v$. The predicate $check(v, s)$ means that the node v exists and has the value s for its variable S. Let $execute(v)$ denote a generic action.

Algorithm 3.1. *Algorithm SSDST*

Predicate $check(v, s) \equiv (v = \bot \lor S.v = s)$

Actions for any node v

$_{ABB}$ $S = B \land check(p_v, A) \land \forall i, 0 \le i < |D_v| : check(D_v[i], B) \longrightarrow execute(v) ; S = A$

$_{BAA}$ $S = A \land check(p_v, B) \land \forall i, 0 \le i < |D_v| : check(D_v[i], A) \longrightarrow execute(v) ; S = B$

Actions ABB and BAA are enabled at node v when the following two conditions are true: (i) either it has no parent, or its parent's S-value is different from its S-value, and (ii) all its children's S-values are the same as its S-value.

For example, given a network of eight nodes starting in a so-called normal starting configuration (Figure 1(a)), the only enabled nodes are of even depth (the root and the children of the root's children). If we assume a synchronous system, the next execution step brings the system into the configuration in Figure 1(b), in which the only enabled nodes are of odd depth. The next configuration is shown in Figure 1(c), followed by the one in Figure 1(d). Then the system returns to the configuration illustrated in Figure 1(a). The cycle repeats forever.

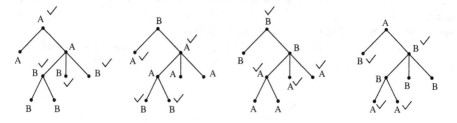

(a) A configuration   (b) After one step   (c) After two steps   (d) After three steps

Fig. 1. Four steps in a synchronous system

3.1   Proof of Correctness for *SSDST*

In this section, we show that Algorithm *SSDST* stabilizes in at most $2h + 2k - 1$ rounds, to the global predicate

k-*Exec*: $\equiv \{\forall$ node v, v has executed macro *execute* at least k times $\}$

and works under the unfair distributed daemon.

We extend the notions of *configuration-string* and *difference-string* to the tree network. We show that in every configuration, during execution of *SSDST*:

- No node is enabled if any of its neighbors is enabled (local mutual exclusion) (Property 1)
- At least one node is enabled (no deadlock); after it executes, a node becomes disabled until all its neighbors execute (Property 2)
- During the first $2h + 2k - 1$ rounds every node executes at least k times (no starvation) (Lemma 1).

We then show that $SSDST$ works under the unfair distributed daemon (Property 3, Section 3.2).

Henceforth, $n > 1$, as the case $n = 1$ is trivial. Let the *configuration tree* be the tree in which every node is represented by its S-value only.

A *normal starting configuration* is a configuration in which each branch of the configuration tree is a prefix of $(AABB)^n$ (the string of length $4n$ obtained by concatenating $AABB$ n times). Starting from a normal starting configuration, the enabled nodes are alternately of even and odd depth (Figure 1). The *binary edge labeling* is the labeling where an edge between nodes with the same S value is labeled 0 and other edges are labeled 1.

Definition 2. *Given a configuration tree C, we let DT_C, the difference tree, be the tree in which every node v is represented by a two-bit string $DT_C(v) = b_0 b_1$ such that:*

$$b_0 = \begin{cases} 1, \text{ if } p_v = \perp \text{ or the link } (p_v, v) \text{ is labeled 1} \\ 0 \text{ otherwise} \end{cases}$$

$$b_1 = \begin{cases} 1, \text{ if } \exists w \in D_v \text{ s.t. the link } (v, w) \text{ is labeled 1} \\ 0 \text{ otherwise} \end{cases}$$

If C is understood, write DT instead of DT_C. Given a binary edge labeling and the S-value of some node, the corresponding configuration tree C is uniquely defined. Given a difference tree DT and the S-value of some node, the corresponding configuration tree C is uniquely defined.

For example, for the configuration in Figure 1(a), the binary edge labeling is given in Figure 2(a) and the difference tree is given in Figure 2(b).

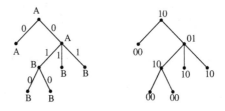

(a) Binary edge labeling   (b) Difference tree

Fig. 2. Some configuration

Given any configuration tree C, a node v is enabled if and only if $DT_C(v) = 10$.

Property 1. For any configuration tree C and for any node v, if node v is enabled to execute, then no neighbor of v is enabled.

Property 2. (i) In any configuration tree C there exists at least one enabled node.
(ii) For any node v, if node v is enabled and is selected to execute, then after the execution is completed, its actions are disabled.

Given a node v and its parent p_v where $S.p_v = a$ and $S.v = b$, the notation "$a \leftarrow b$" denotes that state b does not block state a from being enabled (for p_v

to be enabled in state a, $S.v$ must be b). The notation $a \rightarrow b$ indicates that state a does not block state b from being enabled (for v to be enabled in state b, $S.p_v$ needs to be a).

We use the above notation to define *layers* as follows. We start defining the layers of nodes from node R and going down the tree until we reach the leaf nodes. Node R is placed on some layer. If node v is an internal node on a certain layer, then for any child node $w \in D_v$:

- if $S.v \rightarrow S.w$ then w is one layer higher
- if $S.v \leftarrow S.w$ then w is one layer lower.

We can represent a configuration tree using this notation in a level ordering, where the peak nodes are the enabled nodes. The binary edge labeling is consistent with the orientation of the arrows between a node and its parent, and a node and its children (1 for ↗, 0 for ↘). For example, the sawtooth-like arrangement of the configuration tree in Figure 3(a) is given in Figure 3(b).

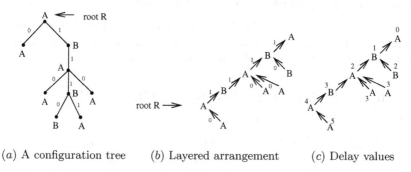

(a) A configuration tree   (b) Layered arrangement   (c) Delay values

Fig. 3. Calculating the delay values

Definition 3 (Node Delay). *For each node v we define $delay[v]$ to be a nonnegative integer characterized as follows: (i) there exists at least one node whose delay is 0, and (ii) if $delay[u] = d$ and node v is a neighbor of node u such that $S.v \rightarrow S.u$ then $delay[v] = delay[u] + 1$, and (iii) if $S.v \leftarrow S.u$ then $delay[v] = delay[u] - 1$.*

The delay of some node is in fact the layer on which the node is arranged in the layered arrangement.

The delay values of the nodes in Figure 3(a) are given in Figure 3(c). An enabled node has all the adjacent arrows pointing towards it. For a tree of height h, for any node v, $delay[v]$ is a value between 0 and $2h$. The number of rounds that a node waits before it becomes enabled cannot exceed its delay value.

Let d_0 be the array of the delay values in the starting configuration and D_0 be the maximal value of d_0 over all nodes: $1 \leq D_0 \leq 2h$.

Lemma 1. *For any node v and any value $t > 0$ node v executes t times within the first $d_0[v] + 2t - 1$ rounds.*

Proof. We define the predicate $\mathcal{P}(q)$ as follows: For any node v, for any $t \geq 1$, node v executes t times within the first q rounds if $q \geq d_0[v] + 2t - 1$. For any $q \geq 1$, Predicate $\mathcal{P}(q)$ holds (induction on q).

Corollary 1. *For any node v and any value $t > 0$ node v executes t times within the first $2h + 2t - 1$ rounds.*

Proof. Follows from Lemma 1: for any node v, $2h \geq d_0[v]$.

3.2   The Unfair Distributed Daemon

In this section we show that Algorithm $\mathcal{SSDST}$ works under the unfair distributed daemon. A sufficient condition to prove that a certain algorithm works under the unfair daemon is to show that a continuously enabled node which is never selected eventually becomes the only enabled node. If a node v is enabled to execute but not selected by the distributed daemon, it remains enabled. Since the unfair daemon must select a non-empty subset of the enabled nodes in every computation step, it will be forced to select v (Property 3).

Property 3. If a node v is enabled to execute but is not selected by the daemon, it remains enabled until it gets selected. Every continuously enabled node will be eventually selected by the unfair distributed daemon.

4   Self-stabilizing Local Mutual Exclusion Algorithm on Rooted Trees $\mathcal{LMET}$

Each node holds three variables: variable S that takes values in the set $\{A, B\}$, a pointer $i \in 0..|D_v| - 1$ to some child of v, and a Boolean variable *request* that is *true* whenever the process requests access to its critical section CS. Thus, the total memory requirement of node v is $2 + \lceil \log(deg) \rceil$ bits per node (*deg* is the node degree).

For some node v, let $S = S.v$ and *request* $=$ *request.v*. The predicate $check(v, s)$ is defined in Section 3.

A protocol solves the local mutual exclusion problem if any configuration of the system running the protocol has two properties ([3]): *(i) safety* - no two neighboring nodes can be simultaneously enabled to execute their critical sections (CS), and *(ii) liveness* - a node requesting to execute its CS will eventually do so.

Algorithm 4.1. *Algorithm $\mathcal{LMET}$*

Actions for any node v

$_{ABB}$ $S = B \wedge check(p_v, A) \wedge \forall i, 0 \leq i < |D_v| : check(D_v[i], B) \longrightarrow$
        if *request* then CS; *request* = *false*
            $S = A$

$_{BAA}$ $S = A \wedge check(p_v, B) \wedge \forall i, 0 \leq i < |D_v| : check(D_v[i], A) \longrightarrow$
        if *request* then CS; *request* = *false*
            $S = B$

Property 1 shows that $\mathcal{LMET}$ has the safety property. Lemma 1 shows that $\mathcal{LMET}$ has the liveness property.

5   Self-stabilizing Min-heap Algorithms for a Rooted Tree

In this section we present two algorithms for min-heap problem in a rooted tree: $\mathcal{AHEAP}$ (Section 5.1), and $\mathcal{HEAP}$ (Section 5.2). Algorithm $\mathcal{AHEAP}$ is implemented in an abstract model. Algorithm $\mathcal{HEAP}$ is implemented in the shared-memory model.

Let x and y be two values to be swapped. Swapping can be done in three steps without using an extra variable, as follows:

1. $x = x + y$ 2. $y = x - y$ 3. $x = x - y$

Alternatively, we could use "$\oplus$, bit-wise exclusive or, instead of addition and subtraction.

5.1   Heap Construction in a Rooted Tree

Algorithm $\mathcal{AHEAP}$ (Figure 5.1) is a particular case of Algorithm $\mathcal{SSDST}$, in which the macro $execute(v)$ is replaced by the macro $heap(v)$ that sets $IV.v$ to the minimal value among itself and its children's IV-values.

Consider an abstract model, different from the shared-memory model, in which a node v, in order to have the heap property locally, can modify the variable $IV.J$ of some child J. Intuitively, since by executing Algorithm $\mathcal{SSDST}$, local mutual exclusion is satisfied in any configuration (see Property 1), a node can synchronize the swap of values with some child. We assume for now that the swap is done in an atomic step (macro $heap$), and we show in Section 5.2 how this is done in the shared-memory model.

Each node, besides the variable IV to be sorted, holds a variable $S \in \{A, B\}$, a pointer $i \in 0 \ldots |D_v| - 1$, and a variable $j \in \{-1, 0, \ldots, |D_v| - 1\}$ that either points to some child of node v that holds a value smaller than node v, or has the value -1 if either node v is a leaf or all its children have larger values. Thus, the total memory requirement of node v is $1 + 2\lceil \log{(deg)} \rceil$ bits per node (deg is the node degree).

For some node v, let $S = S.v$ and $IV = IV.v$. Predicate $check(v, s)$ is defined in Section 3. If all children of v hold values greater than or equal to IV, then $min(v)$ returns the default value -1. Otherwise, $min(v)$ returns the index in the array D_v of a child of node v which holds the minimum value.

The guards $C1$-$C3$ "correct" the variable S of the node to some value in the set $\{A, B\}$ (a result of a fault or arbitrary initialization).

5.2   Heap Construction in the Shared-Memory Model

In Algorithm $\mathcal{HEAP}$ (Figure 5.2), each node v holds, besides the variable IV to be sorted, a variable $S \in \{A, B, X, Y\}$, a pointer i, a variable j, a variable J which is a pointer to some child, and a variable $tmpS \in \{A, B\}$. Variable $tmpS$

Algorithm 5.1. *S-S. Min-Heap in a Rooted Tree in the Abstract Model* $\mathcal{A\_HEAP}$

Macro $heap(v) ::$
$j = min(v)$
if $(j \geq 0)$ then $J = D_v[j]$; $IV.v=IV.v+IV.J$; $IV.J=IV.v-$ $IV.J$; $IV.v=IV.v-IV.J$

Function $min(v) ::$
if $D_v = \bot$ then return -1
else
$\quad j = 0$
$\quad$forall $l \in \{0, |D_v| - 1\}$ do if $(IV.D_v[j] > IV.D_v[l])$ then $j = l$
$\quad$if $(IV.D_v[j] < IV.v)$ then return j else return -1

Heap actions for any node v

$_{ABB}$ $S = B \wedge check(p_v, A) \wedge \forall i, 0 \leq i < |D_v| : check(D_v[i], B) \quad \longrightarrow \quad heap(v)$; $S = A$
$_{BAA}$ $S = A \wedge check(p_v, B) \wedge \forall i, 0 \leq i < |D_v| : check(D_v[i], A) \quad \longrightarrow \quad heap(v)$; $S = B$

---

stores the value of S temporarily while the swap is performed between node v and its child J. Thus, the total memory requirement of node v is $3 + 2\lceil \log{(deg)} \rceil$ bits per node (deg is the node degree).

For any node v, let $S = S.v$, $IV = IV.v$, $J = J.v$, $tmpS = tmpS.v$, $S_p = S.p_v$, $J_p = J.p_v$, $IV_p = IV.p_v$, $S_J = S.(J.v)$, and $IV_j = IV.(J.v)$. The macro $heap'(v, value)$ executes the first step of swapping between node v and the child $J = D_v[j]$, and the value $value$ to be given to variable $S.v$ after the swap is performed is stored in variable $tmpS.v$.

Predicate $check(v, s)$ has been defined in Section 3. Function $min(v)$ is defined in Section 5.1.

In order to perform the swap, nodes v and J_v must change their S-value (from either A or B to either X or Y). Since node v will change its S-value after the swap, the value to-be for $S.v$ and the value of S_J are stored in variables $tmpS.v$, respectively $tmpS.J$, by each node. Node v changes its S-value to X (macro $heap'$) and node J changes its S-value to Y (Guard $S1$). The swap started by node v already in macro $heap'$ is continued by node J in Guard $S1$, and finished by node v in Guard $S2$ (where also node v restores its S). Once the swap is done, the S-values are restored back to A or B, node v in Guard $S2$, node J in Guard $S3$.

In Figure 4, nodes v and J swap their IV-values (a state of is a triple $S; IV; tmpS$).

Fig. 4. Nodes v and J swap their IV values

Algorithm 5.2. *Self-stabilizing Heap in a Rooted Tree in the Shared-Memory Model* $\mathcal{HEAP}$

Macro $heap'(v, tS)$ ::
$j = min(v)$
if $(j \geq 0)$ then $J = D_v[j]$; $tmpS.v = tS$; $IV.v = IV.v + IV.J$; $S.v = X$

Heap actions for any node v

$_{ABB}$ $S = B \wedge check(p_v, A) \wedge \forall i, 0 \leq i < |D_v| : check(D_v[i], B) \quad \longrightarrow \quad heap'(v, A)$
$_{BAA}$ $S = A \wedge check(p_v, B) \wedge \forall i, 0 \leq i < |D_v| : check(D_v[i], A) \quad \longrightarrow \quad heap'(v, B)$

Synchronizing actions for any node v

$_{S1}$ $S \in \{A, B\} \wedge p_v \neq \bot \wedge S_p = X \wedge J_p = v \quad \longrightarrow \quad IV{=}IV_p{-}IV \; ; \; tmpS = S \; ; \; S = Y$
$_{S2}$ $S = X \wedge J \neq \bot \wedge S_J = Y \quad \longrightarrow \quad IV = IV - IV_J \; ; \; S = tmpS$
$_{S3}$ $S = Y \wedge p_v \neq \bot \wedge S_p \neq X \quad \longrightarrow \quad S = tmpS$
$_{C1}$ $S = Y \wedge p_v = \bot \quad \longrightarrow \quad S = tmpS$
$_{C2}$ $S = X \wedge D_v = \bot \quad \longrightarrow \quad S = tmpS$
$_{C3}$ $S = X \wedge D_v \neq \bot \wedge \exists w \in D_v : S.w = X \quad \longrightarrow \quad S = tmpS$

5.3   Proof of Correctness of $\mathcal{A\_HEAP}$

The root node R has level 1. Besides local mutual exclusion, heap-building requires synchronization between neighboring nodes. Each node has a local clock measuring *pseudo-time* such that the comparison between the node and its child with the minimal IV value (and eventual swapping) is done when the two nodes have the same pseudo-time values.

For each configuration, the *pseudo-time* function Ψ is defined from the node to non-negative integers. Ψ is initially computed from the delay values, and is updated at each step.

Ψ_0, the pseudo-time at the initial configuration, is defined as follows:

(i) given node v and its parent p_v, $\Psi_0(v) = \frac{d_0[v] + d_0[p_v] - 1}{2}$, and
(ii) $\Psi_0(R) = max\{\Psi_0(v), v \in child_R\}$, where R is the root.

For example, given the configuration in Figure 3(c), the Ψ_0 values are given in Figure 5(a).

We observe that if a node v is enabled, then $\Psi_0(v) = \Psi_0(w)$ for all $w \in D_v$.

Definition 4. *Let Ψ_j and Ψ_{j+1} be the pseudo-time functions for two consecutive configurations in some execution $C_j \mapsto C_{j+1}$. Then Ψ_{j+1} is computed as follows:*

- if node v has executed during this step then $\Psi_j(v)$ and $\Psi_j(w)$ for all children $w \in D_v$ increase by 1: $\Psi_{j+1}(v) = \Psi_j(v) + 1$ and $\Psi_{j+1}(w) = \Psi_j(w) + 1$.
- if any child of the root R executes, $\Psi(R)$ is updated if necessary, i.e., $\Psi_{j+1}(R) = max_{w \in child_R}\{\Psi_{j+1}(w)\}$
- all other nodes u keep their current pseudo-time values, i.e., $\Psi_{j+1}(u) = \Psi_j(u)$.

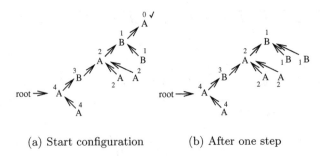

(a) Start configuration   (b) After one step

Fig. 5. Pseudo-time values

For example, given Ψ_0 from Figure 5(a), if the marked node executes, then the next pseudo-time values are the ones in Figure 5(b).

The following relations hold:

(i) $\Psi_0(R) \leq h$

(ii) $\Psi_0(v) \leq i + h - 1$ for $v \neq R$, where $i = level(v)$.

Thus $\Psi_0(v) \leq 2h - 1$, for any node v.

Let $\mathcal{E}(v,t)$ be the predicate: "Node v is enabled if $\Psi(v) = t$."

Observation 1. *(i) If $\mathcal{E}(v,t)$ is true then $\mathcal{E}(v,t+2k+1)$ is false and $\mathcal{E}(v,t+2k)$ is true, for all $k \geq 0$.*
  (ii) If $\mathcal{E}(v,t)$ is false and $t \geq \Psi_0(v)$ then $\mathcal{E}(v,t+2k+1)$ is true and $\mathcal{E}(v,t+2k)$ is false, for all $k \geq 0$.

Property 4. Given a starting configuration C_0, and C_j some configuration after Algorithm $\mathcal{SSDST}$ has executed j steps, then the number of rounds elapsed is $q \leq min\{\forall \text{ nodes } v, \Psi_j(v)\}$.

Proof. A round has elapsed if all nodes enabled in the first configuration of the round have increased their Ψ values by at least one unit; thus the minimum value among all nodes has increased at least by one.

We assume that the values to be placed in min-heap order are distinct. (If necessary, we can add infinitesimal tie-breakers to the values.) Thus they can be arranged in a strict sorted order: $r_1 < r_2 < \ldots < r_n$, and we say that the value r_i has rank i.

Definition 5. *For any given configuration C of Algorithm $\mathcal{A\_HEAP}$, let l_i be the level of the node that holds the value r_i; we call the function $W(C) = \sum_{i=1}^{n} l_i i$ the weighted path length of the configuration C.*

The function W is strictly positive. It increases when a swap is executed between some node v that holds the value r_i and some child $w \in D_v$ that holds the value r_j, where $r_i < r_j$. The value by which W increases is $j - i$.

By Lemma 1, if the heap property does not hold at some node v, node v will execute a swap in finitely many rounds. Since $W(C)$ is an increasing integer function bounded by hn, it must converge in finitely many steps. Thus:

Observation 2. *Function W converges in finitely many of rounds. Let C_* be the configuration after convergence. Then C_* has the heap property.*

Let L_i be the level of the node that holds the value r_i in configuration C_*. Array pos is defined as follows.

Definition 6. *Given j, $1 \leq j \leq n$, and some $t \geq 0$, the value $pos[j, t]$ represents the level of node v that holds the value r_j when $\Psi(v) = t$.*

If initially, the element of r_i is held by the node v situated at level l_i and $\Psi(v) = t_0$, then we assume that for any t, $0 \leq t \leq t_0$, $pos[j, t] = pos[j, t_0]$.

First, we show that once the Ψ value of some node is t, the level $pos[j, t]$ of the element r_j is within a certain range (Property 5). In order to show that Algorithm $\mathcal{A\_HEAP}$ arranges the values as a heap, we show that after $7h/2 - 4$ rounds, $pos[j, t] = L_j$ for all j (Lemma 2).

Property 5. For any $t \geq 0$ and for any j, $1 \leq j \leq n$,

$$min\{L_j, Q[j, t]\} \leq pos[j, t] \leq max\{L_j, P[j, t]\}$$

where $P[j, t] = -t + 2L_j + 3h - 5$ and $Q[j, t] = t + 2L_j + 3 - 4h$, for any j and t.

Proof. Consider the predicates:

$\mathcal{P}(t)$: for any $j \in 1 \dots n$, $pos[j, t] \leq max\{L_j, P[j, t]\}$
$\mathcal{Q}(t)$: for any $j \in 1 \dots n$, $pos[j, t] \geq min\{L_j, Q[j, t]\}$

It can be shown by induction on t that $\mathcal{P}(t)$ holds. The proof that $\mathcal{Q}(t)$ holds is similar.

Lemma 2. *Algorithms $\mathcal{A\_HEAP}$ arranges the values into min-heap order in $7h/2$-4 rounds; thus the stabilization time is $O(h)$ rounds.*

Proof. Follows from Property 5.

5.4 Reduction of Algorithm $\mathcal{HEAP}$ to $\mathcal{A\_HEAP}$

In this section we first show that Algorithm $\mathcal{HEAP}$ reduces to Algorithm $\mathcal{A\_HEAP}$. We can then conclude that, starting from an arbitrary configuration, in at most $4(7h/2 - 4)$ rounds, Algorithm $\mathcal{HEAP}$ arranges the values into min-heap order (Lemma 5).

Definition 7 (Reduction). *Given two different models of communication $\mathcal{M}$ and $\mathcal{M}'$, an algorithm $\mathcal{A}$ in the model $\mathcal{M}$ can be reduced to another algorithm $\mathcal{A}'$ in the model $\mathcal{M}'$ if there exists a one-to-many relation $\mathcal{R}$ from the set of system configurations in the model $\mathcal{M}$ to the set of the system configurations in the model $\mathcal{M}'$ such that the following conditions are true:*

i) *For each configuration of Algorithm $\mathcal{A}$ in the model $\mathcal{M}$ there exists at least one configuration of Algorithm $\mathcal{A}'$ in the model $\mathcal{M}'$.*

ii) *(Lifting property) Given $\mathcal{C}_1$ and $\mathcal{C}_2$ two configurations of Algorithm $\mathcal{A}$ in the model $\mathcal{M}$ such that $\mathcal{C}_1 \longrightarrow \mathcal{C}_2$ is an execution step of Algorithm $\mathcal{A}$, for any configuration $\mathcal{C}_1' \in \mathcal{R}(\mathcal{C}_1)$, if Algorithm $\mathcal{A}'$ in the model $\mathcal{M}'$ starts in $\mathcal{C}_1'$ there exists at least one execution path that starts in $\mathcal{C}_1'$ and ends in some configuration $\mathcal{C}_2' \in \mathcal{R}(\mathcal{C}_2)$.*

If $\mathcal{A}$ accomplishes a task in the model $\mathcal{M}$ and $\mathcal{A}$ reduces to $\mathcal{A}'$, then by Definition 7, $\mathcal{A}'$ accomplishes the same task in the model $\mathcal{M}'$.

We now show that Algorithm $\mathcal{HEAP}$ reduces to Algorithm $\mathcal{A\_HEAP}$. Let $S_v = (s_v, x_v, J_v)$ be the set of all variables of node v in order (S, IV, J) used by Algorithm $\mathcal{A\_HEAP}$ in the abstract model. Let $S_v^{t_v} = (s_v, x_v, t_v, J_v)$ be the set of all variables of node v in order $(S, IV, tmpS, J)$ used by Algorithm $\mathcal{HEAP}$ in the shared-memory model.

Then $\mathcal{R}$ is defined as follows:

$$\mathcal{R}(S_1, \ldots S_n) = \{(S_1^{t_1}, \ldots S_n^{t_n}), t_i \in \{A, B\}, \forall i, 1 \leq i \leq n\}$$

For each state S_i of some configuration $\mathcal{C}_1$ of Algorithm $\mathcal{A\_HEAP}$ in the abstract model, $1 \leq i \leq n$, there exists two possible states S_i^A and S_i^B in the shared-memory model. Thus for each configuration $\mathcal{C}_1$ there exists 2^n configurations in $\mathcal{R}(\mathcal{C}_1)$ of Algorithm $\mathcal{HEAP}$ in the shared-memory model, thus Condition (i) of Definition 7 is satisfied. We are left to show that Condition (ii) of Definition 7 is satisfied (Lemma 3).

Lemma 3. *Given $\mathcal{C}_1$ and $\mathcal{C}_2$, two configurations of Algorithm $\mathcal{A\_HEAP}$ in the abstract model, such that $\mathcal{C}_1 \longrightarrow \mathcal{C}_2$ is an execution step of Algorithm $\mathcal{A\_HEAP}$; for any configuration $\mathcal{C}_1' \in \mathcal{R}(\mathcal{C}_1)$, if Algorithm $\mathcal{HEAP}$ in the shared-memory model starts in $\mathcal{C}_1'$ there exists at least one execution path that starts in $\mathcal{C}_1'$ and ends in some configuration $\mathcal{C}_2' \in \mathcal{R}(\mathcal{C}_2)$.*

Proof. We give a sketch of the proof. A node state contains all the variables stored at that node. The system configuration contains the states of all the nodes. An execution step is a transition from one configuration to another. We break the system configuration into a number of *chunks*. A *chunk* is a set of a node and its descendants in the tree such that the first node in each chunk is enabled, and all the descendants of the first node reachable by a path of disabled nodes are added to the chunk. We build the set of chunks starting from the root in depth-first-search (DFS) order. If the root node is currently disabled, then the root and all nodes reachable from the root reachable by a path of disabled nodes are not part of any chunk. We call the set of those nodes the *null chunk*.

Given a configuration, there is a unique way to break it into chunks. An execution step of Algorithm $\mathcal{A\_HEAP}$ in the abstract model in one chunk affects only the nodes' states in that chunk.

From Property 1 we know that if a non-leaf node is enabled, its children are disabled. So, except for the leaf nodes, every chunk contains at least two nodes.

If the chunk contains at least two nodes, then the last node in the chunk is disabled, so it cannot affect the state of the first node of other chunks.

Instead of considering an execution step between *global* configurations, we consider an execution step between the chunks of a global configuration.

If the starting state of the node is either A or B, then the value to be sorted is its initial value. If some node starting state is either X or Y, then it is possible for some of the three steps of the swap to be applied (see Section 5) and the initial value of that node to be modified accordingly, and that modified value to be sorted. This drawback is caused by arbitrary initialization, and would be encountered even if we had used an extra variable for swapping.

We recall that node $J.v$ is the child of node v that holds the minimal IV value among all node v's children. The variable $J.v$ is $\perp$ if and only if node v is a leaf node ($child.v = \perp$).

For any node v such that $S_v = X$, either S_v remains X and then the node $J.v$ will have its S equal to Y in at most three rounds (by executing Action $S1$), or v changes its S to A or B in at most one round.

For any node v such that $S.v = X \wedge S.(J.v) = Y$ then $IV.v$ gets the value $IV.(J.v)$ and then node v changes its S_v to A or B in at most one round. Node $J.v$ had already stored in $IV.(J.v)$ the old value of $IV.v$ (by executing Action $S1$) and will restore its $S_{J.v}$ from Y to either A or B (depending on the value of $tmpS$) in at most one round. We can then conclude that if $S.v$ is either X or Y, then in at most four rounds $S.v$ is either A or B (Lemma 4).

Lemma 4. *For any node v, if $S.v \in \{X, Y\}$, in at most four rounds $S.v$ becomes either A or B.*

Lemma 5. *Starting from an arbitrary configuration, in at most $4(7h/2 - 4)$ rounds, Algorithm $\mathcal{HEAP}$ arranges the n values in min-heap order.*

Proof. From Lemma 4, each swap takes at most 4 rounds. From Lemma 2, if a swap takes at most 1 round, then heapification takes at most $7h/2 - 4$ rounds. Since the swap takes at most 4 rounds, we obtain a total of at most $14h - 16$ rounds.

6 Conclusion

In this paper, we present the first self-stabilizing algorithm for simultaneously activating non-adjacent processes in a rooted tree, called $\mathcal{SSDST}$. The algorithm is optimal in the space complexity, and asymptotically optimal in the time complexity. We then give two applications of the proposed algorithm for rooted trees, a time and space optimal solution to the local mutual exclusion problem (Algorithm $\mathcal{LMET}$) and a space and (asymptotically) time optimal solution to the min-heap problem (Algorithm $\mathcal{HEAP}$).

All algorithms are self-stabilizing and uniform, and they work under the unfair distributed daemon.

In proving the time complexity of heap-building, we use the notion of *pseudo-time*. Pseudo-time is similar to *logical time* introduced by Lamport [12].

We expect that Algorithm $\mathcal{SSDST}$ can be used to obtain optimal space solutions for other problems in a rooted tree. For example, for broadcasting m messages, a solution based on Algorithm $\mathcal{SSDST}$ stabilizes in at most $2h + 2m - 5$ rounds (the root node executes m times).

References

1. L. Alima. Self-stabilizing max-heap. *Proceedings of the ICDCS Workshop on Self-stabilizing Systems*, pages 94–101, 1999.
2. L. Alima, J. Beauquier, A. Datta, and S. Tixeuil. Self-stabilization with global rooted synchronizers. *Proceedings of the 18-th ICDCS*, pages 102–109, 1998.
3. A. Arora and M. Nesterenko. Stabilization-preserving atomicity refinement. *Journal of Parallel and Distributed Computing*, 62:766–791, 2002.
4. D. Bein, A. Datta, and V. Villain. Snap-stabilizing optimal binary-search-tree. *Proceedings of the 7-th International Symposium on Self-Stabilizing Systems*, 2005.
5. B. Bourgon and A. Datta. A self-stabilizing distributed heap maintenance protocol. *Proceedings of the Second Workshop on Self-stabilizing Systems*, 1995.
6. A. Bui, A. Datta, F. Petit, and V. Villain. State-optimal snap-stabilizing PIF in tree networks. In *Proceedings of the Third Workshop on Self-Stabilizing Systems*, pages 78–85. IEEE Computer Society, 1999.
7. T. Cormen, C. Leiserson, R. Rivest, and C. Stein. *Introduction to Algorithms (second edition)*. MIT Press, 2001.
8. E. W. Dijkstra. Self stabilizing systems in spite of distributed control. *Communications of the Association of the Computing Machinery*, 17:643–644, 1974.
9. S. Dolev, A. Israeli, and S. Moran. Uniform dynamic self-stabilizing leader election. *IEEE Transactions on Parallel and Distributed Systems*, 8(4):424–440, 1997.
10. T. Herman and T. Masuzawa. Available stabilizing heaps. *Information Processing Letters*, 77:115–121, 2001.
11. C. Johnen, L. Alima, A. Datta, and S. Tixeuil. Self-stabilizing neighborhood synchronizer in tree networks. *Parallel Processing Letters*, 12(3 & 4):327–340, 2002.
12. L. Lamport. Time, clocks and the ordering of events in a distributed systems. *Communications of the ACM*, 21:558–565, 1978.
13. S. Ukena, M. Hasegawa, Y. Katayama, T. Masuzawa, and H. Fujiwara. A self-stabilizing max-heap protocol in tree networks. *Electronics and Communications in Japan, Part III: Fundamental Electronic Science (English translation of Denshi Tsushin Gakkai Ronbunshi)*, 86(9):63–72, 2003.

Distance-k Information in Self-stabilizing Algorithms*

Wayne Goddard[1], Stephen T. Hedetniemi[1],
David P. Jacobs[1], and Vilmar Trevisan[2]

[1] Department of Computer Science, Clemson University, SC 29634 USA
{goddard, hedet, dpj}@cs.clemson.edu
[2] Instituto de Matemática, UFRGS, Porto Alegre, Brazil
trevisan@mat.ufrgs.br

Abstract. Many graph problems seem to require knowledge that extends beyond the immediate neighbors of a node. The usual self-stabilizing model only allows for nodes to make decisions based on the states of their immediate neighbors. We provide a general polynomial transformation for constructing self-stabilizing algorithms which utilize distance-k knowledge, with a slowdown of $n^{O(\log k)}$. Our main application is a polynomial-time self-stabilizing algorithm for finding maximal irredundant sets, a problem which seems to require distance-4 information. We also show how to find maximal k-packings in polynomial-time. Our techniques extend results in a recent paper by Gairing et al. for achieving distance-two information.

1 Introduction

Self-stabilization, introduced by Dijkstra [1], is the most inclusive approach to fault tolerance in distributed systems. In a self-stabilizing algorithm, each node maintains its local variables, and can make decisions based on the correct knowledge of its neighbors' states. In a self-stabilizing algorithm, a node may change its local state by making a *move* (an action which causes a change of local state). Algorithms are given as a set of rules of the form "**if** $p(i)$ **then** M", where $p(i)$ is a predicate and M is a move. A node i becomes *privileged* if $p(i)$ is true. When a node becomes privileged, it may execute the corresponding move. We assume a serial model in which no two nodes move simultaneously. A *central daemon* selects, among all privileged nodes, the next node to move. If two or more nodes are privileged, we cannot predict which node will move next. In this paper we say that an algorithm *stabilizes* if no node is privileged. An execution will be represented as a sequence of moves $M_1, M_2, \ldots$, in which M_s denotes the s-th move. One can transform the algorithm to work under other daemons, using established techniques. We refer the reader to [2] for a general treatment of self-stabilizing algorithms.

---

* Research supported by: NSF grant CCR-0222648; CNPq grant 453991/2005-0; and FAPERGS grant 05/2024.1.

P. Flocchini and L. Gąsieniec (Eds.): SIROCCO 2006, LNCS 4056, pp. 349–356, 2006.

A distributed system can be modeled with an undirected graph $G = (V, E)$, where V is a set of n nodes and E is a set of m edges. If $i \in V$, then $N(i)$, its *open neighborhood*, denotes the set of nodes to which i is adjacent, and $N[i] = N(i) \cup \{i\}$ denotes its *closed neighborhood*. Every node $j \in N(i)$ is called a *neighbor* of node i. Throughout this paper we assume G is connected and $n > 1$.

In the usual self-stabilizing model, each node i can read only the variables of its neighbors, that is, those nodes which are a distance of one from i. In this paper, we show how to obtain self-stabilizing algorithms in which a node i can effectively read the contents of variables which are within distance k of i, for any fixed $k \geq 1$, extending results in [3] for achieving distance-two information. This will result in a slowdown of $n^{O(\log k)}$. In Section 3, we obtain a polynomial time self-stabilizing algorithm for finding a maximal irredundant set, a problem which requires distance-4 information.

We assume throughout this paper that all nodes have a unique integer ID. Sometimes we do not distinguish between a node i and its ID. For each $k \geq 1$, we let $N^k[i]$ denote the set of nodes whose distance from i is at most k, and we let $N^k(i) = N^k[i] - \{i\}$. When $k = 1$, these sets correspond, respectively, to the closed and open neighborhoods of i.

A *k-packing* in a graph $G = (V, E)$ is a set $S \subseteq V$ of nodes such that for every pair of distinct nodes, $u, v \in S$, their minimum distance $d(u, v) > k$. A 1-packing is, therefore, a set S having the property that no two nodes in S are adjacent ($d(u, v) > 1$). This is normally called an *independent* set.

Algorithm 1.1 is a well-known and simple self-stabilizing algorithm for finding the characteristic function of a maximal independent set. It is easy to show that this algorithm stabilizes in at most $2n$ moves [6] in the distance-1 model.

Algorithm 1.1. MAXIMAL INDEPENDENT SET

---

local variable: f
ENTER: if $f(i) = 0 \wedge (\forall j \in N(i))(f(j) = 0)$
   then $f(i) = 1$
LEAVE: if $f(i) = 1 \wedge (\exists j \in N(i))(f(j) = 1)$
   then $f(i) = 0$

---

2   Distance-k problems

In [3], it was observed that certain algorithmic problems can be solved more easily on an *extended model* in which each node can instantly see all state information of nodes that are within distance *two*. Having done this, the extended model can be simulated using a conventional self-stabilizing algorithm, provided all nodes have unique IDs. In this paper we show how arbitrary distances greater than two can be achieved. Our idea is to use the technique in [3] recursively.

We now define a class of self-stabilizing algorithm models. For each $k \geq 1$, in the *distance-k self-stabilizing model*, each node i can instantly see all state information of all nodes in $N^k[i]$. We assume that node i can read the ID of j and its state information $f(j)$ for each $j \in N^k[i]$. For brevity, we refer to this as the *distance-k model*. The distance-1 model is the usual self-stabilizing algorithmic model. It will be convenient to assume for now that k is a power of two.

Now let $k = 4$, and consider Algorithm 2.1, which assumes the distance-4 model. If Algorithm 2.1 stabilizes, the set $S = \{i \mid f(i) = 1\}$ is a maximal 4-packing. For if no node is privileged to LEAVE, then S must be a 4-packing, and if no node can ENTER, the 4-packing is maximal. Moreover, the algorithm must always stabilize. Indeed, once a node makes an ENTER move, no node in $N^4(i)$ can ENTER, and so no node in $N^4[i]$ can move again. If a node makes a LEAVE move, its next move must be an ENTER, after which it cannot move. It follows that

Lemma 1. *The distance-4 Algorithm 2.1 finds a maximal 4-packing in at most $2n$ moves.*

Algorithm 2.1. MAXIMAL 4-PACKING IN DISTANCE 4

---

local variable: f
ENTER: if $f(i) = 0 \wedge (\forall j \in N^4(i))(f(j) = 0)$
   then $f(i) = 1$
LEAVE: if $f(i) = 1 \wedge (\exists j \in N^4(i))(f(j) = 1)$
   then $f(i) = 0$

Assume now that we have some distance-$2k$ algorithm $\mathcal{S}_{2k}$, such as Algorithm 2.1, in which every node has a local variable f. We now will describe a way to simulate $\mathcal{S}_{2k}$ using a distance-k algorithm $\mathcal{S}_k$. We will see that the running times of $\mathcal{S}_k$ and $\mathcal{S}_{2k}$ are related to within a factor in $O(n^3)$. In Algorithm $\mathcal{S}_k$, each node i has three local variables:

- The variable f stores the state of node i with respect to $\mathcal{S}_{2k}$, that is, the value of $\mathcal{S}_{2k}$'s local variable.
- The variable σ stores a local copy of $f(j)$ for each $j \in N^k(i)$. We may assume that $\sigma(i)$ is a list of pairs of the form (j, f_j), where j is an ID of a node in $N^k(i)$. We say that $\sigma(i)$ is *correct* if for all $j \in N^k(i)$, $f(j) = f_j$.
- A pointer stores the ID of a member of $N^k[i]$, or has the value NULL. We write $i \rightarrow j$, $i \rightarrow i$, and $i \rightarrow$ NULL to mean, respectively, that i points to j, i points to itself, and i's pointer is NULL.

At each step in the execution of $\mathcal{S}_k$, the values $f(i)$ represent a state with respect to $\mathcal{S}_{2k}$. A node i in the distance-k model can read directly only state information of nodes in $N^k(i)$. However if $j' \in N^{2k}(i)$, then $j' \in N^k(j)$ for some $j \in N^k[i]$. It follows that in the distance-k model, by reading $\sigma(j)$, node i has a *view* of $f(j')$. However, it is possible for this view to be incorrect.

During the execution of $\mathcal{S}_k$, we say that node i is $\mathcal{S}_{2k}$-*alive* if it is privileged for $\mathcal{S}_{2k}$, under the assumption that its view of $\{(j, f(j)) \mid j \in N^{2k}(i)\}$ is correct.

We define

$$minN^k[i] = \min\{j \mid j \in N^k[i] \land j \to j\}, \text{ where } \min\{\emptyset\} = \text{ NULL }.$$

That is, $minN^k[i]$ is the smallest ID, within distance k of i, which is pointing to itself; $minN^k[i]$ is defined to be NULL if no member of $N^k[i]$ points to itself.

Algorithm $\mathcal{S}_k$ is displayed as Algorithm 2.2. When $k = 1$, it is exactly the algorithm described in [3].

Algorithm 2.2. $\mathcal{S}_k$

---

comment: Simulates distance-$2k$ algorithm $\mathcal{S}_{2k}$

local variables: $f, \sigma, \to$
UPDATE-σ: if $\sigma(i)$ is incorrect
  then update $\sigma(i)$
ASK: if i is $\mathcal{S}_{2k}$-alive $\land$ $(\forall j \in N^k[i] : j \to \text{ NULL}) \land \sigma(i)$ is correct
  then $i \to i$
RESET: if $i \not\to minN^k[i] \land \sigma(i)$ is correct
  then $i \to minN^k[i]$
CHANGE: if $\forall j \in N^k[i] : j \to i \land \sigma(i)$ is correct
  then $\begin{cases} \text{if } i \text{ is } \mathcal{S}_{2k}\text{-alive, then update } f(i) \\ i \to \text{ NULL} \end{cases}$

Lemma 2. *If Algorithm $\mathcal{S}_k$ stabilizes, then all pointers are null, $\sigma(i)$ is correct for all i, and no node is $\mathcal{S}_{2k}$-privileged.*

Proof. Assume the algorithm has stabilized. Then no node points to itself, for otherwise the node i pointing to itself having the smallest ID would have all members of $N^k[i]$ pointing to it, and i would be privileged for a CHANGE move. Since no node points to itself, $minN^k[i]$ is *NULL*, and therefore all pointers are *NULL*. All $\sigma(i)$ are correct since no node is privileged for an UPDATE-σ. No node is $\mathcal{S}_{2k}$-privileged, for otherwise it would be privileged to execute ASK.

Lemma 3. *While i is pointing to itself, no node in $N^k(i)$ can execute an ASK or CHANGE.*

Proof. For $j \in N^k(i)$ to execute ASK, i must be NULL. For j to execute CHANGE, i must be pointing to j.

Lemma 4. *If i makes an ASK move, its next move must be a CHANGE move.*

Proof. When i makes an ASK move, all members of $N^k[i]$ are NULL. Suppose its next move is a RESET. Then this means that some $j \in N^k(i)$ is pointing to itself. But this is impossible because $i \to i$. Nor can its next move be an UPDATE-σ, because at the time of the ASK move, $\sigma(i)$ was correct. But this can't change by Lemma 3, nor can its next move be another ASK move because $i \to i$.

Let us say that a move by i is *correct* if $\sigma(j)$ is correct for all $j \in N^k(i)$, and *incorrect* otherwise.

Lemma 5. *If node i makes an ASK move, then its next CHANGE move is correct.*

Proof. Let j be some member of $N^k[i]$. During the interval between the ASK and CHANGE moves, j must have changed its pointer from NULL to i, at which time $\sigma(j)$ was correct, and there must have been a last time during this interval when this occurred. But $\sigma(j)$ must have remained correct, because no member of $N^k[j]$ could have performed a CHANGE while j was pointing to i.

Lemma 6. *If a node i makes a CHANGE move, then its next ASK move is correct.*

Proof. During this interval, all $j \in N^k[i]$ changed their pointers from i to NULL. There is a last time at which j became NULL, prior to the ASK move. At this time, $\sigma(j)$ is correct, and it must remain so up until the ASK move, since no member of $N^k[j]$ could have performed a CHANGE move as long as j's pointer is NULL.

Lemma 7. *Between any two RESET moves made by i, some some $j \in N^k[i]$ must execute an ASK or a CHANGE.*

Proof. This is clear.

For convenience, we define a REAL-CHANGE move as a CHANGE move in which the variable f is assigned. We let $d_i^k = |N^k(i)|$.

Lemma 8. *Consider an interval without a REAL-CHANGE move. Then each node i can make:*

1. *at most one UPDATE-σ move;*
2. *at most one ASK move;*
3. *at most one CHANGE move; and*
4. *$O(d_i^k)$ RESET moves.*

Proof. 8.1 is obvious. To see 8.2, suppose i makes an ASK move. By Lemma 4, its next move must be a CHANGE move. Then by Lemma 5, the CHANGE move is correct. Since this is not a REAL-CHANGE, i is not $\mathcal{S}_{2k}$-privileged. Since no other REAL-CHANGE moves occur, i cannot become $\mathcal{S}_{2k}$-alive again to execute another ASK move. To see 8.3, suppose i makes a CHANGE move, and then makes an ASK move. By Lemma 6, the ASK move is correct. Since no REAL-CHANGE can take place, the σ's remain the same, and if i were to execute another CHANGE move, it would have to be a REAL-CHANGE. Finally, 8.4 follows from Lemma 7.

Lemma 9. *There are at most $O(n^2)$ moves during an interval without REAL-CHANGE moves.*

Proof. This follows immediately from Lemma 8.

Lemma 10. *Each node can make at most one incorrect REAL-CHANGE move.*

Proof. An incorrect REAL-CHANGE move can only occur as a node's first CHANGE move, because subsequent CHANGE moves will be preceded by an ASK move, which by Lemma 5, must be correct.

Lemma 11. *Let (M_i) be a sequence of moves made by Algorithm 2.2 during which no incorrect REAL-CHANGE occurs. Then the subsequence (M_i') of REAL-CHANGE moves is a valid computation of $\mathcal{S}_{2k}$.*

Proof. This is clear.

Lemma 12. *Suppose Algorithm $\mathcal{S}_{2k}$ can execute at most A moves. Then in any interval without an incorrect REAL-CHANGE move, Algorithm $\mathcal{S}_k$ can execute at most $O(An^2)$ moves.*

Proof. By Lemma 11, there can be at most A REAL-CHANGE moves, and by Lemma 9, between any two REAL-CHANGE moves, there are at most $O(n^2)$ moves.

Theorem 1. *In a network with n nodes, a distance-$2k$ algorithm $\mathcal{S}_{2k}$ that stabilizes within A moves can be implemented with a distance-k algorithm $\mathcal{S}_k$ that stabilizes in $O(An^3)$ moves.*

Proof. By Lemma 10 there can be at most n incorrect REAL-CHANGE moves. By Lemma 12, during the intervals without incorrect moves, there can be at most $O(An^2)$ moves. Finally by Lemma 2, the algorithm is correct.

By repeating Theorem 1, we obtain

Theorem 2. *In a network with n nodes, a distance-k algorithm $\mathcal{S}_k$ which stabilizes in A moves can be implemented in the distance-1 model by an algorithm that stabilizes in $O(An^{3\lceil \log_2(k) \rceil})$ moves.*

Corollary 1. *There is a self-stabilizing algorithm to find a maximal 4-packing that stabilizes in $O(n^7)$ moves.*

Proof. This follows by Lemma 1 and Theorem 2.

When we translate, say, a distance-4 algorithm $\mathcal{S}_4$ to a distance-2 algorithm $\mathcal{S}_2$, each node will contain the original variable f used in $\mathcal{S}_4$ in addition to a pointer and a σ. Note that when $\mathcal{S}_2$ is then translated to a distance-1 algorithm $\mathcal{S}_1$, each node will contain these three variables in addition to another pointer and another σ.

A maximal 4-packing can be found by using a single boolean variable in the distance-4 model. However, to find a maximal 3-packing in the distance-4 model, nodes must know the distances of their neighbors. If we assume that in addition to its usual variables, each node displays a list of the IDs of its neighbors, then in the distance-1 model, each node can compute the subgraph induced by its closed neighborhood. In the distance-k model, each node i can compute the subgraph induced by $N^k[i]$, and can compute, for example, the distance $d(i, j)$ for $j \in N^k[i]$. This is illustrated in Algorithm 2.3. This generalizes to a polynomial time self-stabilizing algorithm for maximal k-packing, for any fixed k, and improves upon the maximal k-packing algorithm in [4] that was given without analysis.

Algorithm 2.3. DISTANCE-4 ALGORITHM FOR MAXIMAL 3-PACKING

---

local variable: f
ENTER: if $f(i) = 0 \wedge (\forall j \in N^4(i), d(i,j) \leq 3)(f(j) = 0)$
  then $f(i) = 1$
LEAVE: if $f(i) = 1 \wedge (\exists j \in N^4(i), d(i,j) \leq 3)(f(j) = 1)$
  then $f(i) = 0$

---

3   Maximal Irredundant Sets

Given a set S of nodes, we say a node $s \in S$ has a *private neighbor* with respect to S if there exists some $x \in N[s] - N[S - \{s\}]$. A set S is *irredundant* [5] if every $s \in S$ has a private neighbor with respect to S. Self-stabilizing algorithms have been found for many kinds of related sets, such as maximal independent sets and minimal dominating sets [6], but finding maximal irredundant sets has proven difficult because the problem seems to require distance-4 knowledge.

Let S be a set of nodes, not necessarily irredundant, and let $s \in S$. If s has a private neighbor with respect to S, but s has no private neighbor with respect to $S \cup \{x\}$, we say x *destroys* s. Finally, we say $x \in V - S$ is *safe* if x has a private neighbor with respect to $S \cup \{x\}$, and no $s \in S$ is destroyed by x.

Consider Algorithm 3.1. It is easy to see that if this algorithm stabilizes, then $S = \{i \mid f(i) = 1\}$ is maximal irredundant. For if it is not irredundant, some i is privileged to execute a LEAVE move. And if it is not maximal irredundant, some i can execute an ENTER move. Note also that once a node executes an ENTER, it will never execute a LEAVE. Thus, given a sufficiently powerful model, each node moves at most twice.

Algorithm 3.1. MAXIMAL IRREDUNDANT SET

---

local variable: f
ENTER: if $f(i) = 0 \wedge i$ is safe
  then $f(i) = 1$
LEAVE: if $f(i) = 1 \wedge i$ has no private neighbor
  then $f(i) = 0$

Lemma 13. *Node i can decide if it has a private neighbor from the information in $N^2[i]$.*

Proof. A node x is a private neighbor of i if and only if $x \in N[i]$, but for all $j \in N^2(i)$, $j \in S$ implies $x \notin N[j]$.

Lemma 14. *Node i can decide if it is safe from the information in $N^4[i]$.*

Proof. If node i is not safe, then it must destroy some node $j \in N^2[i]$. However, to know whether such a node j has a private neighbor requires examining $\{f(j') \mid j' \in N^2[j]\}$.

Theorem 3. *There is a self-stabilizing algorithm for finding a maximal irredundant set that stabilizes in $O(n^7)$ moves.*

Proof. By Lemma 13 and Lemma 14 it follows that Algorithm 3.1 can be implemented in the distance-4 model. By our earlier comments, Algorithm 3.1 stabilizes in a linear number of moves. The analysis follows by Theorem 2.

We observe that while Algorithm 3.1 makes a linear number of moves in the distance-4 model, each simulated move may not take constant time, although it will be polynomial.

References

1. Dijkstra, E.W.: Self-stabilizing systems in spite of distributed control. Comm. ACM **17** (11) (1974) 643–644
2. Dolev, S.: *Self-Stabilization.* MIT Press, 2000
3. Gairing, M., Goddard, W., Hedetniemi, S.T., Kristiansen, P., McRae, A.A.: Distance-two information in self-stabilizing algorithms, Parallel Process. Lett., **14** (2004) 387–398
4. Goddard, W., Hedetniemi, S.T., Jacobs, D.P., Srimani, P.K.: Self-stabilizing global optimization algorithms for large network graphs, Int. J. Dist. Sensor Net., **1** (2005) 329–344
5. Haynes, T.W., Hedetniemi, S.T., Slater, P.J.: *Fundamentals of Domination in Graphs,* Marcel Dekker, New York, 1998
6. Hedetniemi, S.M., Hedetniemi, S.T, Jacobs, D.P., Srimani, P.K.: Self-stabilizing algorithms for minimal dominating sets and maximal independent sets, Comput. Math. Appl., **46** (2003) 805–811

Author Index

Lecture Notes in Computer Science

For information about Vols. 1–3958

please contact your bookseller or Springer